Advanced Nanomaterials for Water Remediation (2nd Edition)

Advanced Nanomaterials for Water Remediation (2nd Edition)

Guest Editors

Pedro Manuel Martins
Noelia González-Ballesteros

MDPI
Basel • Beijing • Wuhan • Barcelona • Belgrade • Novi Sad • Cluj • Manchester

Guest Editors
Pedro Manuel Martins
Department of Biology
University of Minho
Braga
Portugal

Noelia González-Ballesteros
Departamento de Química
Inorgánica
Universidade de Vigo
Vigo
Spain

Editorial Office
MDPI AG
Grosspeteranlage 5
4052 Basel, Switzerland

This is a reprint of the Special Issue, published open access by the journal *Nanomaterials* (ISSN 2079-4991), freely accessible at: https://www.mdpi.com/journal/nanomaterials/special_issues/16N5X8BC27.

For citation purposes, cite each article independently as indicated on the article page online and as indicated below:

Lastname, A.A.; Lastname, B.B. Article Title. *Journal Name* **Year**, *Volume Number*, Page Range.

ISBN 978-3-7258-7779-9 (Hbk)
ISBN 978-3-7258-7780-5 (PDF)
https://doi.org/10.3390/books978-3-7258-7780-5

Contents

About the Editors

Pedro Manuel Martins

Pedro Manuel Martins is a researcher at the University of Minho, Portugal, affiliated with the Centre of Molecular and Environmental Biology (CBMA) and the Institute of Science and Innovation for Bio-Sustainability (IB-S). His research focuses on multifunctional materials for environmental applications, particularly photocatalytic and adsorptive membranes, bio-based and hybrid nanomaterials, and advanced manufacturing strategies, including 3D printing, for water remediation and sensing of contaminants of emerging concern. He has coordinated and participated in several national and European research projects, including Marie Skłodowska-Curie and FCT programs, addressing sustainable water treatment and circular-economy-driven materials. He has established strong international collaborations through research stays and joint projects with institutions in Germany, Spain, Algeria, and Portugal. Pedro Martins has authored more than 65 peer-reviewed publications (h-index 28), including contributions to high-impact journals such as *Chemical Reviews*, *Advanced Functional Materials*, and the *Journal of Hazardous Materials*, and has served as a Guest Editor for international scientific journals published by MDPI and Springer Nature. He is actively involved in graduate and postgraduate teaching, PhD and MSc supervision, and scientific management, and currently serves on the Scientific Council of IB-S, where he oversees the operational coordination of the Multifunctional Materials Laboratory.

Noelia González-Ballesteros

Noelia González-Ballesteros is a postdoctoral researcher at the Universidade de Vigo, Spain, affiliated with the Department of Inorganic Chemistry. Her research focuses on the green synthesis and advanced characterization of metallic nanomaterials, with particular emphasis on biosynthesized gold, silver, and platinum nanoparticles and their applications in environmental remediation, photocatalysis, and biomedicine. She has participated in several competitive national and European research projects, including the Spanish projects RTI2018-096376-B-I00 and PID2021-124017-B-I00, as well as the HORIZON Europe Pathfinder project PERSEUS, addressing sustainable nanomaterials, algal biorefineries, and light-activated therapies. Her international experience includes research stays at the International Iberian Nanotechnology Laboratory (Portugal), the University of Parma (Italy), and a two-year postdoctoral fellowship at the University of Minho supported by the Margarita Salas program. She has authored 33 peer-reviewed scientific publications (h-index 16, Scopus), many of them in Q1 journals, and has presented her work at numerous national and international conferences. Her achievements include the Extraordinary PhD Award from the Universidade de Vigo, recognition for best poster presentations at international conferences, and active involvement in open science, peer review for international journals, graduate and postgraduate teaching, and supervision of undergraduate and master's theses.

Preface

This Special Issue focuses on developing advanced nanomaterial-based strategies to improve water and wastewater treatment, aiming to address the ongoing challenge posed by persistent and emerging contaminants. Its scope covers a broad range of materials, processes, and application scenarios, reflecting the multidisciplinary effort needed to address contaminants that are chemically stable, biologically resilient, and increasingly found in complex aquatic environments.

The goal of this Special Issue is to provide a clear, forward-looking overview of how nanotechnology and multifunctional materials can help develop a new generation of water decomposition approaches. The contributions collected explore various mechanisms, including photocatalysis, advanced oxidation, adsorption, and hybrid systems, using materials ranging from inorganic and plasmonic nanostructures to bio-based and polymeric platforms. By integrating material design, process innovation, and device implementation, this Special Issue emphasizes solutions that go beyond traditional lab studies and move towards more practical, real-world applications.

The major motivation for this collection is the growing gap between the complexity of water pollution and the limited effectiveness of conventional treatment technologies. Persistent contaminants often escape removal by existing systems, demanding innovative approaches that are efficient, adaptable, and sustainable. The Guest Editors sought contributions that not only demonstrate high remediation performance but also address eco-friendly synthesis, reusability, and environmental safety as key factors in the development of these technologies. This Special Issue is intended for researchers, engineers, and advanced students in materials science, environmental engineering, chemistry, and related fields, as well as stakeholders interested in innovative and sustainable water treatment solutions. By offering a varied yet consistent collection of studies, it serves as both a reference for current best practices and a source of inspiration for future research on remediating highly resilient water contaminants.

Pedro Manuel Martins and Noelia González-Ballesteros
Guest Editors

nanomaterials

MDPI

Review

Bio-Based Nanomaterials for Groundwater Arsenic Remediation: Mechanisms, Challenges, and Future Perspectives

Md. Mahbubur Rahman [1,*], Md. Nizam Uddin [2,*], Md Mahadi Hassan Parvez [1], Md. Abdullah Al Mohotadi [1] and Jannatul Ferdush [1]

1 Department of Mechanical Engineering, Khulna University of Engineering & Technology, Khulna 9203, Bangladesh; mahadihparvez@gmail.com (M.M.H.P.); abdullahalmohotadi@gmail.com (M.A.A.M.); jferdush555@gmail.com (J.F.)

2 James C. Morriss Division of Engineering, Texas A & M University-Texarkana, 7101 University Ave., Texarkana, TX 75503, USA

* Correspondence: mahbub_rahman@me.kuet.ac.bd (M.M.R.); muddin@tamut.edu (M.N.U.)

Abstract: Arsenic contamination in water poses a significant global health risk, necessitating efficient and sustainable remediation strategies. Arsenic contamination affects groundwater in at least 106 countries, potentially exposing over 200 million people to elevated levels, primarily through contaminated drinking water. Among the most affected regions, Bangladesh remains a critical case study, where widespread reliance on shallow tubewells has resulted in one of the largest mass poisonings in history. Bio-based nanomaterials have emerged as promising solutions due to their eco-friendly nature, cost-effectiveness, and high adsorption capabilities. These nanomaterials offer a sustainable approach to arsenic remediation, utilizing materials like biochar, modified biopolymers, and bio-based aerogels, which can effectively adsorb arsenic and other pollutants. The use of environmentally friendly nanostructures provides a potential option for improving the efficiency and sustainability of arsenic remediation from groundwater. This review explores the mechanisms underlying arsenic remediation using such nanomaterials, including adsorption, filtration/membrane technology, photocatalysis, redox reactions, complexation, ion exchange, and coagulation–flocculation. Despite their potential, challenges such as scalability, stability, and regeneration hinder widespread application. We discuss recent advancements in material design, surface modifications, and hybrid systems that enhance performance. Finally, future perspectives are highlighted, including the integration of these bio-derived systems with smart sensing technologies, sustainable water-treatment frameworks, smart design, and life-cycle integration strategies, particularly for use in resource-constrained regions like Bangladesh and other globally impacted areas.

Keywords: groundwater; arsenic remediation; health hazard; biochar; chitosan; adsorption; kinetic; photocatalysis; hybrid and integrated technologies

1. Introduction

Over 2.5 billion individuals worldwide depend on groundwater for consumption, making the provision of high-quality drinking water a critical global challenge. Despite groundwater being deemed safe, elevated levels of heavy metals such as arsenic may provide significant health risks and hazards to humans. Arsenic contamination in water constitutes a global health issue, particularly in areas dependent on groundwater for potable and agricultural use. Prolonged exposure to arsenic, even at minimal concentrations, is linked to numerous health complications, including dermal lesions, cardiovascular disorders, and malignancies [1–4]. Arsenic is among the ten substances of significant public health concern identified by the World Health Organization (WHO). The WHO's efforts to mitigate arsenic exposure encompass establishing guideline values, evaluating evidence, and offering risk-management recommendations. Approximately 140 million individuals in at least 70 nations have been consuming water with arsenic concentrations exceeding the WHO provisional recommendation of 10 μg/L. Recent statistical modeling indicates that between 94 and 220 million individuals are at risk of exposure to heightened arsenic levels in groundwater [5]. One of the most severely affected countries is Bangladesh, where an estimated 35–77 million people are chronically exposed to arsenic-contaminated groundwater, primarily through shallow tube wells used for drinking and irrigation [6,7]. Arsenic contamination poses a threat not only to human health but also to ecosystems, agriculture, and socio-economic development [8]. It disrupts microbial communities, reduces aquatic biodiversity, and causes sediment toxicity. In agriculture, it contributes to reduced crop yield, phytotoxicity, and arsenic accumulation in food crops such as rice and vegetables. These environmental effects, in turn, lead to economic burdens—including healthcare costs, productivity loss, reduced agricultural income, and mental health stress among affected populations [9–21]. Table 1 summarizes these wide-ranging effects across different domains.

Table 1. Effects of arsenic contamination on different domains.

Category	Effect	Reference
Human Health	Skin lesions, cancers (skin, lung, bladder), cardiovascular disease	[9]
	Neurological effects and cognitive impairment in children	[10]
	Diabetes, hypertension, and metabolic disorders	[11]
	Adverse birth outcomes due to prenatal arsenic exposure	[12]
Environment	Disruption of microbial communities in arsenic-contaminated sites	[13]
	Decreased aquatic biodiversity	[14]
	Sediment arsenic accumulation and benthic toxicity	[15]
Agriculture	Phytotoxic effects: reduced plant growth and yield loss	[16]
	Arsenic accumulation in rice and vegetables	[17]
	Microbial community and enzyme activity decline in agricultural soil	[18]
Socio-economic	Health care costs and productivity loss in arsenic-affected regions	[19]
	Loss of income and agricultural productivity	[20]
	Mental health challenges among chronically exposed populations	[21]

The data and literature surveyed in this review are drawn from scholastic peer-reviewed research articles published mostly over the last ten years and health-related reports published by a renowned and credible organization like the WHO. Targeted keywords such as "arsenic contamination", "bio-based nanomaterials", "groundwater remediation", and "toxicological impact" were selected during the literature search to ensure that the most relevant and up-to-date studies were included in the review. These resources show the alarming rate of increase in As in numerous sources, along with groundwater, and the devastating impact on living organisms on Earth. Additionally, the role of bio-based nanomaterials in mitigating arsenic contamination has been increasingly recognized, offering promising, eco-friendly solutions for arsenic remediation.

1.1. Sources and Environmental Presence

Arsenic is present in two primary forms of oxidation in aqueous environments: arsenite (As^{3+}) and arsenate (As^{5+}). Arsenite (As III) has more toxicity and mobility than arsenate (As V), rendering its removal more challenging. Arsenic contamination in the environment arises from various natural and human-made sources. Natural arsenic sources comprise volcanic eruptions, the weathering of arsenic-laden rocks, and geothermal activity, whereas anthropogenic sources encompass industrial waste, pesticide runoff, and metal smelting. Geogenic processes, including the weathering of arsenic-bearing rocks, microbial reduction of iron oxides, and geothermal emissions, facilitate the spontaneous mobilization of arsenic into groundwater systems. Simultaneously, human activities—especially mining, the application of arsenic-based agrochemicals, industrial waste disposal, and fossil-fuel combustion—have markedly increased arsenic discharge into soils and aquatic systems. The approximate yearly arsenic emission (in kilotons per annum) is shown in Table 2.

Table 2. Global arsenic emissions from various sources [22].

Category	Source	Arsenic (kt/a)
Natural	Windblown dusts	2.6
	Sea salt spray	1.7
	Volcanoes	3.8
	Forest fires	0.19
Anthropogenic	Fossil-fuel combustion	0.81
	Non-ferrous metal production	3.46
	Iron and steel production	0.35
	Cement production	0.27
	Waste disposal	0.12
Biogenic	Continental particulates	0.26
	Continental volatiles	1.3
	Marine	2.3

Figure 1 depicts the worldwide distribution of arsenic concentrations in groundwater, emphasizing areas with contamination levels ranging from below 1 μg/L to exceeding 100 μg/L. Regions including South Asia, certain areas of South America, and the southern United States exhibit notably elevated arsenic concentrations. The map also delineates locations of offshore disposal, land burial, and stockpiling of warfare-related arsenicals, which may present further environmental and public health hazards. This integrated perspective highlights the intricate and diverse origins of worldwide arsenic contamination. As illustrated, countries such as Bangladesh, India, Vietnam, and parts of China and Pakistan experience alarmingly high levels of arsenic in groundwater, often exceeding 100 μg/L—well above the World Health Organization's recommended safe limit of 10 μg/L. Similarly, regions in Argentina and Chile, along with the southwestern United States and parts of Mexico, report elevated concentrations ranging from 50 to 100 μg/L. Moderate contamination is seen across parts of Europe and Central Asia, while some regions in Africa and Oceania are beginning to show signs of emerging arsenic threats. In addition, the map identifies global hotspots for the storage and disposal of warfare-related arsenicals, particularly in areas of the United States, Western Europe, and the South Pacific. These sites raise serious concerns about long-term leaching and environmental toxicity. Collectively, the data reflect a complex interplay between naturally occurring geogenic sources and legacy pollutants, underscoring the urgent need for region-specific arsenic mitigation strategies [23].

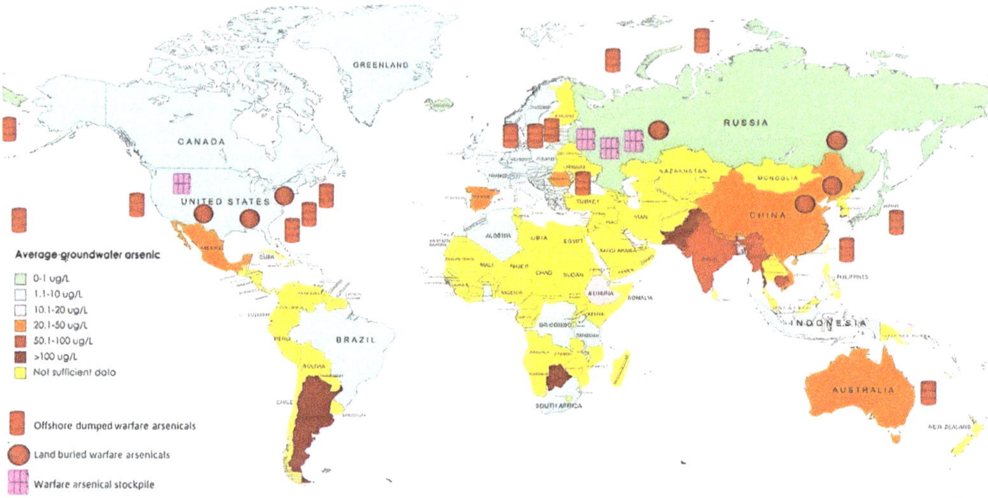

Figure 1. Global distribution of arsenic contamination in groundwater and locations of warfare-related arsenicals [23].

Bangladesh serves as one of the most striking examples of geogenic arsenic contamination, where natural geological processes release arsenic from sediments into shallow aquifers. With millions of people who mostly depend on groundwater from shallow tubewells, where arsenic naturally abounds, Bangladesh stands as a prime example of arsenic contamination on a worldwide scale. It serves as a valuable case for researching natural arsenic mobilisation mechanisms since its pollution is geogenic, unlike that of industrially polluted areas. One of the biggest mass poisonings in history was inadvertently caused by a public health campaign to prevent aquatic illnesses. This unintentional crisis reflects the challenges of balancing pathogen control with geochemical safety in water resource management.Its intricate interplay of public health, socioeconomics, infrastructural issues, and various mitigation attempts creates a rich, real-world backdrop for studying arsenic remediation [6,7].

Figure 2 illustrates the intricate regional distribution of arsenic contamination in groundwater across Bangladesh, based on probabilities derived from concentrations exceeding 10 μg/L, the WHO's threshold for safe drinking water. A color-coded gradient illustrates the following risk levels: red implies places with a very high possibility of contamination (0.99), orange shows moderate to high risk (0.55), yellow represents moderate risk (0.50), and pale yellow designates low-risk zones (0.01). The region's most adversely impacted are predominantly situated in the middle, southern, and southeastern areas of the country. The high-risk zones encompass portions of the Dhaka, Barishal, and Chattogram divisions, but the western and northern regions, including Rajshahi and Rangpur, exhibit markedly lower pollution levels. The Bay of Bengal delineates the southern frontier of the country, providing geographical orientation. This map serves as an essential instrument for pinpointing arsenic hotspots, directing mitigation strategies, and influencing safe water supply planning across Bangladesh.

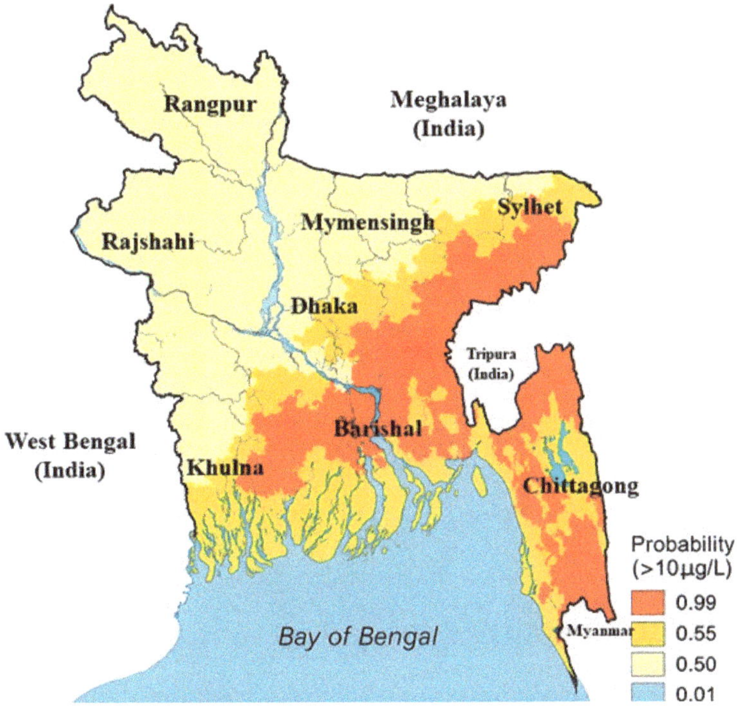

Figure 2. Geospatial distribution of arsenic contamination in tubewell water across Bangladesh (based on figurative data from [7,8,24]).

Worldwide, arsenic contamination in potable water represents a significant public health hazard, impacting between 150 to 220 million people, predominantly in South and Southeast Asia [23]. Importantly, countries like Bangladesh, India, China, Cambodia, and Vietnam also face severe risks, particularly in rural areas lacking centralized water treatment infrastructure. The persistence of arsenic in groundwater across these regions calls for scalable, cost-effective remediation technologies and robust groundwater governance frameworks.

1.2. Role of Bio-Based Nanomaterials in Groundwater Arsenic Remediation

The swift increase in the world population, along with rising energy and raw material use, has heightened apprehension over global warming, environmental pollution, and the exhaustion of natural resources. A revolutionary transition to a circular materials bioeconomy is necessary to maintain contemporary living standards while protecting the Earth. This transition underscores the importance of efficient recycling, upcycling, and the sustainable use of renewable resources [25]. In this context, bio-based nanomaterials have emerged as a highly promising substitute for traditional, fossil-derived materials. These eco-friendly nanostructures, sourced from renewable biological materials like plants, algae, bacteria, fungi, and agricultural waste, provide several benefits, including low production costs, biodegradability, reduced toxicity, and high surface reactivity, rendering them particularly effective for environmental remediation [26]. In contrast to conventional materials, bio-based nanomaterials are environmentally sustainable and biocompatible, while also promoting circular economy principles through the facilitation of greener production methods and the reduction of ecological footprints [27].

Bio-based materials are widely used in various aspects of research and for scientific purposes. Various types of bio-based nanomaterials, such as biochar, chitosan, chitin, and bio-nanocomposites, have demonstrated promising results in contaminated water-treatment applications. The choice of these materials depends on factors such as adsorption capacity, selectivity, reusability, and environmental impact [28,29].

Chitosan and cellulose-derived nanocomposites are efficient biopolymer materials for arsenic adsorption, attributable to their extensive surface area and functional groups that exhibit considerable affinity for arsenic ions. They are biodegradable, renewable, and ecologically friendly [30–34]. Biochar, derived from agricultural byproducts such as corn stalks or rice husks, demonstrates exceptional efficacy in the removal of arsenic and other heavy metals owing to its extensive surface area, porosity, and surface functional groups [35]. Biochar-based composites enhance structural stability and adsorption kinetics, also being applicable to the redox reaction of arsenic-remediation mechanisms, rendering them exceptionally appropriate for field applications in economically disadvantaged regions [36–39]. For instance, agricultural bio-waste-derived biosorbents (orange peel, banana peel, rice husk) and biochar have been utilized for arsenic remediation from As-containing solutions and As-contaminated groundwater, with the results demonstrating efficient adsorption of arsenic at different flow rates and contact durations [40]. Nanomaterials generated from algae, particularly green and brown varieties, possess a high concentration of functional groups like hydroxyl and carboxyl, which enhance arsenic adsorption [40,41]. These naturally derived materials can be transformed into hydrogels or powders, enhancing surface contact with metal ions. Their renewability and minimal toxicity render them optimal for sustainable treatment options [42].

Lignin, a new emerging and underutilized material enriched in functional groups like hydroxyl, methoxy, and phenolic groups, is effective for arsenic adsorption both as a standalone biosorbent and when used in biochar or composite forms. Studies also emphasize that chemical modifications, such as oxidation or metal ion functionalization (e.g., Fe, Zn), significantly enhance lignin's adsorption capacity for As (III) and As (V) through mechanisms including surface complexation and ion exchange [43,44]. Iron oxide supported on bio-derived nanoporous carbon exhibits significant potential for water purification. The magnetic properties facilitate straightforward separation, while the biogenic carbon improves environmental compatibility [45].

Biogenic nanoscale zero-valent iron (nZVI), produced from plant extracts like green tea and eucalyptus, has exhibited remarkable efficacy in arsenic removal. The biogenic nZVI particles exhibit improved dispersion and reduced agglomeration owing to the natural stabilizers included in the plant extracts [46–48]. Rice husk, a plentiful agricultural byproduct, can be treated to obtain nano-silica [49], which functions as an efficient adsorbent for arsenic removal from polluted water. The green synthesis of silver nanoparticles with plant extracts provides an environmentally sustainable method for nanoparticle manufacture [50]. The extensive surface area and functional groups in nano-silica enhance the adsorption of arsenic ions, providing an economical and sustainable method for water filtration. These biosynthesized silver nanoparticles demonstrate substantial antibacterial characteristics and have prospective applications in water-treatment procedures, including the elimination of arsenic and other diseases [50–52]. Table 3 summarizes various bio-based nanomaterials along with their natural sources, nanostructures or forms, targeted arsenic species, and the corresponding remediation mechanisms employed for arsenic removal.

Table 3. Bio-based nanomaterials with their derived sources used in various arsenic-remediation mechanisms.

Bio-Based Nanomaterial	Derived Sources	Possible Nanostructure/Form	As Species Targeted	Remediation Mechanism(s)	Reference
Chitosan	Shrimp shells, fungi	Nanoparticles, beads, films	As (V), As (III)	Adsorption, coagulation/flocculation ion exchange	[30–32]
Cellulose	Plant biomass	Aerogels, membranes	As (V)	Adsorption	[33,34]
Modified Biochar	Rice husk, sawdust	Porous nanosheets, powder	As (III), As (V)	Adsorption, redox reaction	[35–39]
Algae-Based	Green/brown algae	Hydrogel, nanopowder	As (III), As (V)	Complexation, ion exchange	[40–42]
Lignin-Based	Forestry/agricultural waste	Nanoparticles	As (III), As (V)	Adsorption, complexation, ion exchange	[43,44]
Biogenic nZVI	Green tea, eucalyptus	Zero-valent iron nanoparticles	As (V)	Reduction, adsorption	[45–48]
Nano-silica	Algae, agricultural residue	Nanoparticles	As (V)	Adsorption	[50–52]

Recent studies and reports reveal a troubling rise in arsenic levels in groundwater, with severe and wide-ranging impacts on human health. The growing body of evidence underscores the urgent and foremost need for effective mitigation strategies and sustainable technologies to address arsenic contamination on a global scale. Bio-based nanotechnology has opened new avenues for arsenic remediation by offering highly efficient and selective materials that can remove arsenic from water. This review aims to illustrate various mechanisms of arsenic remediation in water, such as adsorption, filtration/membrane technology, photocatalysis, redox reactions, complexation, ion exchange, and coagulation–flocculation, along with their implications, advantages, and disadvantages. This review also discusses challenges related to the scalability, stability, and regeneration of these bio-based systems, followed by advancements in material design, surface modifications, and hybrid systems that enhance performance. The review highlights prospects such as the integration of bio-based nanomaterials with smart sensing technologies, the development of sustainable water-treatment frameworks, smart design, and life-cycle integration for arsenic adsorbents.

2. Mechanisms of Arsenic Remediation

There are different types of bio-based nanomaterials that can be used for arsenic remediation because of their essential properties and mode of action [33,39,46]. Figure 3 shows arsenic remediation from groundwater. The main mechanisms include, but are not limited to, adsorption, filtration/membrane/photo catalysis, redox reactions, complexation, ion exchange, and coagulation–flocculation. An overview of all these arsenic-remediation methods has been presented in a table following the illustrations, detailing each mechanism, along with the nanomaterials used, mode of action, advantages, and disadvantages, to assess their applicability. At the end of the discussion, a succinct summary is also provided, focusing on the applicability in terms of ease of implementation, operational cost, maintenance requirements, reusability, selectivity, environmental impact, scalability, and overall arsenic removal efficiency.

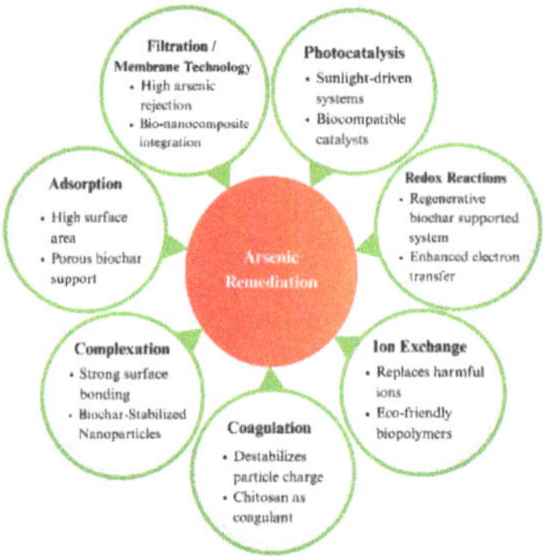

Figure 3. Different mechanisms of arsenic remediation using bio-based nanomaterials.

2.1. Adsorption

Adsorption is a process in which molecules bind (adsorbate) individually to the surface of a solid material (adsorbent) without penetrating its internal structure. Adsorption can take several forms: physical, chemical, and functionalized surface adsorption. The process begins with a transfer of heavy metal ions from the solution to the adsorbent surface, where the ions are adjoined either physically or chemically. The adsorbent must have a large surface area to preferentially and efficiently remove metals [30,36,37]. Chemisorption forms stronger ionic or covalent bonds, while physisorption is characterized by weak van der Waals interactions. Electrostatic attractions, ion exchange, and surface complexation can also form during adsorption, but these primarily occur when more metal ions than binding sites are present on the adsorbent surface [53]. Surface-charged adsorbents attract oppositely charged arsenic species. For example, at particular pH values, iron-based adsorbents develop positive surface charges that attract negatively charged arsenate (As (V)) ions, thereby improving adsorption through electrostatic interaction [54]. During adsorption, the adsorbate is the material that covers the surface of the adsorbent. Arsenic species such as As (III) or As (V) are the adsorbates in arsenic remediation that bind to the surface of adsorbents. A layer of adsorbate, which commonly includes metal ions or other contaminants, accumulates on the surface of the adsorbent as a result of physical or chemical interactions during the adsorption process. Adsorption is seen as a cost-effective, versatile, and simple procedure, with many different adsorbents available from mineral, biological, and organic sources [55]. Some adsorbents may not be effective for every arsenic species or in the presence of competing ions, and the disposal of spent adsorbents can pose environmental challenges [56]. Chitosan and other bio-based materials have been utilized effectively to stabilize magnet-sensitive nanoparticles, thereby greatly improving arsenic removal effectiveness. These stabilized nanomaterials are perfect for practical water-treatment applications, as they have two main benefits: high adsorption performance and simple magnetic recovery [57]. Porous biochar-supported $MnFe_2O_4$ nanocomposites and iron oxide nanoneedle-decorated biochar fibers also show outstanding arsenic removal ability. Fast adsorption kinetics and high arsenic absorption are made possible by their large surface area, well-developed porosity, and high surface reactivity [58]. These materials allow for the rapid and effective binding of arsenic species by combining physical adsorption with surface complexation. They provide a high-performance, scalable, and sustainable solution to arsenic remediation in contaminated water systems. A range of adsorbents can eliminate heavy metals like arsenic from wastewater or untreated water and can include inexpensive and customized engineered products like metal-coated absorbents and nano-adsorbents [59].

Understanding how arsenic attaches to materials during water treatment is essential for designing better and safer water purification systems. This process is explained using two main ideas—adsorption kinetics and isotherms—as shown in Figure 4.

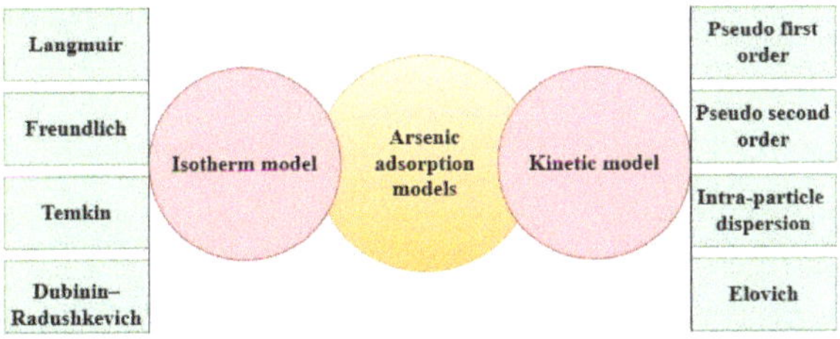

Figure 4. Different types of isotherm and kinetic models for arsenic adsorption.

Kinetics helps us understand how quickly arsenic adheres to a material and how the rate changes over time. Different models are used to describe this. For instance, the pseudo-first-order model is often used when the adsorption occurs quickly at first, usually due to physical interactions. The pseudo-second-order model works better when chemical bonding plays a major role, which is often the case with modified materials. There are also other models, like intraparticle diffusion, which examines how arsenic travels through tiny pores, and the Elovich model, which helps explain adsorption on uneven surfaces. Isotherm models, on the other hand, show us how much arsenic a material can hold once things settle into balance. The Langmuir model assumes that arsenic forms a single layer on a smooth surface, while the Freundlich model allows for more complex, multilayer adsorption on rougher surfaces. The Temkin model considers how heat changes during the process, and the Dubinin–Radushkevich model helps us deduce whether the bonding is physical or chemical [60–64]. Table 4 summarizes the adsorption performance of selected bio-based adsorbents for adsorbates As (III) and As (V) removal, detailing optimum pH, kinetic parameters (k_1, k_2, k_p, β, and q_e), surface area, and maximum adsorption capacities (Q_m) for comparative evaluation. Various bio-based adsorbents have shown strong potential for arsenic remediation, particularly when structurally enhanced to improve efficiency [65–72]. TiO_2-loaded biochar from Chinese medicine dregs shows a surface area of 128.22 m^2/g and a maximum adsorption capacity (Q_m) of 58.456 mg/g at an initial As (III) concentration of 80 mg/L, attributed to adsorption, photocatalysis, and inner-sphere complexation. The adsorption follows pseudo-second-order kinetics (k_2 = 0.0284 g/mg·min), suggesting chemisorption, with additional contributions from pseudo-first-order kinetics (k_1 = 0.0020 min^{-1}) and intra-particle diffusion (k_p = 1.6106 mg/g·$(min)^{1/2}$). The Elovich constant (β = 0.516 g/mg) indicates heterogeneous surface interactions [65]. A chitosan–magnetic graphene oxide nanocomposite exhibits a high surface area (152.38 m^2/g) and an adsorption capacity of 45 mg/g for As (III) at an optimal pH of 7.3, fitting the Langmuir isotherm and pseudo-second-order kinetics, indicating monolayer chemisorption [66]. Likewise, a chitosan–Fe-crosslinked complex (Ch-Fe) shows effective As (III) uptake (13.4 mg/g) at an optimal pH of 9.0. The adsorption process exhibits a pseudo-first-order rate constant of k_1 = 0.0024 min^{-1} with an equilibrium adsorption capacity (q_e) of 2.51 mg/g, and a pseudo-second-order rate constant of k_2 = 0.0042 g/mg·min with q_e = 1.69 mg/g. A notably high intra-particle diffusion constant (k_p = 42.05 mg/g·$(min)^{1/2}$) indicates that pore diffusion plays a dominant role in the overall uptake process [67].

Table 4. Summary of adsorption performance of selected bio-based adsorbents reported in previous studies.

Bio-Based Adsorbent	Adsorbate	Optimum pH	Kinetic Constants k_1: Pseudo 1st Order (/min); k_2: Pseudo 2nd Order (g/mg·min); k_p: Intra-Particle Diffusion (mg/g·(min)$^{1/2}$); β: Desorption Constant (g/mg) (Elovich Model); q_e: Equilibrium Adsorption Capacity (mg/g)	Surface Area (m^2/g)	Maximum Adsorption Capacity (Q_m) (mg/g)	Reference
TiO_2-loaded biochar	As (III)	–	$k_1 = 0.0020$ $k_2 = 0.0284$ $k_p = 1.6106$ $\beta = 0.516$ (For initial concentration 80 mg/L)	128.22	58.456	[65]
Chitosan–magnetic graphene oxide nanocomposite	As (III)	7.3	$k_1 = 0.0767$ $k_2 = 0.0317$ $k_p = 0.5755$	152.38	45	[66]
Chitosan–Fe-crosslinked complex	As (III)	9.0	$k_1 = 0.0024$ ($q_e = 2.51$) $k_2 = 0.0042$ ($q_e = 1.69$) $k_p = 42.05$	–	13.4	[67]
Chitosan-coated bentonite	As (V)	–	$k_1 = 0.0117$ ($q_e = 0.002201$) $k_2 = 4.502$ ($q_e = 0.00834$) $k_p = 1.519 \times 10^9$	–	–	[68]
Control chitosan biosorbent beads (CCBB) and Magnetic chitosan biosorbent beads (MCBB)	As (III), As (V)	6.7	For CCBB, $k_1 = 0.000667$, $k_2 = 0.27$ [As (III)] $k_1 = 0.000975$, $k_2 = 0.17$ [As (V)] For MCBB, $k_1 = 0.00078$, $k_2 = 0.40$ [As (III)] $k_1 = 0.00076$, $k_2 = 0.32$ [As (V)]	38.27 (CCBB) 52.48 (MCBB)	For CCBB 18.87 As (III) 26.13 As (V) For MCBB 73.69 As (III) 79.49 As (V)	[69]
Chitosan quinoxaline Schiff base (CsQ) and cross-linked chitosan quinoxaline Schiff base (CsQG)	As (V)	7 (CsQ) 6 (CsQG)	For CsQ, $k_1 = 0.0027$, $K_2 = 00.064$ For CsQG, $k_1 = 0.0196$, $K_2 = 0.0195$	–	8.811 (CsQ) 31.95 (CsQG)	[70]

Table 4. *Cont.*

Bio-Based Adsorbent	Adsorbate	Optimum pH	Kinetic Constants k_1: Pseudo 1st Order (/min); k_2: Pseudo 2nd Order (g/mg·min); k_p: Intra-Particle Diffusion (mg/g·(min)$^{1/2}$); β: Desorption Constant (g/mg) (Elovich Model); q_e: Equilibrium Adsorption Capacity (mg/g)	Surface Area (m^2/g)	Maximum Adsorption Capacity (Q_m) (mg/g)	Reference
Aluminum-modified food-waste biochar	As (III)	–	$k_1 = 0.00496$ ($q_e = 19.5$) $k_2 = 0.000317$ ($q_e = 20.5$)	–	52.2	[71]
Zero-valent iron/biochar composite	As (III), As (V)	–	–	–	129.24 As (III) 127.15 As (V)	[72]

"–" indicates data not explicitly reported or available in the referenced studies.

Chitosan-coated bentonite exhibits a moderate As (V) adsorption capacity of 1.47 mg/g from contaminated groundwater at an initial concentration of 50.99 μg/L. The adsorption reaches equilibrium within 60 min and follows the Langmuir isotherm, indicating monolayer adsorption onto homogeneous sites. Kinetic data fit the pseudo-second-order model, suggesting chemisorption as the rate-limiting step, although the overall mechanism is a mix of physisorption and chemisorption. Despite a high intra-particle diffusion value ($k_p = 1.519 \times 10^9$ mg/g·(min)$^{1/2}$), the very low equilibrium capacities (q_e = 0.0022–0.0083 mg/g) and small rate constants (k_1 = 0.0117 min^{-1}, k_2 = 4.502 g/mg·min) reflect a slow and low-efficiency adsorption process [68]. Magnetic chitosan biosorbent beads demonstrate effective removal of both As (III) and As (V), with maximum monolayer adsorption capacities of 18.87 mg/g and 26.13 mg/g, respectively, under Langmuir isotherm conditions. The adsorption process follows pseudo-second-order kinetics, indicating chemisorption, and achieves over 99% removal efficiency at pH 6.7. The incorporation of Fe_3O_4 nanoparticles enhances the surface area and exposes functional groups ($-OH$ and $-NH_2$), contributing to improved performance. Additionally, the material shows good reusability with minimal iron leaching, indicating its potential for sustainable arsenic remediation [69]. Finally, Quinoxaline chitosan Schiff base (CsQ) achieves a maximum As (V) adsorption capacity of 8.81 mg/g at pH 7, exhibiting chemisorption-driven uptake with chemical interaction as the rate-limiting step. The material demonstrates a highly porous, wrinkled surface morphology that facilitates arsenic binding, with 98.36% removal efficiency. Compared to its cross-linked counterpart CsQG, CsQ shows superior performance in As (V) removal, attributed to the more accessible functional groups and higher surface activity [70]. Aluminum-modified food waste biochar (Al-FWB) achieves a high As (III) adsorption capacity of 52.2 mg/g, attributed to monolayer chemisorption behavior supported by pseudo-second-order kinetics and Langmuir isotherm fitting. Optimized pyrolysis conditions and aluminum content significantly enhance the sorption performance, while adsorption has been found to be pH-dependent and hindered by competing anions. The process is spontaneous and endothermic, highlighting Al-FWB as a cost-effective and sustainable option for arsenic remediation and food-waste valorization [71]. The zero-valent iron/biochar composite (BC-ZVI), synthesized through a one-step pyrolysis-loading process using $FeCl_3$-treated bamboo, demonstrates exceptional adsorption capacities of 129.24 mg/g for As (III) and 127.15 mg/g for As (V). The high performance can be attributed to its substantial iron loading (up to ~30%), nanoporous structure, and abundant active sites. Effective across a wide pH range and tolerant to competing ions, BC-ZVI also offers magnetic recoverability and reusability, indicating strong potential for practical arsenic remediation in various water matrices [72].

2.2. Filtration/Membrane Technology

Membrane-based treatment, as shown in Figure 5, using nanomaterials has emerged as a highly promising strategy for arsenic removal from water, primarily due to the enhanced filtration efficiency and selectivity offered by nanomaterials [73,74]. The mechanism is relatively simple: arsenic-contaminated water is purified as it passes through the membrane, where arsenic ions are removed through filtration. The four basic filtration processes in filtration/membrane technology include microfiltration, ultrafiltration, nanofiltration, and reverse osmosis with different pore sizes [75].

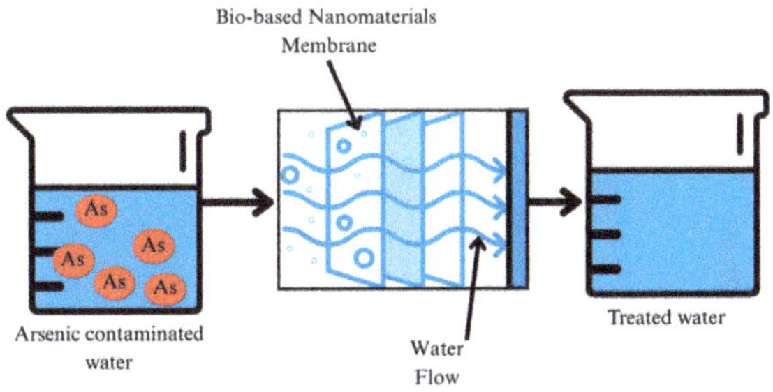

Figure 5. Membrane treatment for arsenic remediation.

Figure 6 illustrates how membrane filtration technologies work together when all the filtration is applied in a sequence of decreasing pore sizes of the filter to remove arsenic from groundwater, with each step offering more precise filtering as the pore size gets smaller. Microfiltration (MF) starts the process by removing large particles, but it does not target arsenic. Ultrafiltration (UF) goes a little further, filtering out viruses and organic matter, though it is still not very effective for arsenic on its own. Nanofiltration (NF), however, is much better at removing arsenate (As (V)). The final stage, reverse osmosis (RO), uses the tiniest pores to remove both forms of arsenic—arsenate and arsenite (As (III))—making it the most powerful option, though it requires more energy and cost. Together, these filtration methods offer a layered approach to cleaning groundwater and ensuring it is safe to drink [76,77]. Advanced membrane processes such as membrane distillation (MD), which operate under atmospheric pressure and controlled thermal conditions, have also proven effective in treating arsenic-contaminated water. MD is especially suited for producing high-quality water while retaining heavy metals like arsenic, commonly found in groundwater and soil water [78]. Despite these advantages, membrane technologies face several challenges. High initial costs associated with membrane fabrication and elevated operational expenses, especially in pressure-driven systems such as reverse osmosis and nanofiltration, pose significant barriers to widespread adoption [75]. Moreover, membrane fouling remains one of the most persistent issues, limiting long-term performance and increasing maintenance needs [79]. The effectiveness of these membranes is influenced by variables such as feed water pH, arsenic concentration, and the presence of competing ions [80,81]. Recent developments have concentrated on the development of organic/inorganic nanocomposites shown to possess improved physical-chemical characteristics for arsenic removal. These include those bio-nanocomposites that provide a thorough approach to arsenic remediation by combining oxidative and adsorptive capacities [76,82]. The use of bio-based polymeric membranes made from natural, renewable materials like chitosan, cellulose, starch, alginate, and lignin makes these environmentally friendly alternatives to synthetic plastics. Because they are biodegradable and biocompatible, they have a much lower environmental impact. The incorporation of nanomaterials such as metal oxides or graphene can improve the membranes' strength, stability, and ability to capture contaminants. These membranes work via mechanisms like physical filtration combined with adsorptive processes such as surface complexation, ion exchange, and redox reactions to remove pollutants, especially arsenic and heavy metals. Chitosan and cellulose membranes stand out for their effectiveness [83]. Bio-nanocomposites such as chitosan-coated magnetic nanoparticles and $MnFe_2O_4$-biochar have shown improved physicochemical qualities by combining oxidation (e.g., As (III) to As (V)) and high-capacity adsorption. Through synergistic processes including surface complexation and redox transformation, these materials offer better arsenic removal efficiency [82,84]. These membranes not only achieve high arsenic uptake but also maintain treated water within safe drinking standards. Nanofiltration membranes have shown impressive performance, with some studies reporting arsenate ion rejection rates as high as 99.80% [85].

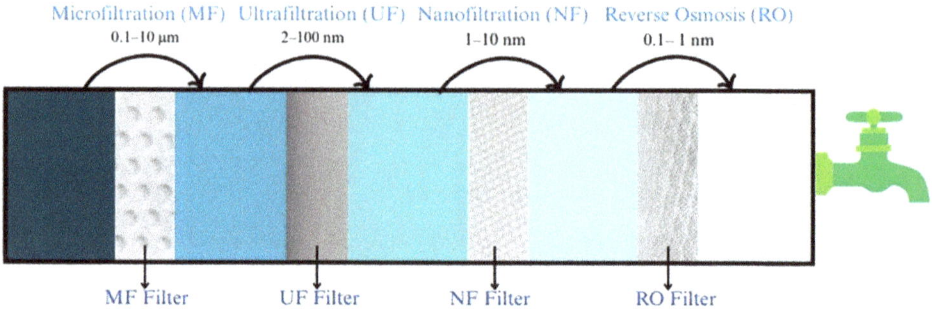

Figure 6. Different categories of filtration for arsenic remediation.

Table 5 shows different membrane filters used in membrane/filtration technology. The pore size has a significant effect on arsenic removal efficiency, signifying that reverse osmosis with the smallest pore size provides better efficiency.

Table 5. Membrane types, pore sizes, and their arsenic removal efficiencies.

Membrane Type	Pore Size	As Species	Initial As Concentration	pH Range	Removal Efficiency (%)	Reference
Microfiltration (MF)	0.1–10 μm	As (III)	20–50 mg/L	7.6–7.9	96%	[79,80]
Ultrafiltration (UF)	2–100 nm	As (V)	100 ng/L	8	>99%	[75,84,86]
Nanofiltration (NF)	1–10 nm	As (V)	0–200 μg/L	6.75	>99%	[80,81]
Reverse Osmosis (RO)	0.1–1 nm	As (III), As (V)	10–1000 μg/L	5.5–8.5	>99%	[75,87]

Removal Efficiency = [(initial concentration − final concentration)/initial concentration] × 100%.

While nanomaterial-enhanced membranes offer substantial improvements in arsenic rejection and operational efficiency, challenges like arsenic speciation, membrane stability, and compatibility with existing infrastructure must be addressed. With further development, these systems have the potential to offer scalable, efficient, and sustainable water purification solutions.

2.3. Photocatalysis

Arsenic remediation is being effectively and sustainably carried out through photocatalysis using bio-based nanomaterials, as shown in Figure 7. Photocatalysis rapidly oxidizes As (III) to As (V) by utilizing TiO_2 or other metal oxides under UV or visible light. This process follows zero-order kinetics and can achieve up to 97% oxidation within minutes [88,89]. This method improves arsenic removal through the processes of both oxidation and adsorption, using the intrinsic properties of nanomaterials, including high surface area, catalytic activity, and environmental compatibility. Photocatalytic methods can be limited by low efficiency under natural sunlight, the need for specific catalysts, and challenges in scaling up for large-scale applications [90].

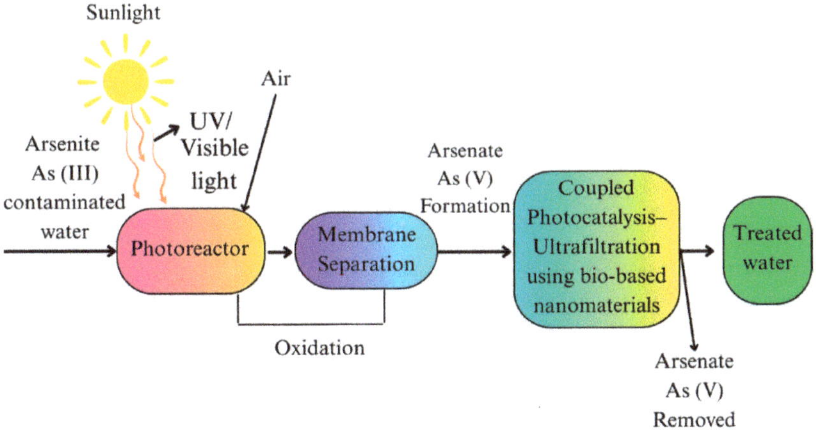

Figure 7. Photocatalysis process for arsenic remediation.

Photocatalytic nanomaterials' dual-function capacity is one of their main advantages. Dual-function remediation of arsenic provides a more effective and efficient solution for arsenic-contaminated environments by using technologies or materials that both transform and immobilize arsenic. Some advanced processes, such as solar light-driven Fe (III)/Fe (II) redox cycles, can convert organic arsenic compounds (like roxarsone) into inorganic arsenate and then immobilize the resulting arsenate through precipitation, thereby significantly reducing both the inorganic and organic arsenic hazards in a single process [91]. They can adsorb both species and oxidize As (III) into As (V), significantly increasing the general removal efficiency. Essentially, this is due to As (III) being more harmful and adaptable than As (V), and its conversion not only lowers toxicity but also enables simpler extraction from water systems. Increasingly, bio-based nanomaterials such as those derived from biopolymers, biochar, and other natural sources are preferred over traditional metal-based photocatalysts. Inherently bio-compatible, biodegradable, and environmentally friendly, these materials are compatible with goals for sustainable development. Moreover, under light exposure, they may generate reactive oxygen species, which are essential for arsenic oxidation. Surface modifications or doping may enhance their optical absorption and charge separation, therefore improving their photocatalytic performance [92]. Photocatalysts activated by ultraviolet (UV) or visible light produce reactive radicals, including hydroxyl (•OH) and superoxide (O_2^{-}•), which drive the oxidation process. Visible light ($\lambda > 420$ nm) can activate modified photocatalysts, enabling the use of sunlight and producing both superoxide and hydroxyl radicals for arsenic remediation [93,94]. Among the most promising candidates, bismuth oxyhalide-based nanomaterials, particularly bismuth oxyiodide (BiOI), have demonstrated outstanding photocatalytic performance under sunlight conditions [95]. Under natural sunlight, these materials efficiently lower arsenic levels to safe levels, therefore showing promise for low-cost, practical, energy-efficient water treatment. Compared to conventional metal oxide photocatalysts like TiO_2, BiOI has a smaller bandgap, which allows for improved light harvesting in the visible spectrum, a significant advantage for solar-driven systems. The hybridization of bio-based photocatalysts with other functional components, e.g., metal oxides and carbon nanotubes, has been shown to greatly improve performance [92]. These hybrids achieve higher removal rates and better long-term stability by using beneficial interactions such as improved charge transfer and stronger pollutant interaction.

By integrating environmental safety, dual-function remediation, and sunlight activation, bio-based nanomaterials present an exciting substitute for photocatalytic arsenic remediation. Their capacity to operate efficiently under sunlight, together with enhanced reusability and low environmental impact, is considered as a next-generation solution for decentralized and sustainable water treatment, particularly for arsenic-contaminated sources.

2.4. Redox Reactions

Bio-based nanomaterials have attracted significant attention as sustainable and efficient instruments for arsenic remediation in redox reactions. Their particular capacity to combine redox transformations with adsorptive removal makes them very effective in dealing with As (III) and As (V) contamination. Among the most significant benefits of bio-based nanomaterials is their capacity to convert As (III) into As (V), a less toxic and more stable form that is more easily adsorbed onto surfaces [46,96]. Because As (III), being neutral in many water conditions, is more difficult to remove than negatively charged As (V), this redox transformation is significant. However, this procedure may be ineffective and time-consuming, notably for converting As (III) to the more easily removable As (V). It often requires the addition of chemicals, which can increase costs and introduce secondary pollution risks [97]. Particularly when stabilized with biochar, iron and copper oxide nanoparticles have demonstrated strong performance in promoting this oxidation. The biochar matrix increases arsenic removal efficiency by performing as both a stabilizer and an extra adsorbent surface [46]. Manganese dioxide (MnO_2), when combined with biochar, also greatly increases the redox conversion of As (III) to As (V) [98]. MnO_2 provides the required surface functional groups that speed electron transfer, thereby enabling oxidative change. By combining redox activity with strong sorption capabilities, these hybrid systems exhibit better stability and multi-functionality than stand-alone metal oxides. Sourced from agricultural or forestry biomass, biochar provides a reasonably priced and environmentally beneficial platform for arsenic removal. Although unaltered biochar can absorb arsenic through weak interactions, surface modification such as loading with metal oxides or changing surface charges greatly increases its redox and adsorption capacity [99]. Its low environmental impact and reusability make it perfect for rural or large-scale arsenic-remediation systems.

Moreover, metal-organic frameworks (MOFs) with redox-active components such as ferrocene have recently surfaced as high-performance materials for targeted arsenic removal [100]. These MOFs provide precise adaptability to allow high-capacity adsorption and selective As (III) oxidation within one structure, something not readily possible with conventional materials. The performance of these materials is greatly affected by environmental factors, including pH and the coexistence of other ions. By lowering competition and electrostatic repulsion, slightly acidic to neutral pH conditions usually improve adsorption and redox processes. Mild alkaline solutions can regenerate many bio-based nanomaterials, especially biochar-based systems, thereby increasing their long-term use in sustainable water-treatment methods [101]. Bio-based nanomaterials offer a highly promising approach to arsenic remediation, as they provide dual-action benefits of adsorption and redox transformation. Their durability, affordability, and functional flexibility qualify them as possible solutions for global water contamination challenges.

2.5. Complexation

Bio-based nanomaterials for arsenic remediation are increasingly interesting because of their efficiency, sustainability, and ability to function through complexation-based methods, as shown in Figure 8. These mechanisms use the distinctive properties of nanomaterials to efficiently remove arsenic through the methods of adsorption, oxidation, electrostatic interactions, and surface complexation. The synthesis of biochar-stabilized metal oxide nanoparticles, such as biochar@Fe and biochar@Cu, is a major development in this field. By preventing nanoparticle aggregation, often considered with bare nanoparticles, these composites help preserve the high surface area necessary for effective arsenic adsorption [96]. With over 95% efficiency in arsenic removal, they provide a strong solution for practical use in water treatment. However, given that the stability of arsenic complexes and their removal from groundwater can be challenging, there is a risk of incomplete removal or secondary contamination in this process [102]. Biodegradable biopolymer chitosan is also used to stabilize magnet-sensitive nanoparticles. Their magnetic characteristics, which enable simple recovery post-adsorption, are a major benefit for scalable uses, causing these chitosan-coated nanomaterials to show improved reusability [57]. This not only increases operational efficiency but also reduces secondary waste production, an often-neglected limitation of conventional treatment methods.

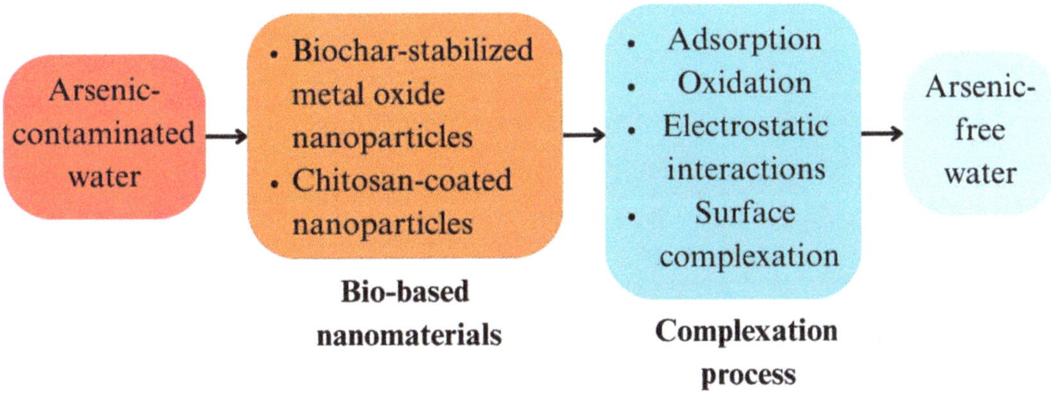

Figure 8. Complexation process for arsenic remediation.

A notable advancement involves $MnFe_2O_4$-biochar nanocomposites, which effectively adsorb both organic and inorganic arsenic species via synergistic mechanisms including electrostatic interactions and surface complexation processes [58]. In contrast to conventional single-metal adsorbents, the presence of bimetallic oxides promotes multi-modal interactions with arsenic, thus expanding the range of arsenic species that can be efficiently removed. Many bio-based nanomaterials, especially those with iron and copper oxides, go beyond physical adsorption to exhibit oxidative properties as well. These nanomaterials are more easily adsorbed and less toxic since they can oxidize As (III) into As (V) [46,96]. The elimination of the need for different pre-oxidation steps and streamlining the treatment process depends on this oxidation-adsorption synergy. The effectiveness of these nanomaterials is largely determined by surface complexation. Functional groups on the surfaces of bio-based materials, such as the hydroxyl, carboxyl, and amino groups, form strong complexes with arsenic ions, improving the selectivity and binding strength of the adsorbents [103]. Sometimes, nanomaterials show photocatalytic activity as well, which allows them to simultaneously oxidize and adsorb arsenic, therefore reducing the process of remediation [104]. Bio-based nanomaterials have significant potential in arsenic cleanup via several complexation-driven processes. Their diverse functionality, sustainability, and compatibility with natural water-treatment conditions establish them as advanced solutions for effective, environmentally friendly arsenic removal in both rural and urban water systems.

2.6. Ion Exchange

Ion exchange is a widely used process in water treatment in which harmful ions in a solution are replaced by other ions of similar charge from a solid material. The process is depicted in Figure 9. In the context of arsenic removal, ion exchange has proven to be an effective technique for extracting toxic metal ions such as arsenic from contaminated water [105].

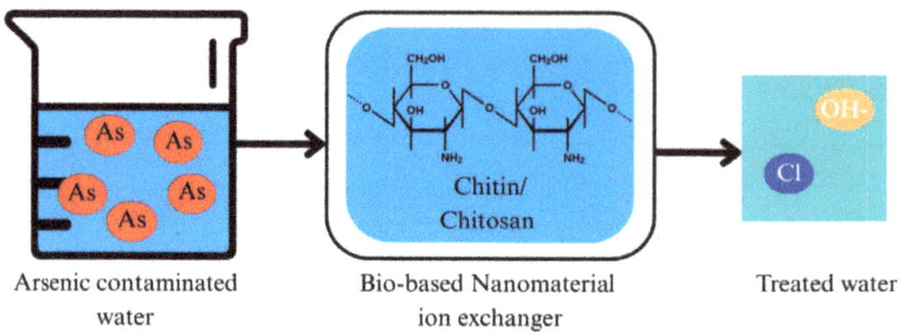

Figure 9. Ion exchange process for arsenic-contaminated water treatment.

Chitin and chitosan, biopolymers derived from the shells of crustaceans like shrimp and crabs, are gaining attention for their potential in arsenic remediation. These materials have natural properties that make them effective in ion exchange due to the presence of amino (-NH_2) and hydroxyl (-OH) groups in their chemical structure. When incorporated into composites or used in their natural form, chitin and chitosan interact with charged particles (ions) in water purification [106]. Chitosan is particularly effective due to its positively charged amino groups, which attract negatively charged ions like arsenate (AsO_4^{3-}). Through ion exchange, arsenate ions in contaminated water are swapped with less toxic ions, such as hydroxide or chloride, which are more readily removed through filtration or precipitation [107].

Chitin, though less soluble than chitosan, also shows potential for ion exchange. In its partially deacetylated form, it exposes amino groups that can interact with contaminants like arsenic. While less commonly used than chitosan, chitin-based materials, especially when modified into nanomaterials, have demonstrated effective metal ion adsorption [108].

2.7. Coagulation–Flocculation

Coagulation–flocculation is a water-treatment process that helps remove suspended particles, colloids, and dissolved contaminants such as arsenic. This is done by adding coagulants, which destabilize the charge of particles, causing them to aggregate into larger clusters, or "flocs" [109]. These flocs can then be easily removed by settling or filtration. Coagulation involves adding a coagulant to the water to neutralize the particles' charge, allowing them to clump together. On the other hand, flocculation is the gentle stirring of the water to encourage the formation of these larger flocs, which can then be removed from the water [110]. The coagulation process is illustrated in Figure 10. Chitosan is commonly used as a coagulant because of its natural flocculating properties. The amino groups in chitosan can bridge between particles, forming larger flocs that are easily removed. This makes chitosan particularly effective in removing colloidal particles, organic pollutants, and heavy metals like arsenic [111].

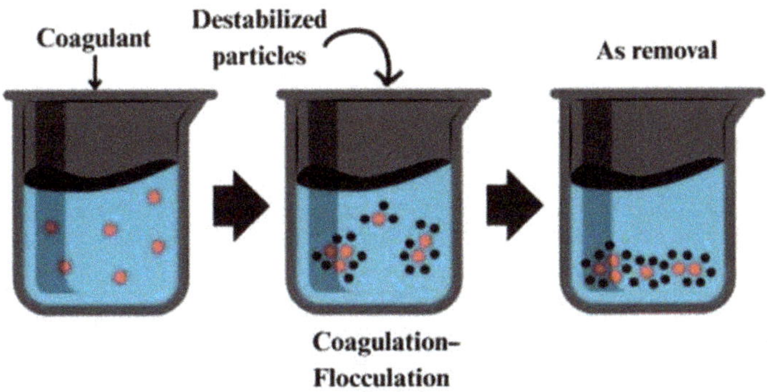

Figure 10. Coagulation–flocculation for arsenic-contaminated water treatment.

When modified into nanomaterials, chitin nanoparticles' effectiveness in particle aggregation improves significantly [112]. Chitin also has potential in coagulation–flocculation, though it is less frequently used than chitosan. Its particles offer several advantages in coagulation–flocculation. They are eco-friendly, biodegradable, and non-toxic, making them a safer alternative to synthetic coagulants like aluminum sulfate, which can be harmful to the environment [103]. These bio-based materials are highly efficient at removing a range of pollutants, including arsenic, due to their natural flocculating abilities and capacity to adsorb metal ions. Additionally, since they are derived from crustacean shells, they are both renewable and cost-effective, utilizing waste materials that would otherwise be discarded.

Nanomaterials are used in several methods to remediate arsenic. Table 6 offers a comparison of these techniques depending on the discussion of mechanisms of arsenic removal using nanomaterials. It highlights the main nanomaterials used in each method and the mode of action. Providing insight into their efficiency, selectivity, and practical application in removing arsenic from water sources, this summary presents a clear overview of how several nanomaterials operate across several remediation methods.

In summary, the mechanism selection for arsenic remediation is influenced by the unique limitations and context-dependent performance of each method. Filtration and membrane-based mechanisms are one of the most efficient and widely recognized for their precision in arsenic removal, with a percentage exceeding 99% through nanofiltration, operational continuity, and minimal chemical usage [85]. Adsorption is notable for its simplicity and cost-effectiveness, making it a promising option for future applications, particularly in resource-limited settings [102]. These advantages establish adsorption methods as strong candidates for future applications, especially in decentralized or rural water-treatment systems [96,113]. The development of bio-based adsorbents, such as chitosan derivatives and metal-modified biochar, has further enhanced the performance and feasibility of this method [114,115]. Ion exchange provides high selectivity for metal ions and the advantage of resin regeneration, but the overall process can be cost-prohibitive, which may restrict its practical application in broader or lower-budget contexts [115]. However, hybrid systems, like photocatalysis–adsorption, are gaining recognition due to their versatility, ease of use, and capacity to overcome the limitations of individual methods, providing more efficient and cost-effective solutions for arsenic removal [90].

Table 6. Different mechanisms of arsenic remediation using nanomaterials.

Method	Mechanism	Nanomaterials Used	Mode of Action	Advantages	Disadvantages	Reference
Adsorption	Binding of arsenic ions to adsorbent surfaces	Chitosan-coated NPs, $MnFe_2O_4$-biochar, iron oxide nanoneedles	Physical/chemical binding, surface complexation, electrostatic interactions	-Cost-effective and simple -High efficiency with materials like chitosan and biochar -Easy regeneration	-Limited selectivity in the presence of competing ions -Potential fouling after multiple uses	[58–60]
Filtration/ Membranes	Physical separation and selective transport using nanoporous structures	Nanofiltration membranes, bio-nanocomposites, TiO_2-coated, iron oxide nanofiber filters	Physical blocking, size exclusion coupled with adsorption	-High selectivity -Can achieve >99% arsenic removal -Good for continuous-flow systems	-Expensive fabrication and operation -Membrane fouling -Not always feasible for rural areas	[73,76,77]
Photocatalysis	Light-driven oxidation combined with adsorption	BiOI, TiO_2, biochar-based composites, carbon nanotube hybrids	Oxidation of As (III) to As (V), followed by adsorption	-Uses solar energy -Effective for As (III) to As (V) conversion	-Depends on light source -Some materials have low visible light efficiency	[88,89]
Redox Reactions	Oxidation-reduction conversion to less toxic and more adsorbable arsenic forms	Biochar@Fe/Cu, MnO_2-biochar, redox-active MOFs (e.g., ferrocene-based)	Redox transformation followed by adsorption	-Converts toxic As (III) to less toxic As (V) -Synergistic adsorption–redox activity	-May require specific pH conditions -Material degradation possible over time	[98–100]
Complexation	Formation of stable complexes between functional groups and arsenic ions	Biochar@Fe/Cu, chitosan-stabilized magnetic NPs, $MnFe_2O_4$ nanocomposites	Surface complexation via hydroxyl, carboxyl, and amino groups, sometimes aided by photocatalysis	-High binding affinity with functional groups -Can be selective and efficient	-Functional group leaching may reduce long-term efficiency -Surface modification can be costly	[104]
Ion Exchange	Replacement of arsenic ions (As^{5+}/As^{3+}) with functional groups (e.g., $-NH_{3+}$, –OH) present on the nanomaterial surface	Chitosan nanoparticles, chitin nanofibers	Amino and hydroxyl groups bind arsenic ions through electrostatic attraction and ligand exchange	-Good for low-concentration arsenic -Reversible process -Eco-friendly materials like chitosan	-Limited capacity -Sensitive to competing anions -Regeneration of chemicals may reduce sustainability	[107,108]
Coagulation–flocculation	Neutralization of surface charges and formation of aggregates that trap arsenic species	Modified chitosan (e.g., with Fe^{3+}, Al^{3+}), chitin-based nanoflocculants	Positively charged biopolymers interact with negatively charged arsenate/arsenite or suspended solids to form flocs	-Fast and scalable -Uses biodegradable coagulants like chitosan -Low-cost option	-Produces sludge -Less effective for dissolved arsenic -Needs post-treatment (filtration/sedimentation)	[109,110,112]

3. Challenges of and Advances in Arsenic Remediation

3.1. Challenges of Arsenic Remediation

Arsenic-contaminated water is a significant environmental and public health issue, particularly in South Asia, with Bangladesh being one of the most severely affected countries, where millions rely on arsenic-laden shallow groundwater for drinking; nanomaterials provide a potential solution [82,115]. However, their practical application presents different kinds of challenges. Knowing and customizing the physicochemical properties of nanomaterials to enhance their selectivity and efficacy in the removal of arsenic is a substantial challenge. Material engineering is highly application-specific due to the complex nature of the removal techniques, which frequently involve surface interactions, redox reactions, and ion transfer. Competing ions, such as phosphates, can also significantly reduce adsorption efficiency, thereby challenging field applications. A further significant issue is the long-term environmental and health effects of these materials, which remain unknown. Since they influence sustainability as well as cost-effectiveness, efficient separation and regeneration of nano-adsorbents still pose significant challenges. Dealing with these challenges requires a diverse effort to maximize material design, develop strong regeneration methods, and evaluate actual performance. Advancing this field requires a balance between invention and safety, therefore ensuring that nanotechnology provides consistent and sustainable water-treatment solutions for arsenic-contaminated water.

3.1.1. Stability and Selectivity

A significant current challenge is ensuring the chemical stability and selectivity of bio-based nanomaterials under actual arsenic-remediation conditions. Although many bio-based materials, such as chitosan-coated nanoparticles and biochar-supported metal oxides, perform well in controlled laboratory environments, their effectiveness usually drops in natural water systems [35,64,116]. Competing anions, which fight for the same active sites on the nanomaterial surface and thereby interfere with arsenic adsorption, are mostly responsible for this reduction. However, some biocompatible nanocomposites may still experience structural or chemical decomposition over time, given their natural biocompatibility and environmentally friendly nature. For example, microbial activity or extended exposure to changing pH and ionic conditions could cause biochar-based nanomaterials to change surface chemistry. Improving the selectivity and long-term reliability during operation of these materials in various environmental matrices by strengthening their surface functionalization and chemical durability is essential [117].

3.1.2. Affordable Scalability

Bio-based nanomaterials are often considered more accessible and sustainable alternatives for arsenic remediation, particularly in resource-limited settings. Many of these materials are made from biodegradable polymers such as chitosan, forestry byproducts, or agricultural waste, so they are affordable at a laboratory scale [118]. However, transitioning to large-scale deployment brings many challenges. This involves the requirement for constant quality in raw biomass substrates, the scaling up of functionalized composite synthesis, and the modification of these materials for application in continuous-flow systems rather than small-scale batch operations. Innovations in low-cost, decentralized treatment systems, as well as standardized synthesis protocols, are required to solve scalability limits [119]. For rural and remote areas harmed by arsenic contamination, modular filter units that use available bio-based nanomaterials could offer a practical way forward. Such systems could be particularly impactful in rural regions of Bangladesh, where affordable, decentralized arsenic treatment is crucial due to widespread groundwater dependence and infrastructure limitations.

3.1.3. Ability to Regenerate and Recycle

The reusability and simplicity of regenerating bio-based nanomaterials are important for the sustainable use of arsenic remediation from groundwater. Although many biopolymer- or biochar-based adsorbents initially show significant removal efficiencies, repeated use can cause surface fouling, structural deterioration, or irreversible arsenic binding, which reduces efficacy [120]. The regeneration of bio-based materials must avoid the use of strong chemicals that could compromise their environmental advantages. Some encouraging methods utilize low-temperature thermal treatments or mild alkaline solutions that restore adsorptive capacity without damaging the material or generating hazardous waste. The addition of magnetic bio-based composites, such as chitosan-coated magnetite nanoparticles, also enables simple recovery and reuse [121,122]. To ensure long-term applicability in arsenic-contaminated water treatment, further research is required to develop green regeneration methods while preserving structural integrity and performance over several treatment cycles.

3.2. Recent Advances in Arsenic Remediation

Bio-based nanomaterials such as chitosan and carbon have all shown enhanced adsorption and electrostatic interactions that are essential for the effective removal of arsenic from both water and sediment. Recent advancements in arsenic remediation have attracted significant attention to these materials [118]. In addition to their chemical and environmental durability, these nanoscale materials exhibit rapid reaction kinetics and exceptionally high removal efficiencies. Particle size is the main element that influences effectiveness. Research indicates that particles with a surface-to-volume ratio of less than 30 nm exhibit significantly greater chemical reactivity and superparamagnetic behavior [123]. This superparamagnetic material is well-suited for advanced separation technologies, as it enables nanoparticles to respond intensely to external magnetic fields while simultaneously losing their magnetism when the field is removed. In addition, their extensive surface area ensures the presence of a greater number of active sites for pollutant binding, thereby expediting the adsorption and reaction processes. For producing clean drinking water, treatment uses nanomaterials to efficiently remove organic and inorganic solutes, microorganisms, and hazardous metal ions from groundwater and contaminated water [124]. Their reactivity is also improved by quantum size effects, which modify the electronic structures of elements at the nanoscale. These properties taken together allow nanomaterials to exhibit quicker kinetics and more efficient arsenic adsorption, resulting in interesting possibilities for sustainable environmental remediation. Bio-based nanomaterials have surfaced as a very promising category of adsorbents because of their biodegradability, sustainability, and capacity to participate in different reactions, including adsorption, oxidation, electrostatic interactions, and surface complexation [99,104]. Significant advancements in material design, surface modification, and hybrid system development have allowed them to improve their performance. In addition to increasing arsenic removal efficiency, these developments address stability, selectivity, and reusability.

3.2.1. Bio- Nano- Materials Design

Material design mostly defines the performance of bio-based nanomaterials. The structural and chemical design of bio-based nanomaterials is a primary factor influencing their adsorption efficiency and selectivity in arsenic remediation. Recent developments highlight the engineering of the structure, composition, and porosity of these materials to maximize their interaction with both As (III) and As (V)—the two common forms of arsenic in water systems [35]. Among the most notable developments are engineered biochar-based composites such Biochar@Fe and Biochar@Cu [96]. These materials are created by soaking biochar with metal oxides to increase their stability and adsorption capacity. While the metal oxides provide active sites for arsenic interaction, biochar serves as a high-surface-area carbon scaffold. These composites reduce nanoparticle aggregation, one of the key challenges in using bare nanoparticles, and demonstrate enhanced adsorption kinetics. Bimetallic systems have helped to bring multimodal interactions into material design [125]. These systems have both adsorptive and oxidative characteristics, therefore allowing simultaneous transformation of As (III) into the more easily removable As (V) and its later acquisition on the adsorbent surface. Different environmental conditions determine the combined impact of the two metals, which broadens the spectrum of arsenic species that can be effectively removed. The addition of hollow nanostructures and core–shell denotes a major development in material engineering. For instance, porous biochar shells encircle a magnetic core, Fe_3O_4, of Fe_3O_4@biochar core–shell structures [58,126]. Such designs are not only efficient in adsorption and redox conversion but also compatible with hybrid remediation and membrane-based systems. Modern arsenic remediation technologies involve materials that are not only efficient but also reusable, stable, and flexible to actual water chemistry. These material design advances thus address both functional efficiency and operational viability and provide several remedial methods—including adsorption, redox reactions, and photocatalysis [93]—making them well-suited for deployment in high-risk areas like Bangladesh, where groundwater arsenic exposure poses a long-term health hazard. Emphasizing porosity, active site availability, and physical stability, these materials promise long-term performance and adaptation to various water chemistry conditions.

3.2.2. Surface Modifications

The adsorption efficiency, selectivity, and mechanism of arsenic removal are all greatly affected by surface chemistry. Recent advances in surface modifications aim to add engineered layers and functional groups to improve surface complexation and binding force [127,128]. The addition of bio-based nanomaterials such as chitosan, biochar, and cellulose enhances arsenic removal by introducing functional groups, including amino ($-NH_2$), carboxyl (-COOH), hydroxyl (-OH), and thiol (-SH), onto the nanomaterial surface is one commonly used approach. Modifying chitosan by combining biochar with $MnFe_2O_4$ improves adsorption, stability, and selectivity, enabling efficient, targeted arsenic binding in contaminated water. By means of ligand exchange, chelation, and electrostatic attraction, these groups interact strongly with arsenic ions, therefore enhancing both capacity and specificity [129]. For example, although the hydroxyl and amino groups are more effective in binding As (V), thiol groups have a strong affinity for As (III). These changes enhance the selectivity of the adsorbent under various pH, ionic strength, and competing ion concentrations, therefore increasing the appropriateness of the materials for real water conditions. Molecular imprinting techniques create recognition sites meant for arsenic ions. The surface modifications reduce nonspecific adsorption, allow for concentrated remediation, and raise the thermodynamic and kinetic properties of the arsenic adsorption process. These developments influence surface-level interactions regulating real efficiency beyond bulk material design.

3.2.3. Hybrid Systems

Hybrid systems are integrated platforms that combine the functions of several components to simultaneously carry out various methods such as adsorption, oxidation, magnetic separation, and even photocatalysis [90]. This multifunctionality is necessary for simplifying arsenic remediation procedures. Chitosan-coated Fe_3O_4 nanoparticles are examples of such systems. A biodegradable biopolymer, chitosan improves the biocompatibility and offers functional amino groups for arsenic binding. The Fe_3O_4 core offers a magnetic response, so it enables easy recovery and reuse. These systems demonstrate great adsorption efficiency with little secondary waste generation. Photocatalytic hybrids comprising TiO_2, ZnO, or graphitic carbon nitride (g-C_3N_4) offer light-activated redox characteristics. Under solar or UV radiation, these materials oxidize As (III) to As (V), a less toxic and more negatively charged species. This enhances subsequent adsorption efficiency, as As (V) exhibits a stronger affinity for the surface functional groups of many bio-based nanomaterials. This eliminates the need for separate oxidation processes, therefore reducing operational complexity and energy consumption. Materials such as $MnFe_2O_4$@biochar [58] or Fe_3O_4@chitosan combine adsorptive capacity with magnetic recovery, thus addressing two main concerns: performance and post-treatment separation. Perfect for scalable water filtration applications, hydrogel-based nanocomposites incorporating bio-nanoparticles into three-dimensional polymer matrices have also been shown to enhance water permeability, adsorption rate, and mechanical flexibility [39,105]. Hybrid systems combine several remedial mechanisms, boost operational scalability, and improve treatment efficiency per cycle. They demonstrate a shift from single-function materials to platforms capable of managing complex and dynamic contamination conditions.

4. Future Perspectives and Research Directions

Arsenic contamination in groundwater is a major health concern, especially in South Asia and other developing countries. Bangladesh, in particular, remains one of the most affected nations, where millions rely on arsenic-contaminated shallow tubewells [7,24]. Natural polymers have given rise to green engineered nanomaterials, which are efficient, affordable, and environmentally benign alternatives. For real-time arsenic detection and focused treatment, future research holds considerable promise, particularly when it comes to combining these materials with smart sensing technologies. The combination of modern digital technologies and chemical research is opening the door to fascinating new developments. Imagine developing intelligent systems that can not only identify arsenic in water in real time but also detect those areas where contamination is growing and take automatic action to treat it. This may be achieved by fusing bio-based nanomaterials with artificial intelligence and smart sensors. The way we monitor water quality could be improved by these AI and IoT-driven systems. The entire process can be sped up and made smarter and more efficient by using them to assist communities in making well-informed decisions, modifying treatment plans as necessary, and sending out timely alerts. As concerns about water safety and resource use grow, future arsenic removal must be sustainable, affordable, and eco-friendly. Using renewable materials, solar energy, and recyclable systems offers a smart, practical path, especially for rural communities. Finding safer and more sustainable ways to clean arsenic from water has become a top priority in environmental research. Additionally, there is increasing interest in creating environmentally friendly technologies that manage harmful substances. Scientists are now combining surface chemistry enhancements with computational modeling to design better materials that make water purification more effective and reliable.

4.1. Integration of Bio-Based Nanomaterials with Smart Sensing Technologies

Bio-based nanomaterials advanced with reactive chemical groups offer a quick, effective, and environmentally sustainable approach to detecting arsenic, which makes them ideal for on-site, real-time monitoring in arsenic-contaminated water treatment. Natural biopolymers such as polydopamine, chitosan, and cellulose were designed with the ability to measure and absorb arsenic with remarkable reliability [130]. Due to their high sensitivity, these modified materials can detect As (III) and As (V) in water. Their exceptional biocompatibility, non-toxic qualities, and natural availability make them particularly suitable for usage in remote or rural locations where conventional water testing equipment might not be accessible [32]. Building on these developments in materials science along with chemical science, the integration of Internet-of-Things (IoT) technology with bio-based sensors is transforming water quality monitoring, particularly in developing nations where arsenic contamination is a major issue [131]. To safeguard access to safe drinking water, these intelligent technologies can identify arsenic in real time and transmit data to central databases instantly, facilitating rapid action and improved decision-making. These days, there are wireless devices that can detect arsenic, send real-time data, and even autonomously initiate treatment operations, including modern biosensing materials such as chitosan-graphene composites or cellulose-based films. In rural and underprivileged regions, this significantly increases the certainty of water safety and minimizes the necessity for manual testing. Moreover, biopolymer sensors with DNA aptamers can accurately detect arsenic, and enzyme-based nanocomposites provide rapid color changes, making them ideal for easy-to-transport arsenic test kits. This technology is unique in that it combines nanomaterials with biosensing techniques that use enzymes or DNA, which increases sensitivity, speed, and accuracy [132]. These biosensing platforms now connect with smartphones and IoT devices, enabling on-the-spot testing, real-time tracking, and data sharing, making sustainable, community-based, and customized water safety solutions accessible.

4.2. Development of Sustainable Arsenic-Contaminated Water-Treatment Frameworks

Future arsenic-contaminated water treatment should focus on reusable, recyclable adsorbents to save resources and support sustainability. Bio-based nanomaterials made from things like rice husks, sawdust, or banana peels are a promising, low-cost option that is easy to find and can be used more than once [133]. To safely dispose of and regenerate arsenic-laden nano-adsorbents, research is essential. Improved regeneration techniques can cut expenses, decrease waste, and increase their useful life. In addition, proper recycling or disposal helps avoid secondary pollution, maintaining the sustainability and environmental friendliness of these bio-based solutions. To improve the efficiency of arsenic removal, integrating bio-based nanomaterials with solar-powered photocatalysis presents a smart and eco-friendly approach. It is a sustainable and energy-efficient solution for safer, cleaner drinking water because it uses sunlight to power the process instead of external power. Photothermal nanocomposites made from natural pigments and biomass-based carbon can soak up sunlight and turn it into heat, making them ideal for powering simple, energy-free water purification [134]. This approach is perfect for off-grid areas, since solar-powered photothermal nanomaterials provide an affordable, sustainable solution for improving arsenic removal in environments with limited resources. Such technologies are especially suitable for rural Bangladesh, where conventional energy and water-treatment infrastructure are limited. Building on this progress, modular, sustainable alternatives that integrate bio-based nanomaterials into membranes or filters should be the focus of water treatment in the future. High flexibility and the ability to remove arsenic make materials like chitosan, alginate, and cellulose-based hydrogels or aerogels useful, effective, and versatile for a range of community-scale applications [135]. Taken together, these innovative materials and technologies offer a complete approach to distributed, sustainable arsenic-remediation systems that are suitable for the demands of the world.

4.3. Smart Design and Life-Cycle Integration for Bio-Based Arsenic Adsorbents

Adding functional groups like amino, thiol, or phosphate to biopolymer-based adsorbents can significantly boost their ability to absorb arsenic, even when other ions are present. When these biopolymers are combined with metal oxides such as Fe_3O_4 or MnO_2, they not only offer a larger surface area but also form stronger bonds with arsenic. This makes them a smart, eco-friendly choice for removing arsenic from complex water sources [136].

In addition, computational modeling techniques like Density Functional Theory (DFT) and Molecular Dynamics (MD) simulations play a crucial role in providing a detailed, atomic-level understanding of adsorption mechanisms. By simulating how molecules interact at the surface of materials, these methods help uncover the fundamental processes that govern material behavior. This deeper insight is essential for the intelligent design of nanomaterials that not only perform better but also exhibit greater durability over time. Such approaches pave the way for developing more efficient, long-lasting materials tailored to specific applications [137].

4.4. Green Synthesis and Life-Cycle Assessment (LCA)

Future research should prioritize eco-friendly fabrication methods for nanomaterials, exploring green synthesis approaches that utilize plant extracts, microorganisms, and low-energy processes. By avoiding toxic solvents and relying on room-temperature techniques, these methods can significantly reduce environmental impacts, making nanomaterial production more sustainable and environmentally responsible [138]. Moreover, LCA tools play a crucial role in evaluating the energy, materials, and emissions involved in the production and use of bio-based nanomaterials. By introducing LCA early in the design process, we can make more informed choices about sustainable materials and ensure that the entire system is designed with environmental impact in mind from the start. This approach helps guide decisions that support both performance and sustainability [139]. To incorporate bio-based nanomaterials into national arsenic-remediation water-treatment programs and secure financing and support for their broad usage, authorities in Bangladesh must work together. For long-term success, it is equally critical to establish trust, educate communities about the efficacy and safety of these materials, and promote involvement in distributed water purification processes.

5. Conclusions

Bio-based nanomaterials offer a feasible and sustainable approach to arsenic remediation, providing a high-performing, affordable, and non-toxic substitute for conventional methods. By employing substances like biochar, bio-based aerogels, and modified biopolymers, these nanomaterials have been observed to be highly effective in eliminating arsenic using several types of techniques, including complexation, adsorption, membrane filtration, photocatalysis, redox reactions, ion exchange, and coagulation–flocculation. The selection of arsenic-remediation mechanisms depends on specific limitations and application contexts. Membrane filtration offers high precision but requires an expensive arrangement; adsorption is cost-effective and enhanced by bio-based materials; ion exchange provides selectivity but is limited by cost; and emerging hybrid systems, like photocatalysis–adsorption, present versatile and efficient alternatives. Practical issues, like limited scalability, long-term stability, and regeneration efficiency, continue to be significant obstacles to wider adoption. Recent developments in hybrid material systems, surface functionalization, and creative design techniques have greatly improved the efficiency and usefulness of bio-based nanomaterials. Looking ahead, the combination of these materials with sustainable arsenic-remediation systems and intelligent sensing technologies has the potential to completely transform arsenic-remediation methods around the world. The safe and universal implementation of bio-based nanotechnologies for access to safe and clean water around the world depends on ongoing interdisciplinary research and development to close current gaps.

Author Contributions: Conceptualization, M.M.R. and M.N.U.; methodology, M.M.R., M.M.H.P. and M.A.A.M.; formal analysis, M.M.R., M.M.H.P. and M.N.U.; investigation, M.M.R., M.M.H.P. and J.F.; resources, M.M.R., and M.N.U.; writing—original draft preparation, all authors; writing—review and editing, all authors; visualization, M.M.R. and M.M.H.P.; supervision, M.M.R. and M.N.U.; project administration, M.M.R. and M.N.U. All authors have read and agreed to the published version of the manuscript.

Funding: This research received no external funding.

Conflicts of Interest: The authors declare no conflicts of interest.

References

1. Shaji, E.; Santosh, M.; Sarath, K.V.; Prakash, P.; Deepchand, V.; Divya, B.V. Arsenic contamination of groundwater: A global synopsis with focus on the Indian Peninsula. *Geosci. Front.* **2021**, *12*, 101079. [CrossRef]
2. Bundschuh, J.; Schneider, J.; Alam, M.A.; Niazi, N.K.; Herath, I.; Parvez, F.; Tomaszewska, B.; Guilherme, L.R.G.; Maity, J.P.; López, D.L.; et al. Seven potential sources of arsenic contamination in Latin America and their environmental and health impacts. *Sci. Total Environ.* **2021**, *780*, 146274. [CrossRef] [PubMed]
3. Li, Y.; Bi, Y.; Mi, W.; Xie, S.; Ji, L. Land-use change caused by anthropogenic activities increase fluoride and arsenic contamination in groundwater and human health risk. *J. Hazard. Mater.* **2021**, *406*, 124337. [CrossRef]
4. Fatoki, J.O.; Badmus, J.A. Arsenic as an environmental and human health antagonist: A review of its toxicity and disease initiation. *J. Hazard. Mater. Adv.* **2022**, *5*, 100052. [CrossRef]
5. World Health Organization. Arsenic. Available online: https://www.who.int/news-room/fact-sheets/detail/arsenic (accessed on 20 March 2025).
6. Argos, M.; Kalra, T.; Rathouz, P.J.; Chen, Y.; Pierce, B.; Parvez, F.; Islam, T.; Ahmed, A.; Rakibuz-Zaman, M.; Hasan, R.; et al. Arsenic Exposure from Drinking Water, and All-Cause and Chronic-Disease Mortalities in Bangladesh (HEALS): A Prospective Cohort Study. *Lancet* **2010**, *376*, 252–258. [CrossRef]
7. Rahaman, M.S.; Mise, N.; Ichihara, S. Arsenic Contamination in Food Chain in Bangladesh: A Review on Health Hazards, Socioeconomic Impacts and Implications. *Hyg. Environ. Health Adv.* **2022**, *2*, 100004. [CrossRef]
8. Flanagan, S.V.; Johnston, R.B.; Zheng, Y. Arsenic in Tube Well Water in Bangladesh: Health and Economic Impacts and Implications for Arsenic Mitigation. *Bull. World Health Organ.* **2012**, *90*, 839–846. [CrossRef]
9. Thakur, B.K.; Gupta, V.; Bhattacharya, P.; Chakraborty, T. Impact of socioeconomic factors on households' willingness to pay for arsenic-free safe drinking water—A case study of Bihar, India. *Groundw. Sustain. Dev.* **2022**, *19*, 100837. [CrossRef]
10. Cervantes, G.I.V.; Esquivel, D.F.G.; Ortega, D.R.; Ayala, T.B.; Chávez, L.A.R.; López-López, H.E.; Salazar, A.; Flores, I.; Pineda, B.; Gómez-Manzo, S.; et al. Mechanisms associated with cognitive and behavioral impairment induced by arsenic exposure. *Cells* **2023**, *12*, 2537. [CrossRef]
11. Kuo, C.-C.; Moon, K.A.; Wang, S.-L.; Silbergeld, E.; Navas-Acien, A. The association of arsenic metabolism with cancer, cardiovascular disease, and diabetes: A systematic review of the epidemiological evidence. *Environ. Health Perspect.* **2017**, *125*, 087001. [CrossRef]
12. Ahmed, S.M.; Branscum, A.; Welch, B.M.; Megowan, M.; Bethel, J.W.; Odden, M.C.; Joya, S.A.; Hasan, M.O.S.I.; Lin, P.I.; Mostofa, G.; et al. A prospective cohort study of in utero and early childhood arsenic exposure and infectious disease in 4-to 5-year-old Bangladeshi children. *Environ. Epidemiol.* **2020**, *4*, e086. [CrossRef]
13. Mohapatra, B.; Saha, A.; Chowdhury, A.N.; Kar, A.; Kazy, S.K.; Sar, P. Geochemical, metagenomic, and physiological characterization of the multifaceted interaction between microbiome of an arsenic contaminated groundwater and aquifer sediment. *J. Hazard. Mater.* **2021**, *412*, 125099. [CrossRef]
14. Clark, A.J.; Labaj, A.L.; Smol, J.P.; Campbell, L.M.; Kurek, J. Arsenic and mercury contamination and complex aquatic bioindicator responses to historical gold mining and modern watershed stressors in urban Nova Scotia, Canada. *Sci. Total Environ.* **2021**, *787*, 147374. [CrossRef]
15. Cai, Y.; Wang, B.; Pan, F.; Fu, Y.; Guo, W.; Guo, Z.; Liu, H. Effects of manganese, iron and sulfur geochemistry on arsenic migration in the estuarine sediment of a small river in Xiamen, Southeast China. *Environ. Pollut.* **2022**, *293*, 118570. [CrossRef]
16. Tang, Z.; Zhao, F.-J. The roles of membrane transporters in arsenic uptake, translocation and detoxification in plants. *Crit. Rev. Environ. Sci. Technol.* **2021**, *51*, 2449–2484. [CrossRef]
17. Zhang, X.; Zhang, P.; Wei, X.; Peng, H.; Hu, L.; Zhu, X. Migration, transformation of arsenic, and pollution controlling strategies in paddy soil-rice system: A comprehensive review. *Sci. Total Environ.* **2024**, *951*, 175500. [CrossRef]
18. Hussain, M.M.; Bibi, I.; Niazi, N.K.; Shahid, M.; Iqbal, J.; Shakoor, M.B.; Ahmad, A.; Shah, N.S.; Bhattacharya, P.; Mao, K.; et al. Arsenic biogeochemical cycling in paddy soil–rice system: Interaction with various factors, amendments and mineral nutrients. *Sci. Total Environ.* **2021**, *773*, 145040. [CrossRef]
19. Gelaye, Y. Public health and economic burden of heavy metals in Ethiopia: Review. *Heliyon* **2024**, *10*, e39022. [CrossRef]

20. Yunus, F.M.; Khan, S.; Chowdhury, P.; Milton, A.H.; Hussain, S.; Rahman, M. A Review of Groundwater Arsenic Contamination in Bangladesh: The Millennium Development Goal Era and Beyond. *Int. J. Environ. Res. Public Health* **2016**, *13*, 215. [CrossRef]
21. Khan, N.; Hoque, S.F.; Mahmud, Z.H.; Islam, M.R.; Alam, M.A.U.; Islam, M.S.; Charles, K.J. Water Quality and Unseen Health Outcomes: A Cross-Sectional Study on Arsenic Contamination, Subclinical Disease and Psychosocial Distress in Bangladesh. *SSM Ment. Health* **2024**, *6*, 100344. [CrossRef]
22. Raju, N.J. Arsenic in the geo-environment: A review of sources, geochemical processes, toxicity and removal technologies. *Environ. Res.* **2022**, *203*, 111782. [CrossRef]
23. Muzaffar, S.; Khan, J.; Srivastava, R.; Gorbatyuk, M.S.; Athar, M. Mechanistic understanding of the toxic effects of arsenic and warfare arsenicals on human health and environment. *Cell Biol. Toxicol.* **2023**, *39*, 85–110. [CrossRef]
24. University College London. Arsenic. Available online: https://blogs.ucl.ac.uk/irdr/tag/arsenic/ (accessed on 20 March 2025).
25. Khan, W.S.; Asmatulu, E.; Uddin, M.N.; Asmatulu, R. *Recycling and Reusing of Engineering Materials, Recycling for Sustainable Developments*; Elsevier: Amsterdam, The Netherlands, 2022; ISBN 9780128224618.
26. Uddin, M.N.; Rab, M.F.; Islam, A.K.M.N.; Asmatulu, E.; Rahman, M.M.; Asmatulu, R. Nanostructured Hybrid Hydrogels for Solar-Driven Clean Water Harvesting from the Atmosphere. *Materials* **2022**, *15*, 7538. [CrossRef]
27. Uddin, M.N.; Desai, F.; Asmatulu, E. Engineered Nanomaterials in the Environment: Bioaccumulation, Biomagnification, and Biotransformation. *Environ. Chem. Lett.* **2020**, *18*, 1073–1083. [CrossRef]
28. Preetha, J.S.Y.; Arun, M.; Vidya, N.; Kowsalya, K.; Halka, J.; Ondrasek, G. Biotechnology Advances in Bioremediation of Arsenic: A Review. *Molecules* **2023**, *28*, 1474. [CrossRef]
29. Saud, A.; Gupta, S.; Allal, A.; Preud'homme, H.; Shomar, B.; Zaidi, S.J. Progress in the Sustainable Development of Biobased (Nano)materials for Application in Water Treatment Technologies. *ACS Omega* **2024**, *9*, 29088–29113. [CrossRef]
30. Anirudhan, T.S.; Sreekumari, S.S. Adsorption characteristics of heavy metal ions and arsenate on chitosan-based composite: Equilibrium and kinetics studies. *J. Hazard. Mater.* **2011**, *190*, 449–460. [CrossRef]
31. Hesami, F.; Bina, B.; Ebrahimi, A.; Amin, M.M. Arsenic removal by coagulation using ferric chloride and chitosan from water. *Int. J. Environ. Health Eng.* **2013**, *2*, 17. [CrossRef]
32. Chauhan, K.; Singh, P.; Sen, K.; Singhal, R.K.; Thakur, V.K. Recent Advancements in the Field of Chitosan/Cellulose-Based Nanocomposites for Maximizing Arsenic Removal from Aqueous Environment. *ACS Omega* **2024**, *9*, 27766–27788. [CrossRef]
33. Sinha, S.; Nigam, S.; Solanki, S.; Batra, L.; Chug, P.; Singh, R. Prospects on Arsenic Remediation Using Organic Cellulose-Based Adsorbents. *Ind. Crops Prod.* **2023**, *201*, 116928. [CrossRef]
34. Motloung, M.T.; Magagula, S.I.; Kaleni, A.; Sikhosana, T.S.; Lebelo, K.; Mochane, M.J. Recent Advances on Chemically Functionalized Cellulose-Based Materials for Arsenic Removal in Wastewater: A Review. *Water* **2023**, *15*, 793. [CrossRef]
35. Sharma, G.; Verma, Y.; Lai, C.W.; Naushad, M.; Iqbal, J.; Kumar, A.; Dhiman, P. Biochar and Biosorbents Derived from Biomass for Arsenic Remediation. *Heliyon* **2024**, *10*, e36288. [CrossRef]
36. Chatzimichailidou, S.; Xanthopoulou, M.; Tolkou, A.K.; Katsoyiannis, I.A. Biochar Derived from Rice by-Products for Arsenic and Chromium Removal by Adsorption: A Review. *J. Compos. Sci.* **2023**, *7*, 59. [CrossRef]
37. Chang Chien, S.W.; Weng, C.M.; Chou, J.S.; Liu, C.C. Application of δ-MnO_2 and Biochar Materials in an Arsenic-Contaminated Groundwater. *Water Environ. Res.* **2022**, *94*, e10811. [CrossRef]
38. Shaikh, W.A.; Alam, M.A.; Alam, M.O.; Chakraborty, S.; Owens, G.; Bhattacharya, T.; Mondal, N.K. Enhanced Aqueous Phase Arsenic Removal by a Biochar-Based Iron Nanocomposite. *Environ. Technol. Innov.* **2020**, *19*, 100936. [CrossRef]
39. Tabassum, R.A.; Shahid, M.; Niazi, N.K.; Dumat, C.; Zhang, Y.; Imran, M.; Bakhat, H.F.; Hussain, I.; Khalid, S. Arsenic Removal from Aqueous Solutions and Groundwater Using Agricultural Biowastes-Derived Biosorbents and Biochar: A Column-Scale Investigation. *Int. J. Phytoremediat.* **2019**, *21*, 509–518. [CrossRef]
40. Chowdhury, S.; Chowdhury, I.R.; Kabir, F.; Mazumder, M.A.J.; Zahir, M.H.; Alhooshani, K. Alginate-based biotechnology: A review on the arsenic removal technologies and future possibilities. *J. Water Supply Res. Technol. AQUA* **2019**, *68*, 369–389. [CrossRef]
41. Saleh, S.M.; Younis, A.M.; Ali, R.; Elkady, E.M. A Novel and Eco-Friendly Algae Amino-Modified Nanoparticles with Significant Environmental Effect for the Removal of As(III) and As(V) from Water. *Environ. Adv.* **2024**, *16*, 100550. [CrossRef]
42. Sandu, T.; Chiriac, A.L.; Zaharia, A.; Iordache, T.V.; Sarbu, A. New Trends in Preparation and Use of Hydrogels for Water Treatment. *Gels* **2025**, *11*, 238. [CrossRef]
43. Park, J.H.; Lee, J.H.; Lee, S.L.; Hwang, S.W.; Seo, D.C. Adsorption Behavior of Arsenic onto Lignin-Based Biochar Decorated with Zinc. *Colloids Surf. A Physicochem. Eng. Asp.* **2021**, *626*, 127095. [CrossRef]
44. Maia, L.C.; Soares, L.C.; Gurgel, L.V.A. A Review on the Use of Lignocellulosic Materials for Arsenic Adsorption. *J. Environ. Manag.* **2021**, *288*, 112397. [CrossRef] [PubMed]
45. Joshi, S.; Sharma, M.; Kumari, A.; Shrestha, S.; Shrestha, B. Arsenic Removal from Water by Adsorption onto Iron Oxide/Nano-Porous Carbon Magnetic Composite. *Appl. Sci.* **2019**, *9*, 3732. [CrossRef]

46. Ainiwaer, M.; Zhang, T.; Zhang, N.; Yin, X.; Su, S.; Wang, Y.; Zeng, X. Synergistic Removal of As(III) and Cd(II) by Sepiolite-Modified Nanoscale Zero-Valent Iron and a Related Mechanistic Study. *J. Environ. Manag.* **2022**, *319*, 115658. [CrossRef]
47. Angai, J.U.; Ptacek, C.J.; Pakostova, E.; Bain, J.G.; Verbuyst, B.R.; Blowes, D.W. Removal of Arsenic and Metals from Groundwater Impacted by Mine Waste Using Zero-Valent Iron and Organic Carbon: Laboratory Column Experiments. *J. Hazard. Mater.* **2022**, *424*, 127295. [CrossRef] [PubMed]
48. Alijani, H.; Shariatinia, Z. Effective Aqueous Arsenic Removal Using Zero Valent Iron Doped MWCNT Synthesized by In Situ CVD Method Using Natural α-Fe_2O_3 as a Precursor. *Chemosphere* **2017**, *171*, 502–511. [CrossRef]
49. Dang, N.T.T.; Nguyen, T.T.A.; Phan, T.D.; Tran, H.; Dang, P.V.; Nguyen, H.Q. Synthesis of Silica Nanoparticles from Rice Husk Ash. *Sci. Technol. Dev. J.* **2017**, *20*, 50–54. [CrossRef]
50. Mukherjee, A.; Sarkar, D.; Sasmal, S. A Review of Green Synthesis of Metal Nanoparticles Using Algae. *Front. Microbiol.* **2021**, *12*, 693899. [CrossRef]
51. Liu, P.; Liang, Q.; Luo, H.; Fang, W.; Geng, J. Synthesis of Nano-Scale Zero-Valent Iron-Reduced Graphene Oxide-Silica Nano-Composites for the Efficient Removal of Arsenic from Aqueous Solutions. *Environ. Sci. Pollut. Res.* **2019**, *26*, 33507–33516. [CrossRef]
52. Zeng, Q.; Zhong, H.; He, Z.; Hu, L. Efficient Removal of Arsenite by a Composite of Amino Modified Silica Supported MnO_2/Fe–Al Hydroxide (SNMFA) Prepared from Biotite. *J. Environ. Manag.* **2021**, *291*, 112678. [CrossRef]
53. Vievard, J.; Alem, A.; Pantet, A.; Ahfir, N.-D.; Arellano-Sánchez, M.G.; Devouge-Boyer, C.; Mignot, M. Bio-Based Adsorption as Ecofriendly Method for Wastewater Decontamination: A Review. *Toxics* **2023**, *11*, 404. [CrossRef]
54. Gupta, A.D.; Rene, E.R.; Giri, B.S.; Pandey, A.; Singh, H. Adsorptive and Photocatalytic Properties of Metal Oxides towards Arsenic Remediation from Water: A Review. *J. Environ. Chem. Eng.* **2021**, *9*, 106376. [CrossRef]
55. Ahmad, H.; Bibi, H.; Murugesan, C.; Ahmad, S.; Kyriakopoulos, G. Sustainable Wastewater Treatment Strategies in Effective Abatement of Emerging Pollutants. *Water* **2024**, *16*, 2893. [CrossRef]
56. Ungureanu, G.; Santos, S.; Boaventura, R.; Botelho, C. Arsenic and Antimony in Water and Wastewater: Overview of Removal Techniques with Special Reference to Latest Advances in Adsorption. *J. Environ. Manag.* **2015**, *151*, 326–342. [CrossRef]
57. Nisticò, R.; Celi, L.R.; Prevot, A.B.; Carlos, L.; Magnacca, G.; Zanzo, E.; Martin, M. Sustainable Magnet-Responsive Nanomaterials for the Removal of Arsenic from Contaminated Water. *J. Hazard. Mater.* **2018**, *342*, 260–269. [CrossRef]
58. Wen, Z.; Xi, J.; Lu, J.; Zhang, Y.; Cheng, G.; Zhang, Y.; Chen, R. Porous Biochar-Supported $MnFe_2O_4$ Magnetic Nanocomposite as an Excellent Adsorbent for Simultaneous and Effective Removal of Organic/Inorganic Arsenic from Water. *J. Hazard. Mater.* **2021**, *411*, 124909. [CrossRef]
59. Chakraborty, R.; Asthana, A.; Singh, A.K.; Jain, B.; Susan, A.B.H. Adsorption of heavy metal ions by various low-cost adsorbents: A review. *Int. J. Environ. Anal. Chem.* **2022**, *102*, 342–379. [CrossRef]
60. Musah, M.; Azeh, Y.; Mathew, J.T.; Umar, M.T.; Abdulhamid, Z.; Muhammad, A.I. Adsorption Kinetics and Isotherm Models: A Review. *Caliphate J. Sci. Technol.* **2022**, *4*, 20–26. [CrossRef]
61. Qiu, H.; Lv, L.; Pan, B.-C.; Zhang, Q.-J.; Zhang, W.-M.; Zhang, Q.-X. Critical Review in Adsorption Kinetic Models. *J. Zhejiang Univ. Sci. A* **2009**, *10*, 716–724. [CrossRef]
62. Azizian, S.; Eris, S. Adsorption Isotherms and Kinetics. In *Interface Science and Technology*; Elsevier: Amsterdam, The Netherlands, 2021; Volume 33, pp. 445–509. [CrossRef]
63. López-Luna, J.; Ramírez-Montes, L.E.; Martinez-Vargas, S.; Martínez, A.I.; Mijangos-Ricardez, O.F.; González-Chávez, M.D.C.A.; Vázquez-Hipólito, V. Linear and Nonlinear Kinetic and Isotherm Adsorption Models for Arsenic Removal by Manganese Ferrite Nanoparticles. *SN Appl. Sci.* **2019**, *1*, 977. [CrossRef]
64. Abiodun, O.-A.O.; Oluwaseun, O.; Oladayo, O.K.; Abayomi, O.; George, A.A.; Opatola, E.; Orah, R.F.; Isukuru, E.J.; Ede, I.C.; Oluwayomi, O.T.; et al. Remediation of Heavy Metals Using Biomass-Based Adsorbents: Adsorption Kinetics and Isotherm Models. *Clean Technol.* **2023**, *5*, 934–960. [CrossRef]
65. Yang, Y.; Zhang, R.; Chen, S.; Zhu, J.; Wu, P.; Huang, J.; Qi, S. Arsenic (III) Removal from Aqueous Solution Using TiO_2-Loaded Biochar Prepared by Waste Chinese Traditional Medicine Dregs. *RSC Adv.* **2022**, *12*, 7720–7734. [CrossRef] [PubMed]
66. Sherlala, A.I.A.; Raman, A.A.A.; Bello, M.M.; Buthiyappan, A. Adsorption of Arsenic Using Chitosan Magnetic Graphene Oxide Nanocomposite. *J. Environ. Manag.* **2019**, *246*, 547–556. [CrossRef] [PubMed]
67. Dos Santos, H.H.; Demarchi, C.A.; Rodrigues, C.A.; Greneche, J.M.; Nedelko, N.; Ślawska-Waniewska, A. Adsorption of As(III) on Chitosan-Fe-Crosslinked Complex (Ch-Fe). *Chemosphere* **2011**, *82*, 278–283. [CrossRef] [PubMed]
68. Yee, J.-J.; Arida, C.V.J.; Futalan, C.M.; de Luna, M.D.G.; Wan, M.-W. Treatment of Contaminated Groundwater via Arsenate Removal Using Chitosan-Coated Bentonite. *Molecules* **2019**, *24*, 2464. [CrossRef]
69. Ayub, A.; Raza, Z.A.; Majeed, M.I.; Tariq, M.R.; Irfan, A. Development of Sustainable Magnetic Chitosan Biosorbent Beads for Kinetic Remediation of Arsenic Contaminated Water. *Int. J. Biol. Macromol.* **2020**, *163*, 603–617. [CrossRef]

70. Abdelwahab, H.E.; Elhag, M.; El Sadek, M.M. Removal of As(V) and Cr(VI) Using Quinoxaline Chitosan Schiff Base: Synthesis, Characterization and Adsorption Mechanism. *BMC Chem.* **2024**, *18*, 225. [CrossRef]
71. Hashimi, S.Q.; Hong, S.-H.; Lee, C.-G.; Park, S.-J. Adsorption of Arsenic from Water Using Aluminum-Modified Food Waste Biochar: Optimization Using Response Surface Methodology. *Water* **2022**, *14*, 2712. [CrossRef]
72. Xu, L.; Zhu, S.; Feng, L.; Zhou, J.; Li, T.; Zhao, Z.; Wang, W. Fresh Biomass Derived Biochar with High-Load Zero-Valent Iron Prepared in One Step for Efficient Arsenic Removal. *J. Clean. Prod.* **2022**, *352*, 131616. [CrossRef]
73. Villena-Martínez, E.M.; Alvizuri-Tintaya, P.A.; Lora-García, J.; Torregrosa-López, J.I.; Lo-Iacono-Ferreira, V.G. Reverse Osmosis Modeling Study of Lead and Arsenic Removal from Drinking Water in Tarija and La Paz, Bolivia. *Processes* **2022**, *10*, 1889. [CrossRef]
74. Algieri, C.; Pugliese, V.; Coppola, G.; Curcio, S.; Calabro, V.; Chakraborty, S. Arsenic Removal from Groundwater by Membrane Technology: Advantages, Disadvantages, and Effect on Human Health. *Groundw. Sustain. Dev.* **2022**, *19*, 100815. [CrossRef]
75. Hamid, N.H.A.; Rushdan, A.I.; Nordin, A.H.; Norrrahim, M.N.F.; Muhamad, S.N.H.; Tahir, M.I.H.M.; Rosli, N.S.B.; Pakrudin, N.H.M.; Roslee, A.S.; Asyraf, M.R.M.; et al. A Review: The State-of-the-Art of Arsenic Removal in Wastewater. *Water Reuse* **2024**, *14*, 279–311. [CrossRef]
76. Siddique, T.A.; Dutta, N.K.; Choudhury, N.R. Nanofiltration for Arsenic Removal: Challenges, Recent Developments, and Perspectives. *Nanomaterials* **2020**, *10*, 1323. [CrossRef]
77. Gokcek, O.B.; Uzal, N. Arsenic Removal by the Micellar-Enhanced Ultrafiltration Using Response Surface Methodology. *Water Supply* **2020**, *20*, 574–585. [CrossRef]
78. Alkhudhiri, A.; Hakami, M.; Zacharof, M.-P.; Abu Homod, H.; Alsadun, A. Mercury, Arsenic and Lead Removal by Air Gap Membrane Distillation: Experimental Study. *Water* **2020**, *12*, 1574. [CrossRef]
79. Xiao, T.; Zhu, Z.; Li, L.; Shi, J.; Li, Z.; Zuo, X. Membrane fouling and cleaning strategies in microfiltration/ultrafiltration and dynamic membrane. *Sep. Purif. Technol.* **2023**, *318*, 123977. [CrossRef]
80. Wang, X.; Liu, W.; Li, D.; Ma, W. Arsenic (V) Removal from Groundwater by GE-HL Nanofiltration Membrane: Effects of Arsenic Concentration, pH, and Co-Existing Ions. *Front. Environ. Sci. Eng.* **2009**, *3*, 428–433. [CrossRef]
81. Boussouga, Y.-A.; Mohankumar, M.B.; Gopalakrishnan, A.; Welle, A.; Schäfer, A.I. Removal of Arsenic (III) via Nanofiltration: Contribution of Organic Matter Interactions. *Water Res.* **2021**, *201*, 117315. [CrossRef]
82. Bhattacharya, S.; Talukdar, A.; Sengupta, S.; Das, T.; Dey, A.; Gupta, K.; Dutta, N. Arsenic contaminated water remediation: A state-of-the-art review in synchrony with sustainable development goals. *Groundw. Sustain. Dev.* **2023**, *23*, 101000. [CrossRef]
83. Morales-Jiménez, M.; Palacio, D.A.; Palencia, M.; Meléndrez, M.F.; Rivas, B.L. Bio-Based Polymeric Membranes: Development and Environmental Applications. *Membranes* **2023**, *13*, 625. [CrossRef]
84. Li, K.; Lu, Y.; Zhou, C.; Liu, Z.; Luo, L.; Zhou, A.; Shao, S. Toward one-step As(III) removal in ultrafiltration with in situ BioMnOx cake layer: Mechanism and feasibility insights. *Water* **2025**, *273*, 123087. [CrossRef]
85. Harisha, R.S.; Hosamani, K.M.; Keri, R.S.; Nataraj, S.K.; Aminabhavi, T.M. Arsenic removal from drinking water using thin film composite nanofiltration membrane. *Desalination* **2010**, *252*, 75–80. [CrossRef]
86. Pookrod, Y.P.; Haller, K.J.; Scamehorn, J.F. Removal of arsenic anions from water using polyelectrolyte enhanced ultrafiltration. *Sep. Sci. Technol.* **2004**, *39*, 811–831. [CrossRef]
87. Pezeshki, H.; Hashemi, M.; Rajabi, S. Removal of arsenic as a potentially toxic element from drinking water by filtration: A mini review of nanofiltration and reverse osmosis techniques. *Heliyon* **2023**, *9*, e14246. [CrossRef] [PubMed]
88. Molinari, R.; Argurio, P. Arsenic Removal from Water by Coupling Photocatalysis and Complexation-Ultrafiltration Processes: A Preliminary Study. *Water Res.* **2017**, *109*, 327–336. [CrossRef] [PubMed]
89. Fontana, K.B.; Lenzi, G.G.; Seára, E.C.R.; Chaves, E.S. Comparison of Photocatalysis and Photolysis Processes for Arsenic Oxidation in Water. *Ecotoxicol. Environ. Saf.* **2018**, *151*, 127–131. [CrossRef]
90. Mahamallik, P.; Swain, R. A Mini-Review on Arsenic Remediation Techniques from Water and Future Trends. *Water Sci. Technol.* **2023**, *87*, 3108–3123. [CrossRef]
91. Chen, N.; Wang, X.; Wan, Y.; Luo, Y.; Huang, Y.; Zhang, L. Simulated Solar Light Driven Fe(III)/Fe(II) Redox Cycle for Roxarsone Degradation and Simultaneous Arsenate Immobilization. *J. Hazard. Mater.* **2020**, *394*, 121635. [CrossRef]
92. Kumar, A.; Sharma, G.; Naushad, M.; Al-Muhtaseb, A.A.H.; García-Peñas, A.; Mola, G.T.; Si, C.; Stadler, F.J. Bio-inspired and biomaterials-based hybrid photocatalysts for environmental detoxification: A review. *Chem. Eng. J.* **2020**, *382*, 122937. [CrossRef]
93. Wang, H.; Murugananthan, M.; Zhang, Y. Graphitic Carbon Nitride Based Photocatalysis for Redox Conversion of Arsenic(III) and Chromium(VI) in Acid Aqueous Solution. *Appl. Catal. B Environ.* **2019**, *248*, 349–356. [CrossRef]
94. Lei, D.; Zhang, W.; Chen, L.; Liu, X.; Li, Y.; Wang, Y.; Liu, J. Simultaneous Removal of Arsenic(III) and Chromium(VI) over $ZnFe_2O_4$ {100}/{111} Z-Scheme Photocatalyst: Facet-Dependent Active Site and Overlooked As(III)/Cr(VI) Complex. *J. Clean. Prod.* **2023**, *383*, 135493. [CrossRef]

95. Yang, Y.; Zhang, C.; Lai, C.; Zeng, G.; Huang, D.; Cheng, M.; Wang, J.; Chen, F.; Zhou, C.; Xiong, W. BiOX (X = Cl, Br, I) photocatalytic nanomaterials: Applications for fuels and environmental management. *Adv. Colloid Interface Sci.* **2018**, *254*, 76–93. [CrossRef] [PubMed]
96. Priyadarshni, N.; Nath, P.; Nagahanumaiah; Chanda, N. Sustainable removal of arsenate, arsenite and bacterial contamination from water using biochar stabilized iron and copper oxide nanoparticles and associated mechanism of the remediation process. *J. Water Process Eng.* **2020**, *37*, 101495. [CrossRef]
97. Moreira, V.R.; Lebron, Y.A.R.; Santos, L.V.S.; de Paula, E.C.; Amaral, M.C.S. Arsenic Contamination, Effects and Remediation Techniques: A Special Look onto Membrane Separation Processes. *Process Saf. Environ. Prot.* **2021**, *148*, 604–623. [CrossRef]
98. Cuong, D.V.; Wu, P.C.; Chen, L.I.; Hou, C.H. Active MnO_2/biochar composite for efficient As(III) removal: Insight into the mechanisms of redox transformation and adsorption. *Water Res.* **2021**, *188*, 116495. [CrossRef]
99. Yamen, R.; Bashir, H.; Bibi, I.; Shaheen, S.M.; Niazi, N.K.; Shahid, M.; Hussain, M.M.; Antoniadis, V.; Shakoor, M.B.; Al-Solaimani, S.G.; et al. A critical review on arsenic removal from water using biochar-based sorbents: The significance of modification and redox reactions. *Chem. Eng. J.* **2020**, *396*, 125195. [CrossRef]
100. Shi, W.; Ma, J.; Gao, F.; Dai, R.; Su, X.; Wang, Z. Metal-organic framework with a redox-active bridge enables electrochemically highly selective removal of arsenic from water. *Environ. Sci. Technol.* **2023**, *57*, 6342–6352. [CrossRef] [PubMed]
101. Singh, P.; Maiti, A. Optimized Synthesis and Characterization of Laterite–Biochar Composite for Arsenic Removal: Examining Colloidal Stability and As(III) Oxidation. *Biochar* **2024**, *6*, 17. [CrossRef]
102. Alka, F.; Shahir, S.; Ibrahim, N.; Ndejiko, M.J.; Vo, D.-V.N.; Manan, F.A. Arsenic Removal Technologies and Future Trends: A Mini Review. *J. Clean. Prod.* **2021**, *278*, 123805. [CrossRef]
103. Zhang, S.; Yi, X.; He, D.; Tang, X.; Chen, Y.; Zheng, H. Recent progress and perspectives of typical renewable bio-based flocculants: Characteristics and application in wastewater treatment. *Environ. Sci. Pollut. Res.* **2024**, *31*, 46877–46897. [CrossRef]
104. Siddiqui, S.I.; Naushad, M.; Chaudhry, S.A. Promising prospects of nanomaterials for arsenic water remediation: A comprehensive review. *Process Saf. Environ. Prot.* **2019**, *126*, 60–97. [CrossRef]
105. Sánchez, J.; Dax, D.; Tapiero, Y.; Xu, C.; Willför, S. Bio-based hydrogels with ion exchange properties applied to remove Cu(II), Cr(VI), and As(V) ions from water. *Front. Bioeng. Biotechnol.* **2021**, *9*, 656472. [CrossRef] [PubMed]
106. De Vargas Brião, G.; de Andrade, J.R.; da Silva, M.G.C.; Vieira, M.G.A. Removal of toxic metals from water using chitosan-based magnetic adsorbents: A review. *Environ. Chem. Lett.* **2020**, *18*, 1145–1168. [CrossRef]
107. Ahmed, M.A.; Mohamed, A.A. The use of chitosan-based composites for environmental remediation: A review. *Int. J. Biol. Macromol.* **2023**, *242*, 124787. [CrossRef] [PubMed]
108. Zia, Q.; Tabassum, M.; Gong, H.; Li, J. A review on chitosan for the removal of heavy metals ions. *J. Fiber Bioeng. Inform.* **2019**, *3*, 103–128. [CrossRef]
109. Kolya, H.; Kang, C.-W. Bio-based polymeric flocculants and adsorbents for wastewater treatment. *Sustainability* **2023**, *15*, 9844. [CrossRef]
110. Yang, R.; Li, H.; Huang, M.; Yang, H.; Li, A. A review on chitosan-based flocculants and their applications in water treatment. *Water Res.* **2016**, *95*, 59–89. [CrossRef]
111. Hsu, C.Y.; Ajaj, Y.; Mahmoud, Z.H.; Ghadir, G.K.; Alani, Z.K.; Hussein, M.M.; Hussein, S.A.; Karim, M.M.; Al-Khalidi, A.; Abbas, J.K.; et al. Adsorption of Heavy Metal Ions Using Chitosan/Graphene Nanocomposites: A Review Study. *Results Chem.* **2024**, *7*, 101332. [CrossRef]
112. Kurniawan, S.B.; Abdullah, S.R.S.; Imron, M.F.; Said, N.S.M.; Ismail, N.I.; Hasan, H.A.; Othman, A.R.; Purwanti, I.F. Challenges and Opportunities of Biocoagulant/Bioflocculant Application for Drinking Water and Wastewater Treatment and Its Potential for Sludge Recovery. *Int. J. Environ. Res. Public Health* **2020**, *17*, 9312. [CrossRef]
113. Shih, M.C. An Overview of Arsenic Removal by Pressure-Driven Membrane Processes. *Desalination* **2005**, *172*, 85–97. [CrossRef]
114. Sun, Y.; Yu, F.; Han, C.; Houda, C.; Hao, M.; Wang, Q. Research Progress on Adsorption of Arsenic from Water by Modified Biochar and Its Mechanism: A Review. *Water* **2022**, *14*, 1691. [CrossRef]
115. Rahman, M.S.; Reza, A.S.; Sattar, G.S.; Siddique, M.A.B.; Akbor, M.A.; Moniruzzaman, M.; Shafiuzzaman, S.M. Mobilization Mechanisms and Spatial Distribution of Arsenic in Groundwater of Western Bangladesh: Evaluating Water Quality and Health Risk Using EWQI and Monte Carlo Simulation. *Chemosphere* **2024**, *366*, 143453. [CrossRef] [PubMed]
116. Burbano, A.A.; Lassalle, V.L.; Horst, M.F.; Méndez, A. The Effect of Carbon Coating on the Arsenite Sorption by Magnetic Carbon Nanocomposites. *Int. J. Environ. Sci. Technol.* **2025**, *22*, 4749–4760. [CrossRef]
117. Feng, W.Y.; Wei, S.D.; Liu, C.B.; Chen, T.; Tang, Y.H.; Ma, J.H.; Yin, K.; Luo, S.L. Efficient removal of arsenic from groundwater using iron oxide nanoneedle array-decorated biochar fibers with high Fe utilization and fast adsorption kinetics. *J. Hazard. Mater.* **2019**, *373*, 141–151. [CrossRef]

118. Nikić, J.; Jazić, J.M.; Watson, M.; Agbaba, J. Application of Nanomaterials in Water Treatment: Arsenic and Natural Organic Matter Removal. *Recent Pat. Nanotechnol.* **2021**, *15*, 197–224. [CrossRef]
119. Huang, Y.; Zhang, J.; Ren, Z.; Xiang, W.; Sifat, I.; Zhang, W.; Zhu, J.; Li, B. Next Generation Decentralized Water Systems: A Water-Energy-Infrastructure-Human Nexus (WEIHN) Approach. *Environ. Sci. Water Res. Technol.* **2023**, *9*, 2446–2471. [CrossRef]
120. Lata, S.; Samadder, S.R. Removal of Arsenic from Water Using Nano Adsorbents and Challenges: A Review. *J. Environ. Manag.* **2016**, *166*, 387–406. [CrossRef]
121. Shaumbwa, V.R.; Liu, D.; Archer, B.; Li, J.; Su, F. Preparation and Application of Magnetic Chitosan in Environmental Remediation and Other Fields: A Review. *J. Appl. Polym. Sci.* **2021**, *138*, 51241. [CrossRef]
122. Alsaiari, N.S.; Alzahrani, F.M.; Katubi, K.M.; Amari, A.; Rebah, F.B.; Tahoon, M.A. Polyethylenimine-Modified Magnetic Chitosan for the Uptake of Arsenic from Water. *Appl. Sci.* **2021**, *11*, 5630. [CrossRef]
123. Chaturvedi, S.; Dave, P.N. Water Purification Using Nanotechnology: An Emerging Opportunity. *Chem. Methodol.* **2019**, *3*, 115–144. [CrossRef]
124. Rahman, M.M.; Khan, K.H.; Parvez, M.M.H.; Irizarry, N.; Uddin, M.N. Polymer Nanocomposites with Optimized Nanoparticle Dispersion and Enhanced Functionalities for Industrial Applications. *Processes* **2025**, *13*, 994. [CrossRef]
125. Lin, Y.; Jin, X.; Khan, N.I.; Owens, G.; Chen, Z. Bimetallic Fe/Ni Nanoparticles Derived from Green Synthesis for the Removal of Arsenic (V) in Mine Wastewater. *J. Environ. Manag.* **2022**, *301*, 113838. [CrossRef] [PubMed]
126. Chutia, G.P.; Phukan, K. Facile Synthesis of Fe_3O_4@biochar@SO_3H as Magnetically Separable Bronsted Acid Nanocatalyst for Biodiesel Production from Different Oil Feedstocks. *Ind. Crops Prod.* **2024**, *215*, 118578. [CrossRef]
127. Uddin, M.N.; Dhanasekaran, P.S.; Asmatulu, R. Synthesis, Characterization, and Applications of Polymer-Based Biomaterials. In *Advances in Nanotechnology*; Bartul, Z., Trenor, J., Eds.; Nova Science Publishers: New York, NY, USA, 2019; Volume 22, pp. 129–168.
128. Uddin, M.N.; Desai, F.; Asmatulu, E. Review of Bioaccumulation, Biomagnification, and Biotransformation of Engineered Nanomaterials. In *Nanotoxicology and Nanoecotoxicology*; Kumar, V., Guleria, P., Ranjan, S., Dasgupta, N., Lichtfouse, E., Eds.; Springer: Cham, Switzerland, 2021; Volume 67, pp. 129–168. [CrossRef]
129. Choi, J.; Kim, B.H. Ligands of Nanoparticles and Their Influence on the Morphologies of Nanoparticle-Based Films. *Nanomaterials* **2024**, *14*, 1685. [CrossRef] [PubMed]
130. Pakizeh, M.; Moradi, A.; Ghassemi, T. Chemical Extraction and Modification of Chitin and Chitosan from Shrimp Shells. *Eur. Polym. J.* **2021**, *159*, 110709. [CrossRef]
131. Zahedi, A.; Liyanapathirana, R.; Thiyagarajan, K. Biodegradable and Renewable Antennas for Green IoT Sensors: A Review. *IEEE Access* **2024**, *12*, 189749–189775. [CrossRef]
132. Yuan, M.; Li, C.; Wang, M.; Cao, H.; Ye, T.; Hao, L.; Wu, X.; Yin, F.; Yu, J.; Xu, F. Low-Cost, Portable, On-Site Fluorescent Detection of As(III) by a Paper-Based Microfluidic Device Based on Aptamer and Smartphone Imaging. *Microchim. Acta* **2023**, *190*, 109. [CrossRef]
133. Paranjape, P.; Sadgir, P. Heavy Metal Removal Using Plant Origin Biomass and Agricultural Waste-Derived Biomass from Aqueous Media: A Review. *Water Conserv. Sci. Eng.* **2023**, *8*, 9. [CrossRef]
134. Tang, Z.; Ma, D.; Chen, Q.; Wang, Y.; Sun, M.; Lian, Q.; Shang, J.; Wong, P.K.; He, C.; Xia, D.; et al. Nanomaterial-Enabled Photothermal-Based Solar Water Disinfection Processes: Fundamentals, Recent Advances, and Mechanisms. *J. Hazard. Mater.* **2022**, *437*, 129373. [CrossRef]
135. Varamesh, A.; Abraham, B.D.; Wang, H.; Berton, P.; Zhao, H.; Gourlay, K.; Minhas, G.; Lu, Q.; Bryant, S.L.; Hu, J. Multifunctional Fully Biobased Aerogels for Water Remediation: Applications for Dye and Heavy Metal Adsorption and Oil/Water Separation. *J. Hazard. Mater.* **2023**, *457*, 131824. [CrossRef]
136. Wang, J.; Zhuang, S. Removal of Various Pollutants from Water and Wastewater by Modified Chitosan Adsorbents. *Crit. Rev. Environ. Sci. Technol.* **2017**, *47*, 2331–2386. [CrossRef]
137. Wang, B.; Xuan, J.; Yang, X.; Bai, Z. Synergistic DFT-Guided Design and Microfluidic Synthesis of High-Performance Ion-Imprinted Biosorbents for Selective Heavy Metal Removal. *Colloids Surf. A Physicochem. Eng. Asp.* **2021**, *626*, 127030. [CrossRef]
138. Uddin, M.N.; Subeshan, B.; Rahman, M.M.; Asmatulu, R. Bioinspired Electrospun Nanocomposites: An Emerging Technology of Atmospheric Fog Water Generator. In Proceedings of the International Conference on Mechanical, Industrial and Energy Engineering, Khulna, Bangladesh, 19–21 December 2020.
139. Del Pilar Rodríguez-Rojas, M.; Bustos-Terrones, V.; Díaz-Cárdenas, M.Y.; Vázquez-Vélez, E.; Martínez, H. Life Cycle Assessment of Green Synthesis of TiO_2 Nanoparticles vs. Chemical Synthesis. *Sustainability* **2024**, *16*, 7751. [CrossRef]

nanomaterials

MDPI

Article

Natural Sunlight Driven Photocatalytic Degradation of Methylene Blue and Rhodamine B over Nanocrystalline Zn_2SnO_4/SnO_2

Maria Vesna Nikolic [1,*], Zorka Z. Vasiljevic [1], Milena Dimitrijevic [1], Nadezda Radmilovic [2], Jelena Vujancevic [3], Marija Tanovic [1] and Nenad B. Tadic [4]

[1] Institute for Multidisciplinary Research, University of Belgrade, 11030 Belgrade, Serbia; zorkav@imsi.bg.ac.rs (Z.Z.V.); milena.dimitrijevic@imsi.bg.ac.rs (M.D.); mtanovic@imsi.bg.ac.rs (M.T.)
[2] Institute of Nuclear Sciences of Vinca, National Institute of the Republic of Serbia, University of Belgrade, 11351 Belgrade, Serbia; nadezdas@vin.bg.ac.rs
[3] Institute of Technical Sciences of SASA, 11000 Belgrade, Serbia; jelena.vujancevic@itn.sanu.ac.rs
[4] Faculty of Physics, University of Belgrade, 11000 Belgrade, Serbia; nenad.tadic@ff.bg.ac.rs
* Correspondence: mariavesna@imsi.bg.ac.rs

Abstract: The natural sunlight driven photocatalytic degradation of organic pollutants is a sustainable solution for water purification. The use of heterojunction nanocomposites in this process shows promise for improved photodegradation efficiency. In this work, nanocrystalline Zn_2SnO_4/SnO_2 obtained by the solid-state synthesis method was tested as a heterojunction photocatalyst material for the degradation of methylene blue (MB) and Rhodamine B (RhB) dyes as single and multicomponent systems in natural sunlight. Characterization of the structure and morphology of the synthesized nanocomposite using X-ray diffraction (XRD), Fourier transform infrared spectroscopy (FTIR), field emission scanning electron microscopy (FESEM), transmission electron microscopy (TEM) combined with energy dispersive X-ray spectroscopy (EDS), and photoluminescence (PL) spectroscopy confirmed the formation of Zn_2SnO_4/SnO_2 and heterojunctions between Zn_2SnO_4 and the SnO_2 nanoparticles. A photodegradation efficiency of 99.1% was achieved in 120 min with 50 mg of the photocatalyst for the degradation of MB and 70.6% for the degradation of RhB under the same conditions. In the multicomponent system, the degradation efficiency of 97.9% for MB and 53.2% for RhB was obtained with only 15 mg of the photocatalyst. The degradation of MB occurred through N-demethylation and the formation of azure intermediates and degradation of RhB occurred through sequential deethylation and fragmentation of the xanthene ring, both in single and multicomponent systems.

Keywords: Zn_2SnO_4; SnO_2; heterojunction; photocatalysis; methylene blue; Rhodamine B; natural sunlight

1. Introduction

Photocatalysis, an advanced oxidation process involving the photocatalytic degradation of pollutants in waste water into harmless products such as CO_2 and water has been the focus of much scientific research due to the continuous and increasing pollution of water and the demand for a clean water supply [1,2]. Organic dyes, applied for coloring in different industrial processes, are a common pollutant [3,4]. The photocatalysis process involves a chemical reaction accelerated by light in the presence of a catalyst material [2].

Photocatalysis degrades organic pollutants into water and carbon dioxide as well as other inorganic substances that are commonly regarded as safe [4].

The photocatalysis process depends on the catalyst material used and the light source. Semiconducting metal oxides have been widely investigated as catalysts in photocatalysis due to their suitable band gap, abundance, versatility, non-toxicity, stability, and photocatalytic activity [5]. Metal oxides commonly investigated and applied for the photocatalytic degradation of organic dyes include TiO_2 [1] and ZnO [6]. When the surface of a semiconducting catalyst material dispersed in polluted waste water is illuminated with a light source with an energy equivalent or higher than the band gap energy of the semiconducting metal oxide catalyst, this leads to the activation of electrons from the valence band (VB) and their migration to the conduction band (CB) [2]. Positive holes remaining in the VB react with water molecules to form hydroxyl radicals. These radicals degrade the molecules of the organic dye pollutants into intermediates, and in some cases, into carbon dioxide (CO_2) and water (H_2O). Excited electrons in the VB can react with dissolved oxygen species, leading to the formation of superoxide radicals that also decompose pollutant species adsorbed on the catalyst semiconductor into intermediate products. Over the years, research has focused on improving the photocatalytic efficiency of metal oxide catalyst materials by modifying them with dopants, forming composites and heterostructures [3,5,7]. Thus, ZnO has been combined with SnO_2, which showed an improved degradation efficiency of the methylene blue (MB) and Rhodamine B (RhB) organic dyes under UV light [8]. Semiconductor heterojunction catalysis can combine semiconductors with other semiconductors, metals, carbon, or form multicomponent heterojunctions in order to improve the photocatalytic performance, especially in the visible light region [9]. Semiconductor-semiconductor heterojunctions, as heterostructure catalysts, represent a combination of two metal semiconductors with different properties (including ionization potential and electron affinity) [10,11]. When a heterostructure is irradiated by light, electrons are excited from the VBs to CBs of both catalysts. Photogenerated electrons are then injected from the catalyst with a higher CB edge to the catalyst with the lower CB edge, with simultaneous hole injections in the opposite direction, leading to increased charge separation and reduced electron-hole pair recombination [11]. Thus, ZnO/SnO_2 heterostructures treated by mechanical milling showed an improved photocatalytic degradation of MB in both UV and natural sunlight due to an increase in surface defects and their synergy with heterojunction ZnO/SnO_2 particles [12]. ZnO/Zn_2SnO_4 nanocomposites have shown improved photodegradation efficiency of MB and RhB in visible light, as the synergistic effect of the ZnO and Zn_2SnO_4 resulted in an improved lifetime of the photogenerated charge carriers and recombination reduction in photogenerated electron-hole pairs [13].

Wide band-gap semiconductors such as ZnO, SnO_2, and TiO_2 have shown good photocatalytic activity in ultraviolet (UV) light irradiation [2]. However, recent research has focused on achieving good photocatalytic properties in visible light, which represents a higher percentage of the solar spectrum. Research has also focused on applying metal oxide nanostructures for the photocatalysis of organic dyes such as MB in natural sunlight [12,14].

Zinc-tin-oxide is an environmentally friendly multicomponent oxide that can have two forms: rhombohedral perovskite $ZnSnO_3$ and n-type inverse spinel orthorhombic Zn_2SnO_4 [15]. Zn_2SnO_4 has high electron mobility, good stability, and a wide band gap, but a relatively unfavorable recombination rate for achieving efficient photocatalysis [16]. The composition, structure and morphology have an influence on the resulting properties of zinc tin oxides, including photocatalytic degradation of organic polluting dyes such as MB and RB [17]. Thus, high porosity was achieved in Zn_2SnO_4 synthesized with the aid of different carbon-based templates achieving 96% degradation of MB in 25 min [18]. The optical band gap value of zinc tin oxides depends on the particle size and morphology of the synthesized particles. An interesting feature is that photoluminescence studies of zinc tin oxides have shown the presence of energy levels that can be associated with oxygen vacancies. These sub-gap states enable absorption of photons whose energy is lower than the measured optical band gap and enable photocatalytic activity in visible light besides photocatalytic activity in the UV light region. Urbach energies have been determined for different zinc tin oxide morphologies and compositions, and they also infer possible absorption below the optical band gap [17].

Both Zn_2SnO_4 and SnO_2 are wide band gap semiconductors. An improved photocatalytic degradation of MB in UV light was achieved by Zn_2SnO_4/SnO_2 nanocomposites obtained by sonochemical synthesis combined with high-temperature calcination [19]. The CB edge of Zn_2SnO_4 is more negative than the CB edge of SnO_2, making SnO_2 a better electron acceptor. At the same time, the VB edge of SnO_2 is more positive than the VB edge of Zn_2SnO_4, leading to a more effective separation of photogenerated electron-holes and a smaller electron-hole recombination [20]. A one-pot hydrothermal method applied for the synthesis of the Zn_2SnO_4/SnO_x composite achieved improved photocatalytic degradation of methyl orange (MO) organic dye under UV light irradiation [21], while a hydrothermally synthesized Zn_2SnO_4/SnO_2 nanocomposite achieved an enhanced and stable photocatalytic degradation of RhB under UV light irradiation [22]. The molar ratio of Zn:Sn also influenced the resulting photocatalytic activity toward MB degradation under UV light irradiation [23]. Different composite Zn_2SnO_4/SnO_2 microstructures obtained using varied synthesis procedures have also been analyzed and include multi-shelled and hybrid structures [24] as well as hollow engineered microboxes [25], achieving an improved photocatalytic degradation of RhB under simulated sunlight. Hollow cubes of Zn_2SnO_4/SnO_2 showed a better photocatalytic degradation of MB and MO than solid ones in simulated sunlight [26].

The aim of this work was to analyze the use of nanocrystalline Zn_2SnO_4/SnO_2 obtained by a simple method of solid-state synthesis as a heterojunction photocatalyst material for the degradation of single and multicomponent systems of organic dyes in an aqueous medium in natural sunlight. Multicomponent systems containing a mixture of dyes or other active components may have different effects than single-component systems [8] and can show the competitive effects of different dyes in the mixtures [27]. The analysis of these differences is significant from the viewpoint of wastewater treatment processes in the textile industry, especially under natural sunlight [28]. The purpose was to analyze and elucidate the dye decomposition process and degradation mechanism in single-component systems of MB and RhB as well as the interaction of degradation and nanoparticle components in a multicomponent system of MB and RhB organic dyes under the simplest conditions of natural sunlight.

2. Materials and Methods

Nanocrystalline Zn_2SnO_4/SnO_2 was obtained by the solid-state synthesis of starting ZnO and SnO_2 nanopowders (both from Sigma Aldrich, St. Louis, MO, USA, particle size < 100 nm) consisting of mixing/homogenizing the two powders in a 1:1 M ratio with an agate mortar and pestle. Subsequent calcination was performed at 1050 °C for 5 h in a chamber furnace in air, followed by grinding with an agate mortar and pestle. This calcination temperature was selected to ensure Zn_2SnO_4 formation in line with previous TG/DTA analysis [29].

The structure of the obtained Zn_2SnO_4/SnO_2 nanocrystalline composite powder was analyzed using X-ray diffraction (XRD) spectroscopy on a Rigaku Ultima IV diffractometer

(Tokyo, Japan) in the range of 10–90° with a step of 0.02° and Fourier transform infrared (FTIR) spectroscopy measured on a Perkin Elmer Spectrum Two (Waltham, MA, USA) in the range of 400–4000 cm^{-1} with a resolution of 8 cm^{-1}. A Shimadzu UV-2600 device with an ISR2600 Plus Integrating sphere attachment (Kyoto, Japan) was used to measure the UV–Vis diffuse reflectance spectrum. The morphology was observed by field emission scanning electron microscopy (FESEM) on a Tescan MIRA3 XM (Brno, Czech Republic) and transmission electron microscopy (TEM) on an FEI Talos F200X microscope (Thermo Fischer Scientific, Waltham, MA, USA) combined with an energy dispersive X-ray spectroscopy (EDS) system. 3D and 2D photoluminescence (or fluorescence) spectra were acquired using a Fluorolog FL3-221 with a 450 mW Xe lamp (Jobin Yvon Horiba, Paris, France) and FluorEssence 3.5 software (Horiba Scientific, Kyoto, Japan) with the following settings for 3D: excitation range, 325–350 nm; emission range, 446–640 nm; increment 4 nm; slit (band pass) 2 nm for establishing excitation and emission spectra, and 2D spectra settings with an increment of 2 nm for acquiring an emission range of 350–460 nm (400–700 nm) for the excitation of 325 nm (385 nm). The emission detector signal was scaled by the reference quantum counter signal (S1/R1).

The photocatalytic activity of Zn_2SnO_4/SnO_2 was evaluated by monitoring the photodegradation of aqueous solutions of methylene blue (MB) and Rhodamine B (RhB) individually and combined in binary dye mixtures under direct natural sunlight in July and August 2022 between 11.00 a.m. and 01.00 p.m. (average daily temperature 30 ± 2 °C). The experiments were conducted on fully sunny days with no clouds, and the light intensity was measured as 800–1000 W/m^2 using a Voltcraft PL-110SM solar radiation measuring instrument. We varied the photocatalyst amount from 10 to 50 mg and applied it to 50 mL 10 ppm aqueous solutions of MB, RhB, or a 50 mL equimolar mixture of 10 ppm MB and RhB. Before irradiation, the solutions were mixed in the dark on a magnetic mixer for 60 min to achieve adsorption and desorption equilibrium between the photocatalyst and pollutant. Photolysis of the solutions was also analyzed under the same conditions. Absorption was monitored on a Shimadzu UV-2600 spectrophotometer (Kyoto, Japan) at 663 nm (MB) and 554 nm (RhB).

The degradation efficiency was determined as:

$$\%\ \text{degradation efficiency} = \frac{(C_0 - C_t)}{C_0} \times 100 \tag{1}$$

where C_0 is the concentration of the dye solution at the start of the exposure to natural sunlight (t = 0) and C_t is the concentration of the dye solution at time t.

Kinetic analysis of the dye degradation process was performed using the simplified Langmuir–Hinshelwood (L-H) kinetic model [14,24,27]:

$$-\ln\frac{C_t}{C_0} = -\ln\frac{A_t}{A_0} = kt \tag{2}$$

where the normalized temporal concentration changes of the dye during the photocatalytic process can be assumed to be proportional to the normalized maximum absorbance change of the dye at time t (A_t) and 0 (A_0) derived from the change in the absorption profile at irradiation time t. The pseudo-first-order kinetic constant (k) can be determined from the fitted linear relation.

Mass spectrometry (MS) analysis was performed on a TSQ Quantum Access Max mass spectrometer equipped with a HESI source (Thermo Fisher Scientific, Waltham, MA, USA). The ion source settings were as follows: spray voltage, 3500 V; vaporizer temperature, 300 °C; sheath gas, N_2; pressure, 27 AU; ion sweep gas pressure, 1.0 AU;

auxiliary gas (N_2) pressure, 10 AU; capillary temperature, 275 °C; skimmer offset, 0 V. Data were acquired in positive mode including full scanning in the m/z range from 100 to 1000 for qualitative analysis and product ion scanning mode for the quantitative analysis. Collision-induced fragmentation experiments were performed using Ar as the collision gas, with the collision energy set at 10 eV. Samples were introduced into the mass spectrometer with a syringe pump and continuous flow injection at a flow rate of 10 μL min^{-1} for a period of 3 min. Xcalibur software Analyst version 1.4 (Thermo Fisher Scientific) was used for data acquisition and processing.

3. Results and Discussion

3.1. Structure, Morphology, and Optical Properties

The measured XRD pattern of the synthesized Zn_2SnO_4/SnO_2 powder is shown in Figure 1a and confirms the formation of two phases: cubic spinel Zn_2SnO_4 (JCPDS 24-1470) and tetragonal SnO_2 (JCPDS 88-0287). The FESEM image of the Zn_2SnO_4/SnO_2 powder (Figure 1b) showed relatively uniform nanoparticles of Zn_2SnO_4 and SnO_2 with a similar shape and size. The average particle size was estimated as ~150 nm. Rietveld refinement of the measured XRD pattern using the GSAS II software package [30] showed good agreement between the measured and refined patterns (Figure 1a) and enabled the determination of structural parameters (unit cell parameters, atomic positions, crystallite size, and microstrain), as shown in Figure 1c. The crystal structures drawn using the determined crystalline lattice parameters for Zn_2SnO_4 ($Fd\bar{3}ms$) and SnO_2 ($P4_2/mnm$) using the VESTA software package [31] are shown in Figure 1d. The Zn ions are shown in pale grey and the Sn ions are pale purple, while the oxygen ions are shown as red balls. For Zn_2SnO_4, we obtained the inversion parameter of 0.81, which is in the range previously determined for zinc stannate [29,32,33] and confirms the presence of randomly distributed Zn^{2+} and Sn^{4+} ions on both tetrahedral and octahedral sites in the cubic spinel lattice (as shown in Figure 1d). The ideal cation symmetry for the cubic spinel space groups of Zn_2SnO_4 was distorted, as the oxygen parameter was determined to be 0.2591 (rather than 0.25), as noted previously for zinc stannate with larger tetrahedra and smaller octahedra [29]. The crystallite sizes of 108 nm for Zn_2SnO_4 and 88 nm for SnO_2 were similar to values previously obtained for this composite [29].

The measured FTIR spectrum of Zn_2SnO_4/SnO_2 shown in Figure 2 confirms the presence of metal-oxide bonds in the nanocomposite. Thus, Sn−O stretching vibrations can be noted in the range 430–620 cm^{-1} (501 and 553 cm^{-1}), and Sn−O−Sn band vibrations were present in the range 645–700 cm^{-1} (645 and 698 cm^{-1}), as noted before for Sn present in both SnO_2 and Zn_2SnO_4 [29]. Zn-O vibration bands were anticipated in the range 150–420 cm^{-1}, and the tail of one peak was possibly present (marked in Figure 2), as previously noted [29].

The morphology and structure of the Zn_2SnO_4/SnO_2 powder were further analyzed by TEM, high-resolution TEM (HRTEM), high-angle annular dark-field scanning (HAADF) TEM, and EDS. The TEM images shown in Figure 3a,b enabled us to have a closer look at the powder particles, confirming that the SnO_2 and Zn_2SnO_4 nanoparticles were similar in size and shape. HAADF combined with EDS mapping, as shown in Figure 3c, further confirmed the presence of a homogenous mixture of the two components of the nanocomposite powder. Analysis of the periodic lattice fringes for the two particles shown in the HRTEM image in Figure 3d using fast Fourier transform (FFT) enabled us to determine a lattice spacing of 0.24 nm for the (200) plane of tetragonal SnO_2 and 0.50 nm for the (111) plane of cubic spinel Zn_2SnO_4, with corresponding FFT images for SnO_2 (Figure 3e) and Zn_2SnO_4 (Figure 3f). Further analysis of the TEM images of Zn_2SnO_4/SnO_2, as shown in Figure 3g,h, and the HRTEM image shown in Figure 3i, revealed a lattice spacing of 0.33 for the (110) plane of tetragonal SnO_2 and a lattice spacing of 0.30 for the (220) plane of cubic spinel Zn_2SnO_4. The SnO_2 plane and the neighboring Zn_2SnO_4 plane formed a heterointerface with no interlayer gaps between them, which indicated the presence of Zn_2SnO_4 and SnO_2 heterojunctions, as noted before for Zn_2SnO_4/SnO_2 microcubes [34]. Combining two individual oxides in a nanocomposite that forms heterojunctions between SnO_2 and Zn_2SnO_4 has previously been noted for this type of nanocomposite, which has led to improved sensing and photocatalytic properties [35–37]. The close proximity and increased contact between Zn_2SnO_4 and SnO_2 particles in the synthesized heterostructures resulted in the formation of heterojunctions in the interfacial region and led to faster electron transfer [38,39].

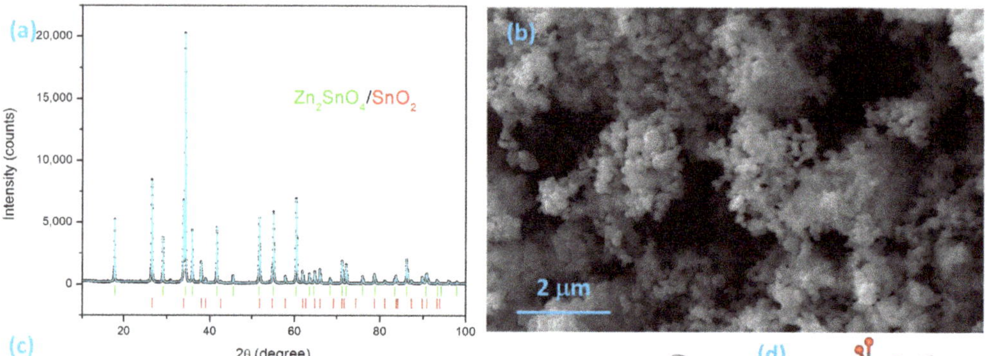

(c)

Zn_2SnO_4 (70.8 wt.%) a = 8.6724(6) Å, Crystallite size 108 nm, Microstrain 1323(26)						
Atom	Site	x	y	z	U_{aniso}	Occu.
Zn1	8a	0.125	0.125	0.125	0.022	0.81
Sn1	8a	0.125	0.125	0.125	0.022	0.19
Zn2	16d	0.500	0.500	0.500	0.008	0.61
Sn2	16d	0.500	0.500	0.500	0.008	0.39
O	32e	0.2591(7)	0.2591	0.2591	0.001	1.00
O-Zn1, Sn1 distance 2.015(4); O-Zn2,Sn2 distance 2.09196(24)						

SnO_2 (29.1 wt.%) $a = b$ = 4.7419(6) c = 3.1892 Å, Crystallite size 88 nm, Microstrain 1541(4)						
Atom	Site	x	y	z	U_{aniso}	Occu.
Sn	2a	0	0	0	0.001	1
O	4f	0.3087(26)	0.3078	0.3078	0.003	1
Sn-O distances 2.064(9) and 2.050(5)						

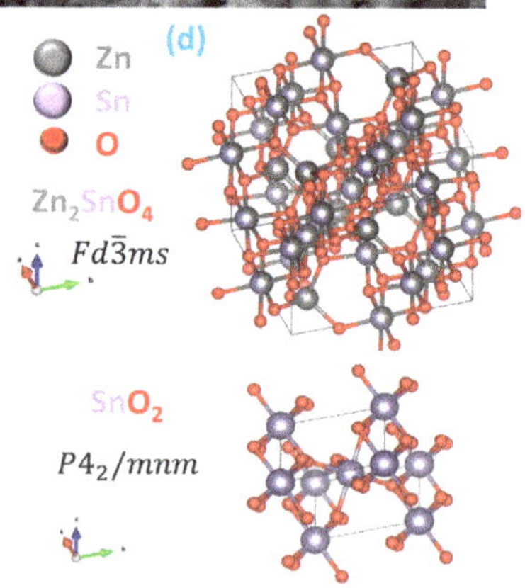

Figure 1. XRD pattern (points) measured for the Zn_2SnO_4/SnO_2 powder and Rietveld refinement (pale blue line) with the identified phases marked in green—Zn_2SnO_4 and red—SnO_2 (**a**), FESEM image of the Zn_2SnO_4/SnO_2 powder (**b**), structural parameters determined from the Rietveld refinement (**c**), and crystal structure drawn by VESTA [31] for Zn_2SnO_4 and SnO_2 (**d**).

Generally, the surface area of materials obtained by solid-state synthesis is not very high, and this was confirmed by the BET surface area determined from the measured nitrogen adsorption/desorption isotherms (type II) of 4.9 m^2/g (Zn_2SnO_4/SnO_2 obtained by solid-state synthesis with a shorter calcination time) [29]. This value was slightly lower but comparable with some previous results obtained for this composite such as 9.9 m^2/g obtained for microcubes [34] or 13.6 m^2/g obtained for a specific cube-type morphology [20]. The mercury porosimetry measurements in [29] of the solid-state obtained

Zn_2SnO_4/SnO_2 indicated a stable pore system with high porosity values obtained for both intrusion/extrusion runs. The average pore diameters were similar and close to the average particle size. Taking into account the results obtained by the nitrogen adsorption/desorption measurements and the mercury porosimetry measurements, we concluded that the analyzed nanocomposite was macroporous with very little micro and mesopores, similar to that expected for the Zn_2SnO_4/SnO_2 powder synthesized in this work. Even though a high BET specific surface area was not obtained by solid-state synthesis, the macroporous microstructure is also beneficial for photocatalysis.

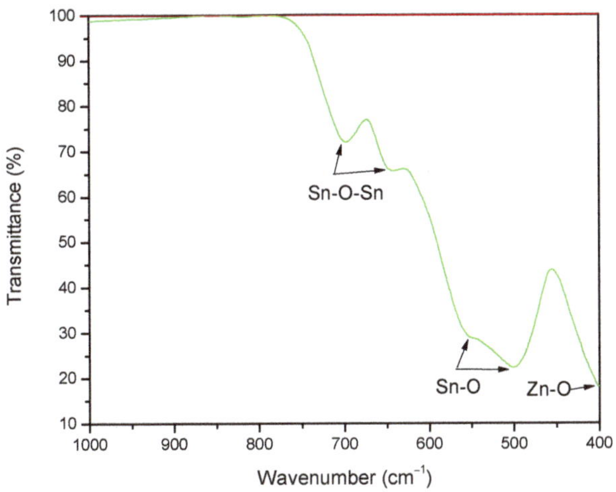

Figure 2. FTIR spectrum measured for the Zn_2SnO_4/SnO_2 powder.

Kubelka–Munk transformation of the measured diffuse reflectance spectrum of the Zn_2SnO_4/SnO_2 powder enabled the calculation of the equivalent absorption coefficient used for the estimation of the optical band gap, as shown in Figure 4 [40]. The direct band gap of the analyzed Zn_2SnO_4/SnO_2 powder was estimated from the Tauc plot to be 3.87 eV (inset in Figure 4a). A direct band gap was assumed, taking into account that both SnO_2 and Zn_2SnO_4 have direct optical energy gaps. SnO_2 has a direct optical energy gap estimated to be 3.68 eV [41], though for nanoparticles and thin films, it has been found to be between 3.4 and 4.6 eV [42,43]. Zn_2SnO_4 also has a direct optical energy gap, which has been found to be between 3.2 and 4.1 eV, with 3.6 eV determined for bulk samples, but both higher and lower for nanoparticles and films with different morphologies [17,21,44,45]. The absorption edge for Zn_2SnO_4/SnO_2 was estimated to be 337 nm (3.68 eV), as shown in Figure 4a. The estimated direct band gap value from the Tauc plot confirmed this value, as shown in the inset in Figure 4a. The estimated band gap value of 3.68 eV was slightly higher than the value of 3.63 eV determined by Li et al. [20] for a cube-shaped Zn_2SnO_4/SnO_2 heterojunction composite powder. The absorption edge of pure SnO_2 powder (used in the synthesis) was lower than the absorption edge of Zn_2SnO_4/SnO_2. A redshift in the absorption edge of Zn_2SnO_4/SnO_2 compared with the absorption edge of pure SnO_2 is an indication of a narrower band gap caused by the formation of heterojunctions between Zn_2SnO_4 and SnO_2, as previously noted [39]. The Tauc plot shown in the inset enabled the estimation of the direct band gap of 3.84 eV. Ye et al. [39] concluded that heterojunction formation between Zn_2SnO_4 and SnO_2 in a heterostructured nanocomposite can be confirmed by a band gap value higher than the band gap of pure Zn_2SnO_4 and lower than the band gap of pure SnO_2. Wang et al. [21] found that heterojunctions formed at the interface between SnO_2

and Zn_2SnO_4 promoted a better separation of photoinduced electrons (e^-) and holes (h^+), leading to an extended absorption range and improved photocatalytic activity.

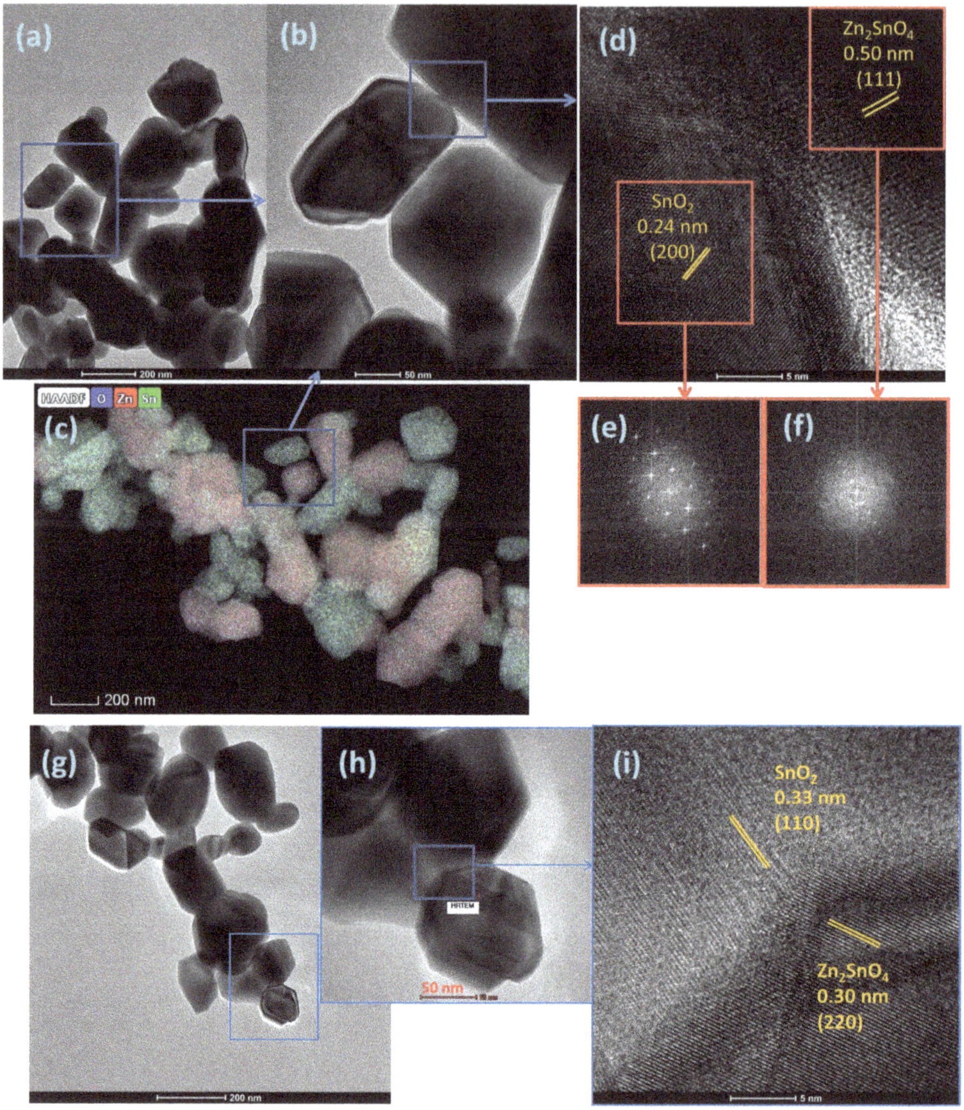

Figure 3. TEM (**a**,**b**,**g**,**h**), HAADF TEM (**c**), HRTEM (**d**,**i**), and FFT (**e**,**f**) images for the Zn_2SnO_4/SnO_2 powder; blue rectangles mark the magnified areas, while the red squares mark the areas used for FFT analysis.

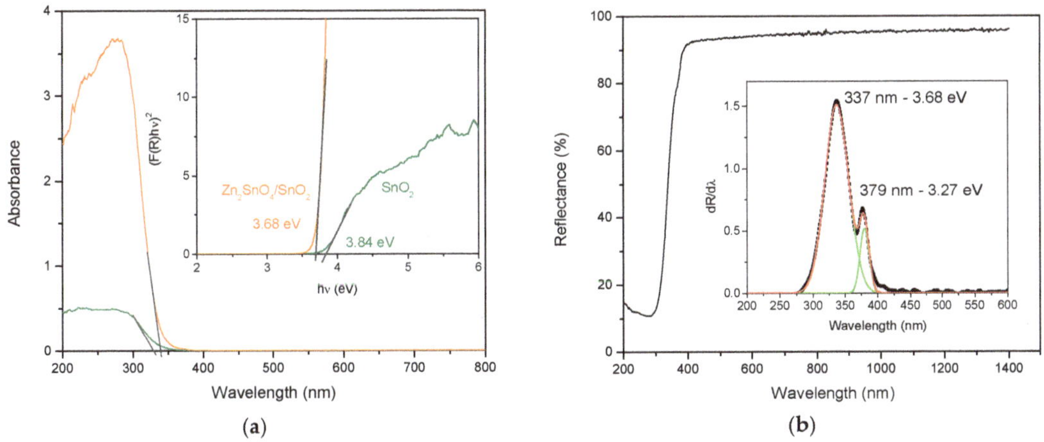

Figure 4. Zn_2SnO_4/SnO_2 (orange) and SnO_2 (green) powder absorbance obtained by Kubelka–Munk approximation (**a**) from the measured diffuse reflectance spectra (**b**), insets (**a**)—Tauc plot for estimation of direct band gap and (**b**)—first derivative of the measured diffuse reflectance spectra of Zn_2SnO_4/SnO_2 (black points) and the calculated transition energy values (the red line represents the sum of fitted peaks, while the green lines represent the Gaussian fits of the two noted peaks with maxima at 337 and 379 nm, respectively).

The first derivative of the measured reflectance ($dR/d\lambda$) has been used to enable the identification of additional transition energies as well as estimate the band gap [40,46]. The presence of additional energy levels in the band gap of doped metal oxides or metal oxide nanocomposites has led to enhanced sensing properties or improved electrocatalytic activity [47,48]. In the case of Zn_2SnO_4/SnO_2, there was one prominent peak at 337 nm accompanied by a smaller but noticeable shoulder at 379 nm, showing the presence of additional transitions within the band gap (Figure 4b).

The photoluminescence (PL) spectra of Zn_2SnO_4/SnO_2 measured under an excitation of 325 and 385 nm are shown in Figure 5a,b, while the photoluminescence 3D spectrum of Zn_2SnO_4/SnO_2 measured in the emission range of 446–640 nm for the excitation range of 325–350 nm is shown in Figure 5c, and the contour plot of this spectrum is shown in Figure 5d. PL emission spectra provide information on the migration and recombination of photo-induced hole-electron pairs [16]. The emission intensity decreased with an increase in excitation wavelength, as shown in Figure 5c,d with a maximum intensity at around 575 nm for an excitation of 325 nm. After multi-peak Gaussian fitting, the spectrum obtained under an excitation of 325 nm (Figure 5a) was deconvoluted into a green and an orange emission band located at 530 nm (2.33 eV) and 597 nm (2.07 eV), respectively. The spectrum obtained under the excitation of 385 eV (Figure 5b) was deconvoluted into a violet, blue, and green emission band located at 434 nm (2.85 eV), 482 (2.57 eV), and 560 nm (2.21 eV), respectively. Similar emission bands have previously been observed in Zn_2SnO_4/SnO_2 nanocomposites [22] as well as in Zn_2SnO_4 [49] and SnO_2 [50]. The determined emission bands can be ascribed to different luminescent centers due to defect energy levels originating from oxygen vacancies, tin vacancies, or oxygen interstitials [13,49,50].

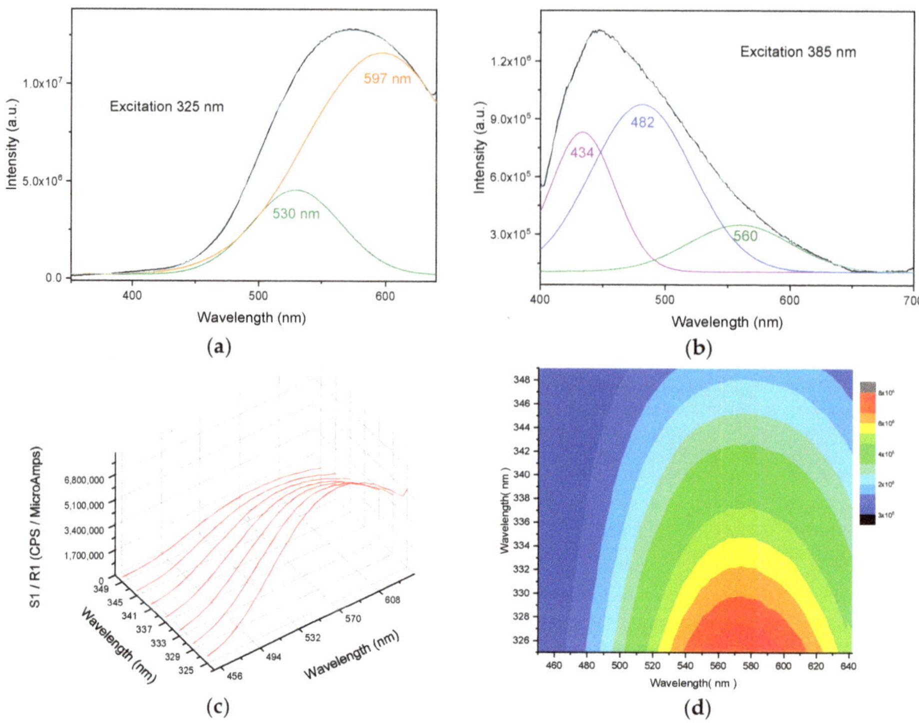

Figure 5. PL spectra of Zn_2SnO_4/SnO_2 with an excitation wavelength of 325 nm (**a**) and 385 nm (**b**) (the black points represent the measured spectra, the pale blue lines represent the sums of fitted peaks, and the green, orange, purple and blue lines represent the Gaussian fits of the individual peaks). 3D PL spectrum (**c**) and contour plot (**d**) under an excitation range of 325–350 nm and emission range of 446–640 nm.

3.2. Photodegradation of Dye in Natural Sunlight

3.2.1. Photodegradation of MB

Figure 6a shows that the degradation of MB dye (10 ppm = 10 mg/L) under natural sunlight irradiation using 50 mg of the photocatalyst resulted in a noticeable and significant reduction in the absorption peaks of MB and in an almost white solution (as shown in the inset), indicating the almost complete degradation of the dye after 120 min (2 h) of irradiation in natural sunlight. MB is a blue cationic thiazine dye with absorption peaks at 246 nm, 291 nm, and 663 nm and a shoulder at 615 nm, three mesomeric structures where the positive charge is located either on nitrogen or sulfur atoms [51]. The reduced forms (Leucomethylene blue and MBH_2^+) obtained using reducing agents are colorless and stable in aqueous solutions. With an increase in the irradiation time, we also noted a slight blueshift in the strongest absorption band starting at 663 nm, which reduced to 636 nm after 120 min. The occurrence of a blueshift has been previously noted for photocatalytic MB degradation with other photocatalysts such as PANI/TiO_2 composites [52]. This is associated with the occurrence of N-demethylated derivatives of MB during the degradation process. Variation in the Zn_2SnO_4/SnO_2 concentration from 10 to 50 mg (Figure 6b) showed that a decrease in the number of active sites on the catalyst surface reduced the photocatalytic activity and the characteristic absorption peak at 663 nm. The blueshift remained, but was less with a decrease in catalyst load (650 nm–10 mg, 649 nm–20 mg, and 638 nm–50 mg). The degradation efficiency of 99.1% was achieved after 120 min of irradiation in natural sunlight for 50 mg of the photocatalyst, which was reduced to 91.1% for 30 mg, 80.8% for 20 mg, and 74.8% for 10 mg of the photocatalyst, as shown in Figure 6c. The photodegradation rate followed the pseudo-first-order kinetic model (Equation (2)) as shown in Figure 6d, with R^2 values between 0.91 and 0.96. The determined pseudo-first-order kinetic constant k values showed an increase with the increase in catalyst load, with 0.01151 for 10 mg and 0.03828 for 50 mg.

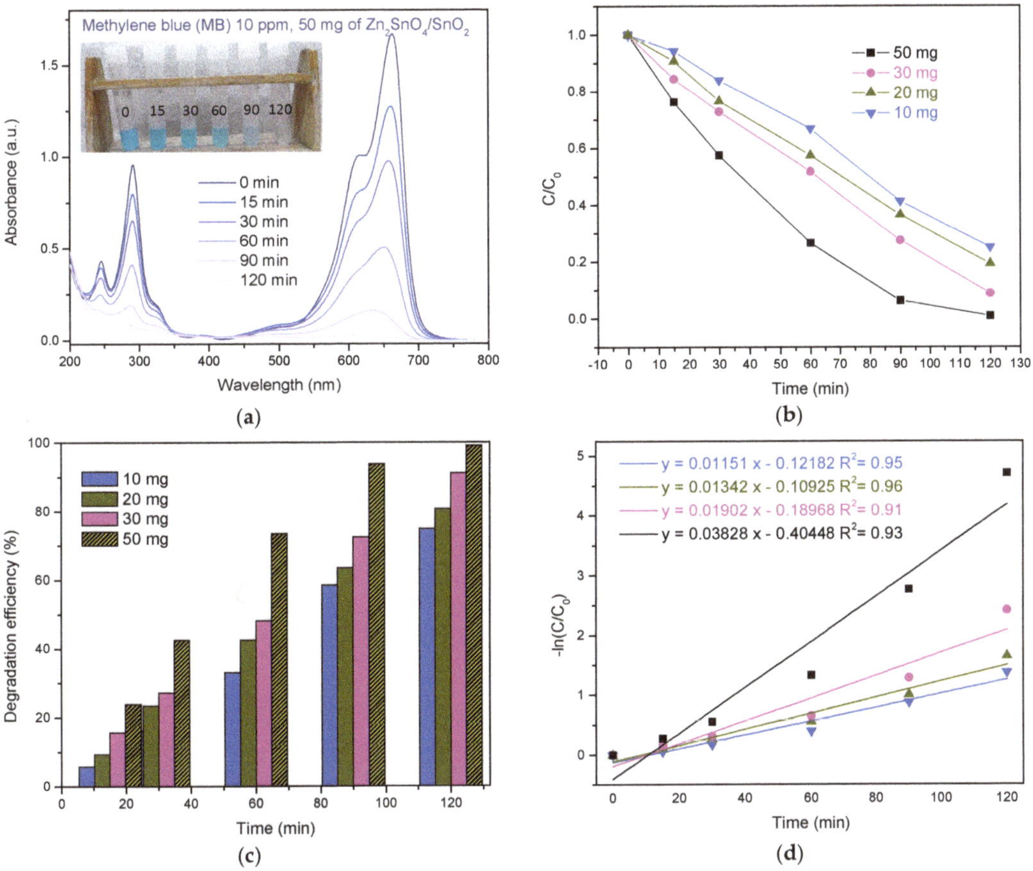

Figure 6. UV–Vis absorption spectra of the photocatalytic degradation of MB (10 ppm) under natural sunlight irradiation in the presence of 50 mg of Zn_2SnO_4/SnO_2 (**a**), C/C_0 at different time intervals (**b**), degradation efficiency (**c**), and the pseudo-first-order kinetic model (**d**) for different photocatalyst loads (10, 20, 30, and 50 mg).

Methylene blue (MB) degradation has been previously achieved by this type of composite, but mostly in UV light, as shown in Table 1, and shows a significant influence of the composite morphology on the degradation efficiency and pseudo-kinetic constant. Rovisco et al. [17] also showed that the morphology and phase had a significant influence on the photocatalytic activity of zinc-stannate in both UV and visible light for the degradation of MB. Li et al. [26] attributed the improved photocatalytic activity of Zn_2SnO_4/SnO_2 hollow cubes compared with solid ones to the presence of cavities within the cubes, enabling higher light absorption, a lower recombination rate of photogenerated electrons, and a highly efficient separation of photoelectrons from vacancies. In this case, the presence of heterojunctions between Zn_2SnO_4 and SnO_2 nanoparticles can bring about improved photocatalytic activity due to the improved separation of photoinduced electrons and holes [21]. This was reflected in the high value obtained for the pseudo-kinetic constant for 50 mg of photocatalyst as well as the relatively high values for lower catalyst loads, showing that the complete degradation of MB would occur for durations longer than 120 min, but comparable with the results shown in Table 1.

Table 1. Photocatalytic degradation efficiency (DE) and pseudo-first-order kinetic constant (k) determined for different Zn_2SnO_4/SnO_2 nanocomposite morphologies applied as photocatalysts for MB.

Sample	Conditions	DE (%)	k (min^{-1})	Ref.
Zn_2SnO_4/SnO_2 nanocubes Zn:Sn = 1:1	100 mg-100 mL 2.5 ppm, UV-light	91.6% 360 min	0.007236	[23]
Zn_2SnO_4/SnO_2 nanocubes Zn:Sn = 3:2	100 mg-100 mL 2.5 ppm, UV-light	99.4% 360 min	0.012244	[23]
Zn_2SnO_4/SnO_2 nanoparticles	100 mg-100 mL 2 ppm UV-light	98.6% 220 min	0.01763	[19]
Zn_2SnO_4/SnO_2 nanocubes	100 mg-100 mL 2 ppm UV-light	85.1% 220 min	0.008893	[19]
Zn_2SnO_4/SnO_2 spheres	5 mg-50 mL 10 ppm UV light	60% 140 min	0.0049	[53]
Zn_2SnO_4/SnO_2 urchin-like hollow spheres	5 mg-50 mL 10 ppm UV light	97% 140 min	0.0209	[53]
Zn_2SnO_4/SnO_2 hollow cubes	100 mg-100 mL, 10 ppm sim. sunlight	97.1% 150 min	0.0241	[26]
Zn_2SnO_4/SnO_2 solid cubes	100 mg-100 mL 10 ppm sim. sunlight	83.2% 150 min	0.0114	[26]
Zn_2SnO_4/SnO_2 heterojunctions	50 mg-50 mL 10 ppm natural sunlight	99.1% 120 min	0.03828	This work

3.2.2. Photodegradation of RhB

The degradation of RhB in natural sunlight using the Zn_2SnO_4/SnO_2 heterojunction nanocomposite was less efficient than the degradation of MB under the same conditions. Though they are both cationic dyes, MB is a blue cationic thiazine dye, while RhB is a pink dye part of the triphenylmethane family. Figure 7a shows the absorption spectra measured for RhB degradation in natural sunlight using 50 mg of Zn_2SnO_4/SnO_2 as the photocatalyst. Though there was discoloration of the solution, which changed from bright pink to a lighter, more milky pink, as shown in the inset, 120 min was not sufficient for the complete degradation of RhB. We also noted a slight blueshift of the characteristic peak at 554 nm, which shifted to 547 nm after 120 min of natural sunlight irradiation. This blueshift was attributed to the formation of N-deethylated intermediates of RhB and has been previously noted for both the Zn_2SnO_4/SnO_2 nanocomposite [22] and other photocatalysts such as $NaBiO_3$ [54]. The photocatalytic activity was lower for smaller catalyst loads, as shown in Figure 7b. The highest degradation efficiency of 70.6% was obtained for the highest catalyst load of 50 mg, and it decreased with a decrease in catalyst load, with 10 mg achieving a degradation efficiency of only 22.6% after 120 min in natural sunlight, as shown in Figure 7c. The photodegradation rate also followed the pseudo-first-order kinetic model (Equation (2)), as shown in Figure 7d, with R^2 values between 0.92 and 0.99. The determined pseudo-first-order kinetic constant k values increased with the increase in catalyst load, with 0.00216 for 10 mg and 0.01018 for 50 mg. Table 2 shows a comparison between the results obtained in this work for the Zn_2SnO_4/SnO_2 nanocomposite heterojunctions with the literature data obtained for different Zn_2SnO_4/SnO_2 nanocomposite morphologies and light sources (simulated UV or UV–Vis light). The kinetic rate we obtained was not the highest, but also not the lowest, indicating that for a longer irradiation time in natural sunlight, we could expect the complete degradation of RhB, possibly after 240 min.

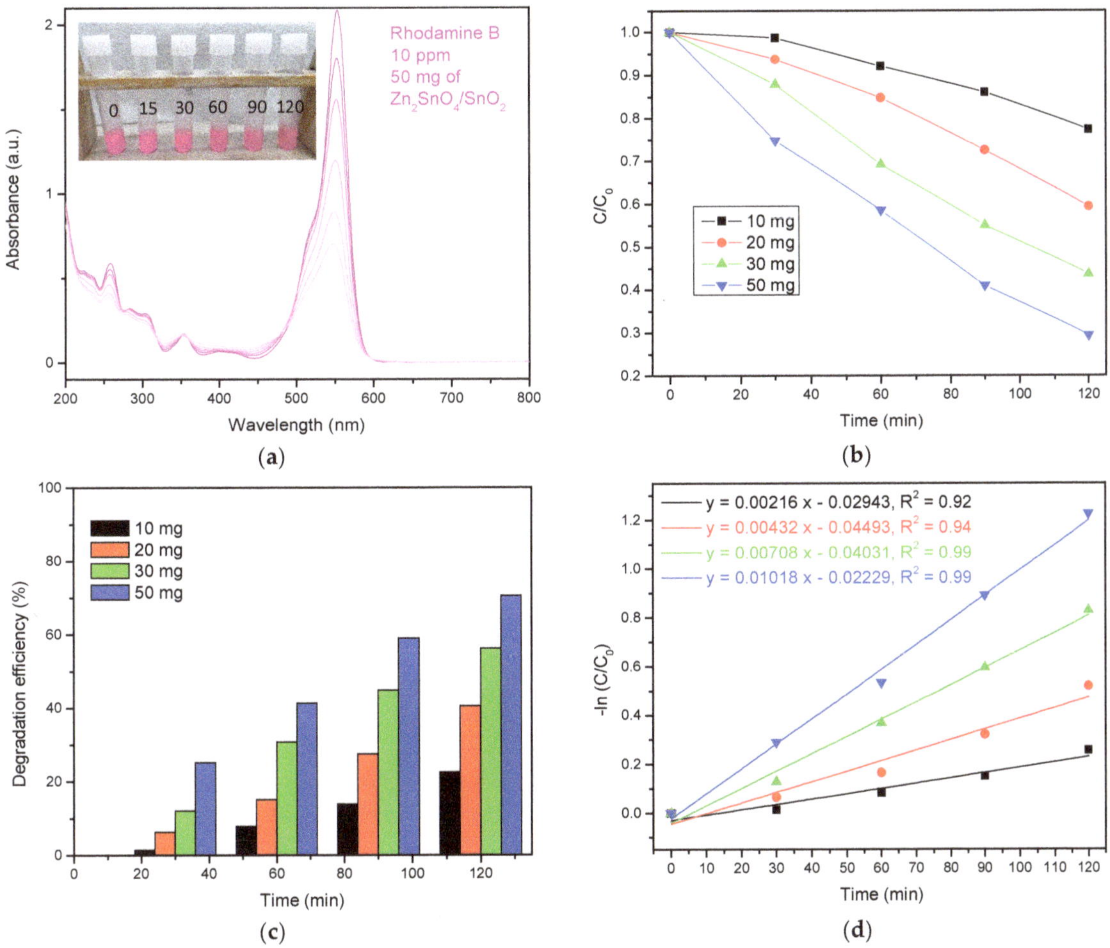

Figure 7. UV–Vis absorption spectra of the photocatalytic degradation of RhB (10 ppm) under natural sunlight irradiation in the presence of 50 mg of Zn_2SnO_4/SnO_2 (**a**), C/C_0 at different time intervals (**b**), degradation efficiency (**c**), and pseudo-first-order kinetic model (**d**) for different photocatalyst loads (10, 20, 30, and 50 mg).

Table 2. Photocatalytic degradation efficiency (DE) and pseudo-first-order kinetic constant (k) determined for different Zn_2SnO_4/SnO_2 nanocomposite morphologies applied as photocatalysts for RhB.

Sample	Conditions	DE (%)	k (min^{-1})	Ref.
Zn_2SnO_4/SnO_2 double shell nanocubes	50 mg-50 mL 6.4 ppm, simulated sunlight	95%, 120 min	0.0214	[24]
Zn_2SnO_4/SnO_2 yolk shell nanocubes	50 mg-50 mL 6.4 ppm, simulated sunlight	85%, 240 min	0.0074	[24]
Zn_2SnO_4/SnO_2 hollow microbox	20 mg-50 mL 10 ppm simulated sunlight	83%, 45 min	-	[25]
Zn_2SnO_4/SnO_2 nanocomposite	40 mg-100 mL 5 ppm, UV-C light	62% 60 min 94.5%, 240 min	0.0111	[22]
Zn_2SnO_4/SnO_2 microspheres consisting of nanobelts	80 mg on stainless steel mesh 100 mL, 10 ppm, simulated UV–Vis light	93.4% 100 min	0.028	[55]
Zn_2SnO_4/SnO_2 microspheres—nanoparticle-based nanosheets	80 mg on stainless steel mesh 100 mL, 10 ppm simulated UV–Vis light	99.9% 100 min	0.036	[55]
Zn_2SnO_4/SnO_2 heterojunctions	50 mg-50 mL 10 ppm natural sunlight	70.6%, 120 min	0.0108	This work

3.2.3. Photodegradation of MB+RhB

The measured absorption curves showing the degradation of MB+RhB under natural sunlight using 50 mg of Zn_2SnO_4/SnO_2 are presented in Figure 8a, indicating the complete degradation of MB and the significant degradation of RhB. A blueshift in both characteristic peaks for MB and RhB (at 663 and 554 nm) was also noted, indicating the presence of both N-demethylated N-deethylated intermediates of MB and RhB, respectively, in the mixed dye solution. Widening of the absorption bands was also noticeable after longer irradiation times. As expected from the results obtained for the degradation of MB and RhB as single dye aqueous solutions, MB degraded faster compared with RhB. The starting purple solution completely changed color after 120 min of irradiation, resulting in a milky pale pink color, as shown in Figure 8b. Change in the catalyst amount (10 and 30 mg) showed a paler purple for 10 mg and a darker shade of pink after 120 min for 30 mg of Zn_2SnO_4/SnO_2. Figure 8c,e, and Figure 8d,f shows the change in dye concentration ratios and the degradation efficiency of MB and RhB in MB+RhB, respectively, with the catalyst load varying between 5 and 50 mg. The degradation efficiency of MB was higher, achieving 97.9% with only 15 mg of photocatalyst. The degradation efficiency of 84.1% was obtained for RhB in MB+RhB with 50 mg of the photocatalyst. Analysis of the first-order reaction kinetics showed that the determined pseudo-first-order kinetic constant k values were overall higher for the same catalyst load applied to the single dye solutions (Figure 8g,h). Thus, 30 mg of Zn_2SnO_4/SnO_2 degraded RhB, and 20 mg of Zn_2SnO_4/SnO_2 degraded MB with a similar kinetic rate to 50 mg of Zn_2SnO_4/SnO_2 in single dye solutions. Thus, we can conclude that Zn_2SnO_4/SnO_2 nanocomposite heterojunctions perform better in a mixed MB+RhB dye solution. These changes could be attributed to adsorption competition, light shielding, or some effects of dye–dye interactions. Dlugosz et al. [8] studied the degradation of single, bi, and multiple mixed dye solutions using the ZnO-SnO_2 nanocomposite and determined that the presence of components of the same nature did not affect the dye decomposition rate, except for quinolone yellow (dye with a neutral character), while Tang et al. [27] noted a 5% reduction in degradation rate for malachite green (cationic dye). MB and RhB degradation did not change in the mixed dye solutions.

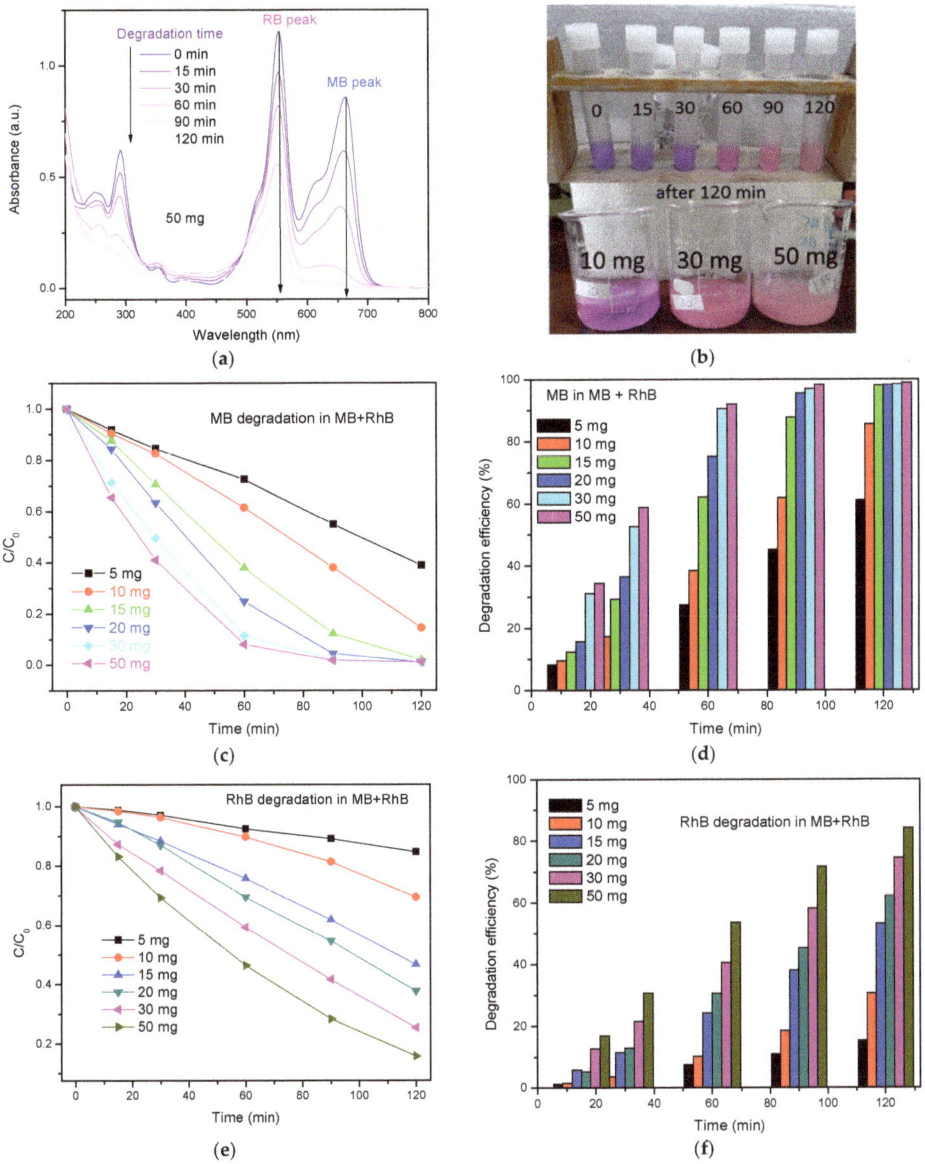

Figure 8. *Cont.*

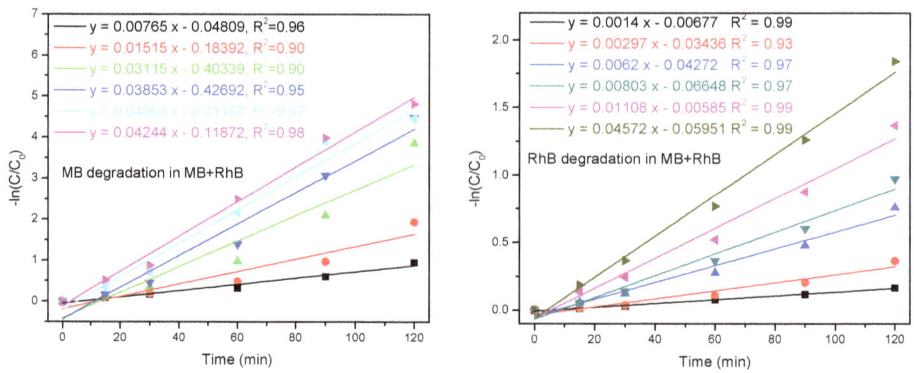

(**g**) (**h**)

Figure 8. UV–Vis absorption spectra of the photocatalytic degradation of MB+RhB (10 ppm) under natural sunlight irradiation in the presence of 50 mg of Zn_2SnO_4/SnO_2 (**a**), color change in the solution after 120 min using 50 mg of Zn_2SnO_4/SnO_2 and final solutions after irradiation for 120 min using 10, 30, and 50 mg Zn_2SnO_4/SnO_2 (**b**), C/C_0 at different time intervals for MB (**c**) and RhB (**e**) and degradation efficiency for MB (**d**) and RhB (**f**) in MB+RhB, and the pseudo-first-order kinetic model for MB (**g**) and RhB (**h**) in MB+RhB for different photocatalyst loads (5–50 mg).

3.2.4. Zn_2SnO_4/SnO_2 Pollutant Degradation Mechanism

Irradiation of the Zn_2SnO_4/SnO_2 nanocomposite with natural sunlight initiates electron excitation from valence bands (VBs) to conduction bands (CBs) for both components. Then, the electron (e^-) transfer occurs from Zn_2SnO_4 with a higher CB edge to SnO_2 with the lower CB edge (as shown in Figure 9). Simultaneous hole (h^+) transfer also occurs from SnO_2 to Zn_2SnO_4, which leads to more effective charge carrier separation, thus reduced charge carrier recombination increases the carrier lifetime [11,22]. Photogenerated electrons in the CBs can react with O_2 absorbed on the catalyst (Zn_2SnO_4/SnO_2 nanocomposite) surface or dissolved in water and create superoxide anion radicals (O_2^-). At the same time, photogenerated holes can interact with OH^- or H_2O and create hydroxyl radicals ($OH\cdot$) [22]. Both superoxide anion and hydroxide radicals are responsible for the degradation of organic pollutants (MB, RhB, MB+RhB) into CO_2 and H_2O [17]. The proposed degradation mechanism is shown in Figure 9.

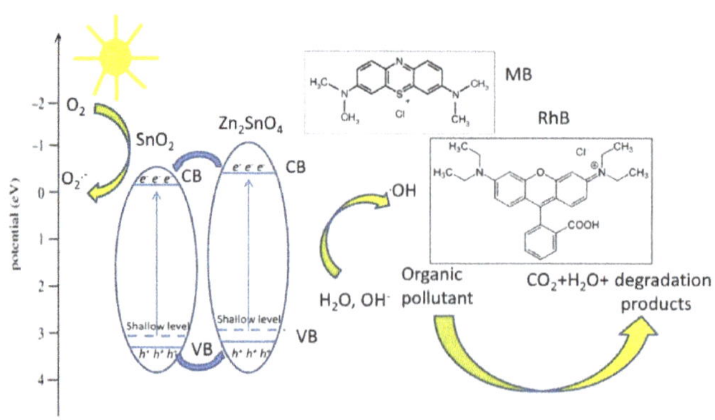

Figure 9. Degradation mechanism scheme for MB, RhB, and MB+RhB degradation under natural sunlight by Zn_2SnO_4/SnO_2.

3.3. MS Analysis—Degradation Mechanism

In order to elucidate the degradation pathways of MB, RhB, and their mixture under photocatalytic conditions in greater detail, an electrospray ionization mass spectrometry (ESI-MS) analysis was carried out in positive mode. The obtained spectra indicated a stepwise degradation process that occurs through the formation of intermediate products and can be classified into three main mechanisms: N-demethylation, ring opening, and complete fragmentation into smaller organic residues. The degradation process was monitored at time intervals of 0, 15, 30, 60, 90, and 120 min, which allowed for an observation of the kinetics and progressive fragmentation of the dye molecules. The results provide deeper insights into the individual stages of degradation, simultaneously confirming the spectral shifts and changes in absorption observed in UV–Vis analysis, thereby further contributing to the understanding of the photocatalytic degradation mechanism of these dyes and their mixture.

3.3.1. MB Degradation Pathway

The ESI-MS spectrum of the initial MB solution (MB 0) showed a dominant molecular ion peak at m/z 284 [56–61], confirming the presence of intact MB. After 15 min of irradiation (MB 15), peaks mainly appeared at m/z 256 and m/z 270 corresponding to Azure A, and Azure B, respectively [57]. The peak at m/z 223 corresponded to the N-demethylation products, as suggested by the observed blueshift in the UV–Vis spectrum (Figure 4). The peak at m/z 297 can be explained by multiple hydroxylation of the MB molecule [58,60]. The spectrum after 30 min of irradiation did not show any significant changes, but as the irradiation continued (MB 60), the formation of peaks at m/z 188, 172, and 158 [59] revealed oxidative fragmentation and ring-opening processes, aligning with the diminishing absorption intensity in the UV–Vis measurements. The increase in the intensity of products formed by oxidation and hydroxylation of the MB molecule, particularly in the final stages (MB 90 and MB 120), led to the conclusion that MB had been completely degraded. Scheme 1 presents the presumed degradation mechanism of MB during photodegradation in the presence of Zn_2SnO_4/SnO_2, while the possible structures of degradation products detected at various UV exposure times are provided in Table 3. The heatmaps shown in Figure 10, which track the relative intensities of the peaks in the mass spectra over time (up to 30 min—Figure 10a, and from 60 min onward—Figure 10b), illustrate the variation in the abundance of intermediates during MB photodegradation. These data demonstrate that as the initial intensity of intact MB (m/z 284) gradually decreases, early intermediates, such as N-demethylation products (m/z 223), Azure A (m/z 256), and Azure B (m/z 270), form rapidly, but their intensity decreases as the reaction progresses, confirming that they are transient degradation products [59]. The intermediate product at m/z 277 exhibited oscillatory dynamics, suggesting a continuous transformation. The formation of oxidative products, especially the peaks at m/z 340 and m/z 396, became more pronounced in the later stages, supporting the hypothesis that MB does not degrade solely via demethylation but also through advanced oxidative transformations that lead to the final fragmentation and eventual mineralization of the initial molecule. However, additional validation is required. The observed evolution of the peak intensities indicates that MB degradation occurs through a sequence of interconnected reactions from the initial demethylation and hydroxylation processes, through the formation of intermediate oxidized products, to complete fragmentation and mineralization.

Table 3. Exposure time and ion masses of the degradation products of MB and their possible structures.

Ion *m/z*	Exposure Time (min)	(Possible) Structure
188.17	60	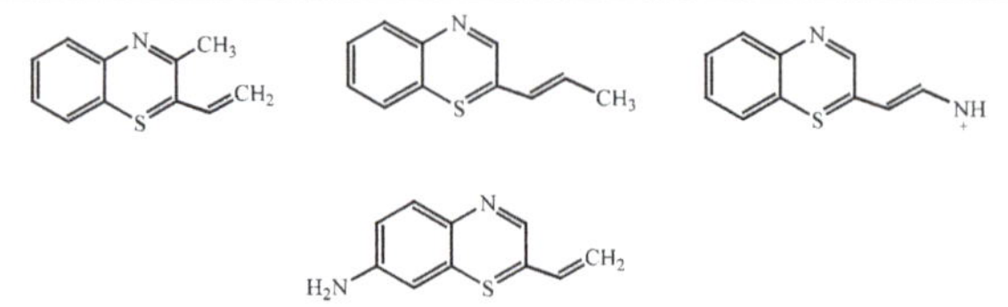
172.19	60	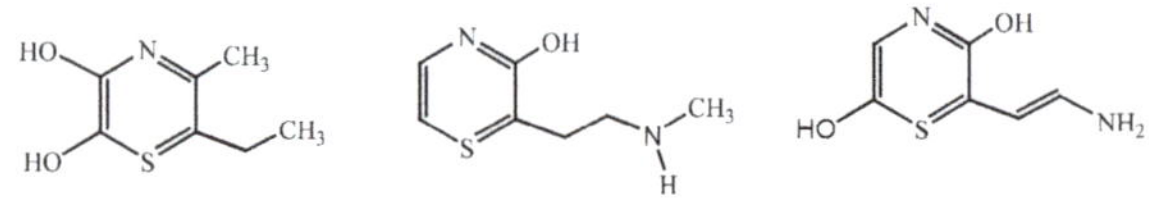
158.18	60	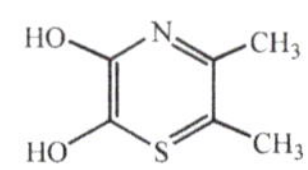
143.16	60	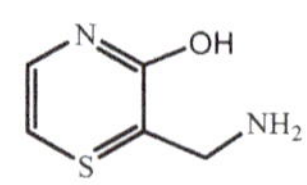
127.90	90	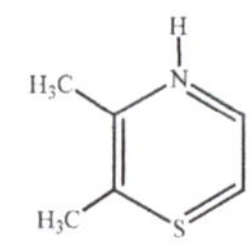
188.17	60	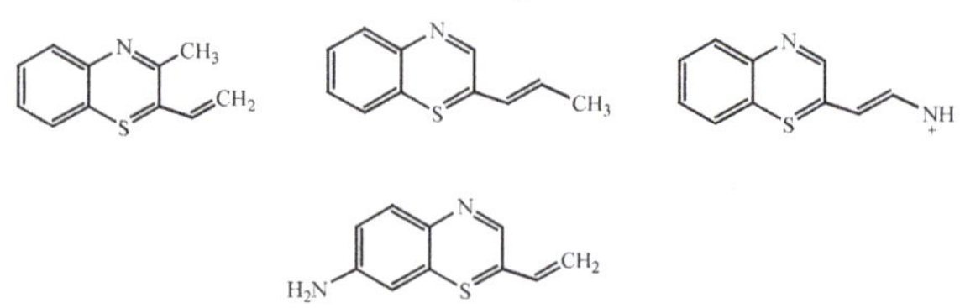
172.19	60	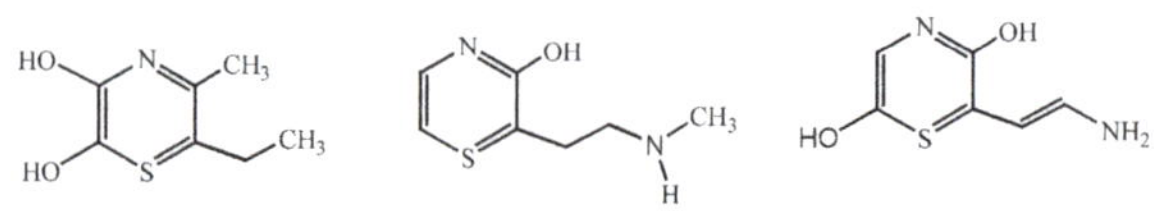

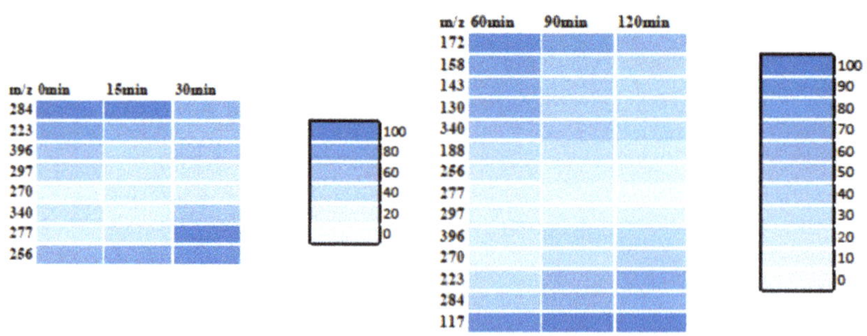

(**a**) (**b**)

Figure 10. Heatmaps illustrating the change in intensity of the intermediary peaks during MB degradation in the first 30 min (**a**) and the 60–120 min interval (**b**) in the presence of Zn_2SO_4/SnO_2 under natural sunlight.

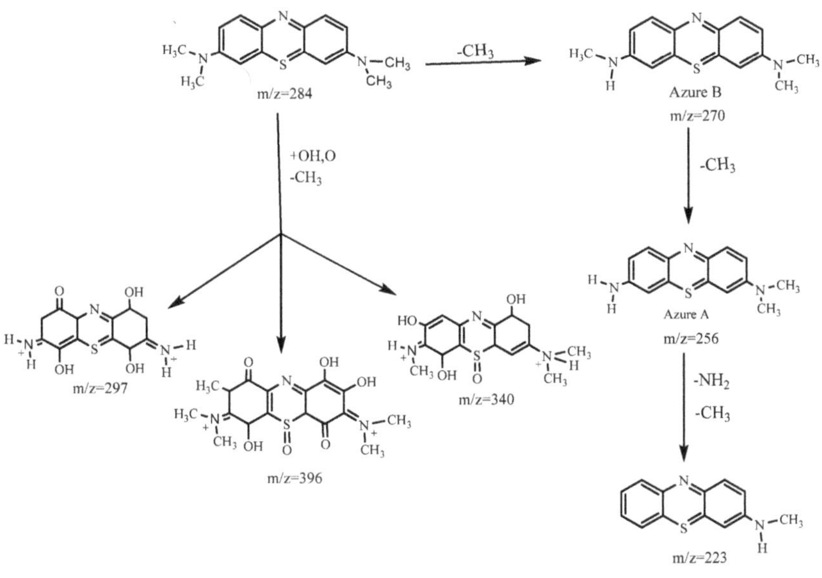

Scheme 1. Presumed degradation mechanism of MB during photodegradation in the presence of Zn_2SnO_4/SnO_2.

3.3.2. RhB Degradation Pathway

Based on the heatmap of m/z intensity values from the MS spectra of RhB photodegradation (time intervals: 0, 15, 30, 60, 90, and 120 min) shown in Figure 11, it is possible to hypothesize a photodegradation pathway for Rhodamine B that essentially consists of two notable processes: sequential N-deethylation [54,61–63] and fragmentation of the xanthene ring [54]. In addition to these two mechanisms, dealkylation, deamination, decarboxylation, dehydration, chromophore cleavage or disruption of conjugation, and rupture of the cyclic structure were also observed. The molecular ion (Rhb0) at m/z 443 [54,61–64] increased during the first 15 min and then decreased significantly by 60 min, eventually being present at less than approximately 6%. This trend indicates that the main molecular ion of RhB is rapidly consumed, thereby initiating the degradation process. It was observed that the main photodegradation products were the N-deethylated intermediates at m/z 415 [54,61–63] and m/z 397 (the 28 mass unit difference between m/z 443 and m/z 415 corresponded to the loss of one ethyl group, which is characteristic of the first step of N-deethylation). The subsequent transition to the ion at m/z 397 indicates an additional transformation via a decarboxylation mechanism [61]. The subsequent deethylation step, which leads to the formation of more stable products, is presumed to be responsible for the ions observed at m/z 340, which displayed a continuous increase (rising during the first 30 min and maintaining high values up to 120 min), indicating the formation of a more stable fragment resulting from the further degradation of the preceding intermediates. Lower fragments, m/z 221, likely arise as a consequence of further cleavage of the aromatic system [61]. These peaks confirm a gradual fragmentation toward smaller molecules. We hypothesize that the ions at m/z 453, m/z 459 [65], and m/z 481 are predominantly transient signals in the early stages of degradation characteristic of short-lived intermediates that rapidly transform or may represent the presence of an impurity. The changes in the MS peak intensities clearly indicate that the photodegradation of Rhodamine B follows a mechanism in which the initial rapid decrease in the parent ion (m/z 443) is accompanied by the simultaneous formation of N-deethylation intermediates (*m*/*z* 415 and *m*/*z* 397) along with a further deethylation step (*m*/*z* 340 and lower fragments). This pattern, confirmed by the temporal changes in intensity, is largely consistent with the results presented in previous studies [54,61–65]. Based on the current mass spectrum analysis, the possible photodegradation mechanism of RhB is proposed in Scheme 2.

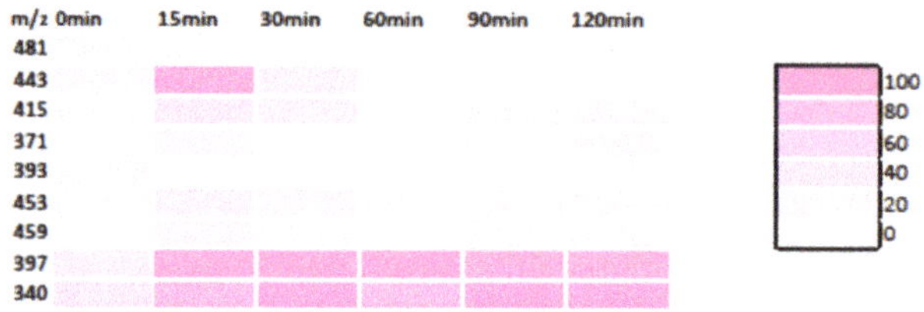

Figure 11. Heatmap illustrating the change in intensity of intermediary peaks during RhB degradation in the presence of Zn_2SO_4/SnO_2 for 0–120 min under natural sunlight.

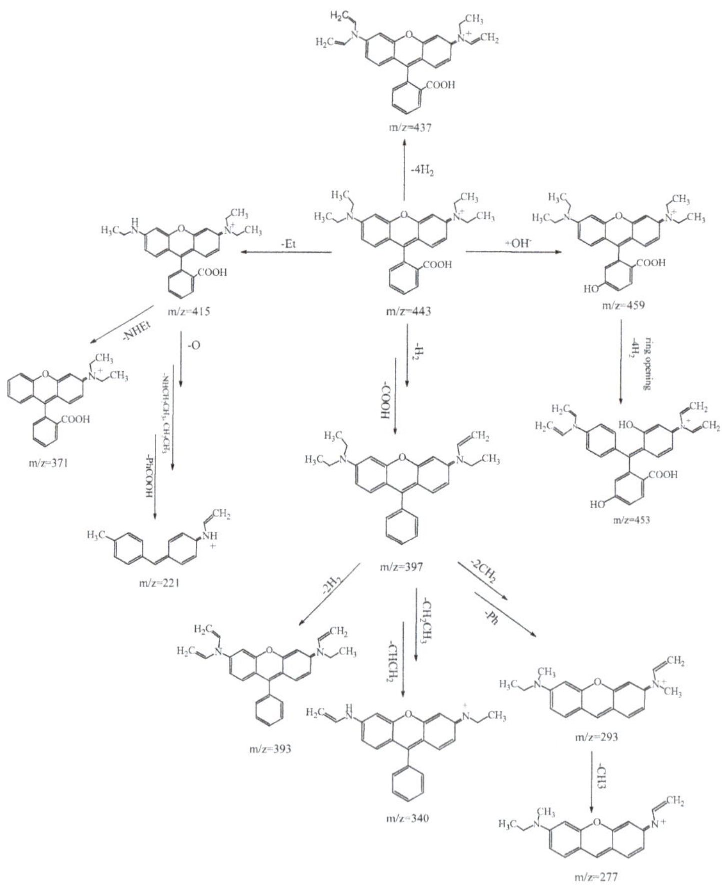

Scheme 2. Presumed degradation mechanism of RhB during photodegradation in the presence of Zn_2SO_4/SnO_2.

3.3.3. Degradation Pathway of the MB/RhB Mixture

Based on the heatmap shown in Figure 12, we assumed that both dyes degraded through characteristic pathways: MB primarily via N-demethylation and the formation of "azure" intermediates, and RhB predominantly through sequential N-deethylation followed by fragmentation of the xanthene ring. More stable ions (e.g., m/z 340 for RhB and m/z 223/256 for MB) initially dominated and then gradually decreased as photodegradation proceeded toward smaller, final fragments. We further assumed that fluctuations in the intensity of ionic peaks arose from competing reactions (oxidation, deamination, decarboxylation, and ring cleavage) as well as potential interactions between the two dyes when present together in solution. Overall, these observations led us to assume that the photodegradation of the MB/RhB mixture occurred through a series of interconnected steps demethylation/deethylation, oxidative fragmentation, and ring cleavage, resulting in the gradual loss of the characteristic chromophoric structures and the formation of smaller, potentially mineralized products.

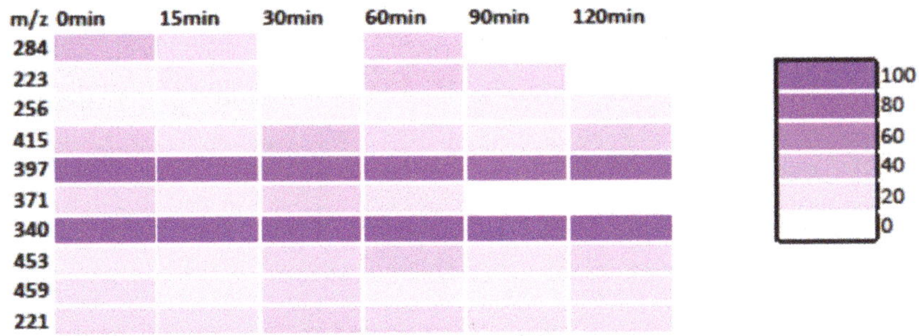

Figure 12. Heatmap illustrating the degradation ion profile of the MB/RhB mixture (0–120 min) in the presence of Zn_2SO_4/SnO_2 under natural sunlight.

4. Conclusions

Solid-state synthesis was applied to obtain a Zn_2SnO_4/SnO_2 nanocomposite with ~70% Zn_2SnO_4 and ~30% SnO_2. Detailed structural and morphological characterization confirmed the formation of the cubic spinel Zn_2SnO_4 (crystallite size 108 nm) and SnO_2 (crystallite size 88 nm) heterojunction nanocomposite powder. The estimated band gap value from the absorption edge was 3.68 eV, and the first derivative of the measured reflectance enabled the identification of additional transition energy that can be attributed to the occurrence of heterojunctions between Zn_2SnO_4 and SnO_2. Analysis of the photoluminescence spectra showed emissions that can be attributed to luminescent centers due to defect energy levels originating from oxygen vacancies, tin vacancies, or oxygen interstitials. The photocatalytic activity of this nanocomposite was evaluated on single-component organic dyes RhB and MB as well as a multicomponent system containing a mixture of RhB and MB under natural sunlight irradiation. A high degradation rate of 99.1% (k = 0.03828 min^{-1}) was achieved for MB after 120 min irradiation, while 70.6% (k = 0.01018 min^{-1}) of RhB degraded in the same time using 50 mg of the photocatalyst material. In the case of the multicomponent system, 15 mg of the photocatalyst material achieved a degradation efficiency of 97.9% for MB. The degradation pathway was described for MB, RhB, and MB+RhB, showing that in the mixed (multicomponent) dye system, it occurred through a series of interconnected steps consisting of demethylation/deethylation, oxidative fragmentation, and ring cleavage, resulting in the gradual loss of the characteristic chromophoric structures and the formation of smaller, potentially mineralized products.

Author Contributions: Conceptualization: M.V.N.; Investigation: M.V.N., Z.Z.V., M.D., N.R., J.V., M.T., and N.B.T.; Methodology: M.V.N. and Z.Z.V.; Formal analysis: M.V.N., N.R., M.T., and M.D.; Visualization: M.V.N. and M.D.; Writing—original draft: M.V.N. and M.D.; Writing—review and editing: M.V.N., Z.Z.V., M.D., N.R., and N.B.T. All authors have read and agreed to the published version of the manuscript.

Funding: This work was funded by the Ministry for Science, Innovations, and Technological Development of the Republic of Serbia, contracts 451-03-136/2025-03/200053 (M.V.N., Z.Z.V., M.D., M.T.), 451-030-136/2025-03/200017 (N.R.), and 451-030-136/2025-03/200175 (J.V.).

Data Availability Statement: Data available on request from the authors.

Acknowledgments: UHPLC/MS2 analysis was performed at the Department of Plant Physiology at the Institute for Biological Research "Siniša Stanković"—National Institute of the Republic of Serbia, University of Belgrade. The authors thank Danijela Mišić, Uroš Gašić, and Valentina Ćurić for their professional help.

Conflicts of Interest: The authors declare no conflicts of interest.

References

1. Byrne, C.; Subramanian, G.; Pillai, S.C. Recent advances in photocatalysis for environmental applications. *J. Environ. Chem. Eng.* **2018**, *6*, 3531–3555. [CrossRef]
2. Son, B.T.; Long, N.V.; Hang, N.T.N. Fly ash-, foundry sand-, clay- and pumice-based metal oxide nanocomposite as green photocatalysts. *RSC Adv.* **2021**, *11*, 30805. [CrossRef]
3. Chiu, Y.H.; Chang, T.F.M.; Chen, C.Y.; Sone, M.; Hsu, Y.J. Mechanistic insights into photodegradation of organic dyes using heterostructure photocatalysts. *Catalysts* **2019**, *9*, 430. [CrossRef]
4. Iqbal, M.A.; Akram, S.; Khalid, S.; Lal, B.; Hassan, S.U.; Ashraf, R.; Kezembayeva, G.; Mushtaq, M.; Chinibayeva, N.; Hosseini-Bandegharaei, A. Advanced photocatalysis as a viable and sustainable wastewater treatment process: A comprehensive review. *Environ. Res.* **2024**, *253*, 118947. [CrossRef]
5. Yadav, S.; Shakya, K.; Gupta, A.; Singh, D.; Chandran, A.R.; Aanappalli, A.V.; Goyal, K.; Rani, N.; Saini, K. A review on degradation of organic dyes by using metal oxide semiconductors. *Environ. Sci. Pollut. Res.* **2023**, *30*, 71912–71932. [CrossRef]
6. Samadi, M.; Zirak, M.; Naseri, A.; Kheirabadi, M.; Ebrahimi, M.; Moshfegh, A.Z. Design and tailoring of one dimensional ZnO nanomaterials for photocatalytic degradation of organic dyes: A review. *Res. Chem. Intermed.* **2019**, *45*, 2197–2254. [CrossRef]
7. Din, M.I.; Khalid, R.; Najeeb, J.; Hussain, Z. Fundamentals and photocatalysis of methylene blue dye using various nanocatalytic assemblies—A critical review. *J. Clean. Prod.* **2021**, *298*, 126567. [CrossRef]
8. Dlugosz, O.; Staron, A.; Brzoza, P.; Banach, M. Synergistic effect of sorption and photocatalysis on the degree of dye removal in single and multicomponent systems on ZnO-SnO_2. *Environ. Sci. Pollut. Res.* **2022**, *29*, 27042–27050. [CrossRef] [PubMed]
9. Wang, H.; Zhang, L.; Chen, Z.; Hu, J.; Li, S.; Wang, Z.; Liu, J.; Wang, X. Semiconductor heterojunction photocatalysts: Design, construction, and photocatalytic performance. *Chem. Soc. Rev.* **2014**, *43*, 5234–5244. [CrossRef]
10. Uddin, T.; Nicolas, Y.; Olivier, C.; Toupance, T.; Servant, L.; Müller, M.M.; Kleebe, H.J.; Ziegler, J.; Jaegermann, W. Nanostructured SnO_2-ZnO heterojunction photocatalysis showing enhanced photocatalytic activity degradation of organic dyes. *Inorg. Chem.* **2012**, *51*, 7764–7773. [CrossRef]
11. Kumar, N.; Jung, U.; Jung, B.; Park, J.; Naushad, M. Zinc hydroxystannate/zinc-tin oxide heterojunctions for the UVC assisted photodegradation of methyl orange and tetracycline. *Environ. Pollut.* **2023**, *316*, 120353. [CrossRef]
12. Markovic, S.; Stankovic, A.; Dostanic, J.; Veselinovic, L.; Mancic, L.; Skapin, S.D.; Drazic, G.; Jankovic-Castvan, I.; Uskokovic, D. Simultaneous enhancement of natural sunlight and artificial UV-driven photocatalytic activity of a mechanically activated ZnO/SnO_2 composite. *RSC Adv.* **2017**, *7*, 42725–42737. [CrossRef]
13. Le, T.H.; Khiem, T.N.; Quynh, H.T.T.; Tuan, P.V. Effect of the composition ratio of ZnO/Zn_2SnO_4 composite on its visible photocatalytic activities in the photodegradation of methylene blue and rhodamine B organic dyes. *Phys. Scr.* **2025**, *100*, 1065986. [CrossRef]
14. Vasiljevic, Z.Z.; Dojcinovic, M.P.; Vujancevic, J.D.; Jankovic-Castvan, I.; Ognjanovic, M.; Tadic, N.B.; Stojadinovic, S.; Brankovic, G.O.; Nikolic, M.V. Photocatalytic degradation of methylene blue under natural sunlight using iron titanate nanoparticles prepared by a modified sol-gel method. *R. Soc. Open Sci.* **2020**, *7*, 200708. [CrossRef] [PubMed]

15. Rovisco, A.; Branquinho, R.; Martins, J.; Oliveira, M.J.; Nunes, D.; Fortunato, E.; Martins, R.; Barquinha, P. Seed-layer free zinc tin oxide tailored nanostructures for nanoelectronic applications: Effect of chemical parameters. *ACS Appl. Nano Mater.* **2018**, *1*, 3986–3997. [CrossRef] [PubMed]
16. Jia, T.; An, J.; Yu, D.; Li, J.; Fu, F.; Wang, K.; Wang, W. Continuously improved photocatalytic performance of $Zn_2SnO_4/SnO_2/Cu_2O$ composites by structural modification. *Nanomaterials* **2019**, *9*, 1390. [CrossRef] [PubMed]
17. Rovisco, A.; Branquinho, R.; Deuermeier, J.; Freire, T.; Fortunato, E.; Martins, R.; Barquinha, P. Shape effect of zinc tin oxide nanostructures on photodegradation of methylene blue and rhodamine B under UV and visible light. *ACS Appl. Nano Mater.* **2021**, *4*, 1149. [CrossRef]
18. Silvestri, S.; de Oliveira, J.F.; Foletto, E.L. Degradation of methylene blue using Zn_2SnO_4 catalysts prepared with pore-forming agents. *Mater. Res. Bull.* **2019**, *117*, 56–62. [CrossRef]
19. Junploy, P.; Thongtem, S.; Thongtem, T.; Phurnangrat, A. Photocatalytic activity of Zn_2SnO_4-SnO_2 nanocomposites produced by sonochemistry in combination with high temperature calcination. *Superlattice. Microst.* **2014**, *74*, 173–183. [CrossRef]
20. Li, B.; Luo, L.; Xiao, T.; Hu, X.; Lu, L.; Wang, J.; Tang, Y. Zn_2SnO_4-SnO_2 heterojunction nanocomposites for dye sensitized solar cells. *J. Alloys Compd.* **2011**, *509*, 2186–2191. [CrossRef]
21. Wang, J.; Li, H.; Meng, S.; Zhang, L.; Fu, X.; Chen, S. One-pot hydrothermal synthesis of highly efficient SnO_x/Zn_2SnO_4 composite photocatalyst for the degradation of methyl orange and gaseous benzene. *Appl. Catal. B* **2017**, *200*, 19–30. [CrossRef]
22. Keles, E.; Yildirim, M.; Öztürk, T.; Yildirim, O.A. Hydrothermally synthesized UV light active zinc stannate: Tin oxide (ZTO: SnO_2) nanocomposite photocatalysts for photocatalytic applications. *Mater. Sci. Semicond. Proc.* **2020**, *110*, 104959. [CrossRef]
23. Junploy, P.; Phuruangrat, A.; Plubphon, N.; Thongtem, S. Photocatalytic degradation of methylene blue by Zn_2SnO_4-SnO_2 system under UV visible radiation. *Mater. Sci. Semicond. Proc.* **2017**, *66*, 56–61. [CrossRef]
24. Sun, L.; Han, X.; Jiang, Z.; Ye, T.; Li, R.; Zhao, X.; Han, X. Fabrication of cubic Zn_2SnO_4/SnO_2 complex hollow structures and their sunlight-driven photocatalytic activity. *Nanoscale* **2016**, *8*, 12858. [CrossRef] [PubMed]
25. Sun, L.; Li, S.; Su, Y.; He, D.; Zhang, Z. Surface-driven disorder engineered Zn_2SnO_4/SnO_2 hollow microboxes with enhanced solar-driven photocatalytic activity. *Appl. Surf. Sci.* **2019**, *463*, 474–480. [CrossRef]
26. Li, Z.; Zhao, Y.; Hua, L.; Xie, J.; Zhou, Y. Hollow SnO_2/Zn_2SnO_4 cubes with porous shells towards n-butylamine sensing and photocatalytic applications. *Vacuum* **2020**, *182*, 109693. [CrossRef]
27. Tang, M.; Snoussi, Y.; Bhakta, A.K.; El Garah, M.; Khalil, A.M.; Ammar, S.; Chehimi, M.M. Unusual, hierarchically structured composite of sugarcane pulp bagasse biochar loaded with Cu/Ni bimetallic nanoparticles for dye removal. *Environ. Res.* **2023**, *232*, 116232. [CrossRef]
28. Yusuff, A.S.; Popoola, L.T.; Aderibigbe, E.I. Solar photocatalytic degradation of organic pollutants in textile industry wastewater by ZnO/pumice composite photocatalyst. *J. Environ. Chem. Eng.* **2020**, *8*, 103907. [CrossRef]
29. Nikolic, M.V.; Labus, N.; Pavlovic, V.P.; Markovic, S.; Lukovic, M.D.; Tadic, N.; Vujancevic, J.; Vlahovic, B.; Pavlovic, V.B. Nanocrystalline Zn_2SnO_4/SnO_2: Crystal structure and humidity influence on complex impedance. *J. Electroceram.* **2020**, *45*, 135–147. [CrossRef]
30. Toby, B.H.; Von Dreele, R.B. GSAS II the genesis of an open-source all-purpose crystallography software package. *J. Appl. Crystallogr.* **2013**, *46*, 544–549. [CrossRef]
31. Momma, K.; Izumi, F. VESTA 3 for three-dimensional visualization of crystal, volumetric and morphology data. *J. Appl. Crystallogr.* **2011**, *44*, 1271–1276. [CrossRef]
32. Gracia, L.; Beltran, A.; Andres, J. A theoretical study on the pressure induced phase transitions in the inverse spinel structure Zn_2SnO_4. *J. Phys. Chem. C* **2011**, *115*, 7740–7746. [CrossRef]
33. Bao, L.; Zang, J.; Wang, G.; Li, X. Atomic-scale imaging of cation ordering in inverse spinel Zn_2SnO_4 nanowires. *Nano. Lett.* **2014**, *14*, 6505–6509. [CrossRef] [PubMed]
34. Wang, K.; Liu, D.; Deng, P.; Liu, L.; Lu, S.; Sun, Z.; Ma, Y.; Wang, Y.; Li, M.; Xia, B.Y.; et al. Band alignment in Zn_2SnO_4/SnO_2 heterostructure enabling efficient CO_2 electrochemical reduction. *Nano Energy* **2019**, *64*, 103954. [CrossRef]
35. Li, H.; Ma, Q.; Guo, J.; Zhang, Q.; Wang, X.; Wang, Y. Enhanced surface electron migration of porous and hollow SnO_2/Zn_2SnO_4 heterostructures for efficient triethylamine sensing performance. *Appl. Surf. Sci.* **2022**, *597*, 153752. [CrossRef]
36. Hu, J.; Xiong, X.; Guan, W.; Tan, C. Hollow mesoporous SnO_2/Zn_2SnO_4 heterojunction and RGO decoration for high-performance detection of acetone. *ACS Appl. Mater. Interfaces* **2022**, *14*, 55249–55263. [CrossRef]
37. Dong, S.; Cui, L.; Tian, Y.; Xia, L.; Wu, Y.; Yu, J.; Bagley, D.M.; Sun, J.; Fan, M. A novel and high-performance double Z-scheme photocatalyst ZnO-SnO_2-Zn_2SnO_4 for effective removal of the biological toxicity of antibiotics. *J. Hazard. Mater.* **2020**, *399*, 123017. [CrossRef]
38. Lu, J.; Xie, Y.; Luo, F.; Fu, H.; Huang, X.; Liu, Y.; Liu, H. Heterostructures of mesoporous hollow Zn_2SnO_4/SnO_2 microboxes for high-performance acetone sensors. *J. Alloys Compd.* **2020**, *844*, 155788. [CrossRef]

39. Ye, W.; Li, J. Zn_2SnO_4/SnO_2 heterojunctions regulated by the mineralizer sodium hydroxide promotes highly sensitive detection of formaldehyde vapor at room temperature. *Ceram. Int.* **2024**, *50*, 30541–30551. [CrossRef]
40. Vasiljevic, Z.Z.; Lukovic, M.D.; Nikolic, M.V.; Tasic, N.B.; Mitric, M.; Aleksic, O.S. Nanostructured Fe_2O_3/TiO_2 thick films: Analysis of structural and electronic properties. *Ceram. Int.* **2015**, *41*, 6889–6897. [CrossRef]
41. Arlinghaus, F. Energy bands in stannic oxide (SnO_2). *J. Phys. Chem. Solids* **1974**, *35*, 931–935. [CrossRef]
42. Akgul, F.A.; Gumus, C.; Er, A.O.; Farha, A.H.; Akgul, G.; Ufuktepe, Y.; Liu, Z. Structural and electronic properties of SnO_2. *J. Alloys Compd.* **2013**, *579*, 50–56. [CrossRef]
43. Ahmed, A.S.; Azam, A.; Shafeq, M.; Chaman, M.; Tabassum, S. Temperature dependent structural and optical properties of tin oxide nanoparticles. *J. Phys. Chem. Solids* **2012**, *73*, 943–947. [CrossRef]
44. Alpuche Aviles, M.A.; Wu, J. Photoelectrochemical study of the band structure of Zn_2SnO_4 prepared by the hydrothermal method. *J. Am. Chem. Soc.* **2009**, *131*, 3216–3224. [CrossRef]
45. Jaculine, M.M.; Raj, C.J.; Das, S.J. Hydrothermal synthesis of highly crystalline Zn_2SnO_4 nanoflowers and their optical properties. *J. Alloys Compd.* **2013**, *577*, 131–137. [CrossRef]
46. Marotti, R.E.; Giorgi, P.; Machado, G.; Dalchiele, E.A. Crystallite size dependence of band gap energy for electrodeposited ZnO grown at different temperatures. *Sol. Energy Mater. Sol. Cells* **2006**, *90*, 2356–2361. [CrossRef]
47. Flak, D.; Braun, A.; Vollmer, A.; Rekas, M. Effect of titania on the electronic structure and transport properties of FSS-made Fe_2O_3 nanoparticles for hydrogen sensing. *Sens. Actuators B* **2013**, *187*, 347–355. [CrossRef]
48. Rajic, V.; Stojkovic Simatovic, I.; Veselinovic, L.; Belosevic Cavor, J.; Novakovic, M.; Popovic, M.; Skapin, S.D.; Mojovic, M.; Stojadinovic, S.; Rac, V.; et al. Bifunctional catalytic activity of $Zn_{1-x}Fe_xO$ toward the OER/ORR: Seeking an optimal stoichiometry. *Phys. Chem. Chem. Phys.* **2020**, *22*, 22078–22095. [CrossRef] [PubMed]
49. Ali, M.B.; Bark-Bouaifel, F.; Elhouichet, H.; Sieber, B.; Addad, A.; Boussekey, L.; Ferid, M.; Boukherroub, R. Hydrothermal synthesis, phase structure, optical and photocatalytical properties of Zn_2SnO_4 nanoparticles. *J. Colloid Interface Sci.* **2015**, *457*, 360–369. [CrossRef]
50. Mishra, R.K.; Kushwaha, A.; Sahay, P.P. Influence of Cu doping on the structural, photoluminescence and formaldehyde sensing properties of SnO_2 nanoparticles. *RSC Adv.* **2014**, *4*, 3904. [CrossRef]
51. Soltani, T.; Entezari, M.H. Photolysis and photocatalysis of methylene blue by ferrite bismuth nanoparticles under sunlight irradiation. *J. Mol. Catal. A Chem.* **2013**, *377*, 197–203. [CrossRef]
52. Wang, F.; Min, S.; Han, Y.; Feng, L. Visible-light-induced photocatalytic degradation of methylene blue with polyaniline-sensitized composite photocatalysts. *Superlattice. Microst.* **2010**, *48*, 170–180. [CrossRef]
53. Lee, J.W.; Nam, S.H.; Yu, J.H.; Kim, D.I.; Jeong, R.H.; Boo, J.H. Morphological modulation of urchin-line Zn_2SnO_4/SnO_2 hollow spheres and their applications as photocatalysts and quartz crystal microbalance measurements. *Appl. Surf. Sci.* **2019**, *474*, 78–84. [CrossRef]
54. Yu, K.; Yang, S.; He, H.; Sun, C.; Gu, C.; Ju, Y. Visible light-driven photocatalytic degradation of Rhodamine B over $NaBiO_3$: Pathways and mechanism. *J. Phys. Chem. A* **2020**, *113*, 10024–10032. [CrossRef]
55. Li, Z.; Yang, H.; Zhang, L.; Liu, R.; Zhou, Y. Stainless steel mesh-supported three-dimensional hierarchical SnO_2/Zn_2SnO_4 composite for the applications in solar cell, gas sensor and photocatalysis. *Appl. Surf. Sci.* **2020**, *502*, 144113. [CrossRef]
56. Small, J.M.; Hintelmann, H. Methylene blue derivatization then LC–MS analysis for measurement of trace levels of sulfide in aquatic samples. *Anal. Bioanal. Chem.* **2007**, *387*, 2881–2886. [CrossRef]
57. Molla, A.; Sahu, M.; Hussain, S. Under dark and visible light: Fast degradation of methylene blue in the presence of Ag–In–Ni–S nanocomposites. *J. Mater. Chem. A* **2015**, *3*, 15616–15625. [CrossRef]
58. Martin, N.; Leprince-Wang, Y. HPLC-MS and UV–Visible coupled analysis of methylene blue photodegradation by hydrothermally grown ZnO nanowires. *Phys. Status Solidi* **2021**, *218*, 2100532. [CrossRef]
59. Gupta, V.K.; Manisha, S.; Raj, K.V. Hydrothermal modification and characterization of bentonite for reactive adsorption of methylene blue: An ESI-MS study. *J. Environ. Chem. Eng.* **2015**, *3*, 2172–2179. [CrossRef]
60. Nguyen, T.B.; Doong, R.A.; Huang, C.P.; Chen, C.W.; Dong, C.D. Activation of persulfate by CoO nanoparticles loaded on 3D mesoporous carbon nitride (CoO@ meso-CN) for the degradation of methylene blue (MB). *Sci. Total Environ.* **2019**, *675*, 531–541. [CrossRef]
61. Rovisco, A.; Morais, M.; Branquinho, R.; Fortunato, E.; Martins, R.; Barquinha, P. Microwave-assisted synthesis of Zn_2SnO_4 nanostructures for photodegradation of Rhodamine B under UV and sunlight. *Nanomaterials* **2022**, *12*, 2119. [CrossRef]
62. Rasalingam, S.; Wu, C.M.; Koodali, R.T. Modulation of pore sizes of titanium dioxide photocatalysts by a facile template free hydrothermal synthesis method: Implications for photocatalytic degradation of rhodamine B. *ACS Appl. Mater. Interfaces* **2015**, *7*, 4368–4380. [CrossRef]

63. Zhong, H.E.; Shaogui, Y.A.N.G.; Yongming, J.U.; Cheng, S.U.N. Microwave photocatalytic degradation of Rhodamine B using TiO_2 supported on activated carbon: Mechanism implication. *J. Environ. Sci.* **2009**, *21*, 268–272. [CrossRef]
64. Siwinska-Ciesielczyk, K.; Swigon, D.; Rychtowski, P.; Moszynska, D.; Zgola-Grzieskowiak, A.; Jesionowski, T. The performance of multicomponent oxide systems based on TiO_2, ZrO_2 and SiO_2 in the photocatalytic degradation of Rhodamine B: Mechanism and kinetic studies. *Colloids Surf. A* **2020**, *586*, 124272. [CrossRef]
65. Wang, G.L.; Wang, J.; Zhou, L.P.; Cai, X.; Xu, M.; Lin, J.; Muddassir, M.; Sakiyama, H. A multi-functional Cd (II)-based coordination polymer for the highly sensitive detection of nitrofurazone and photocatalytic efficiency of Rhodamine B. *Inorg. Chim. Acta* **2021**, *527*, 120566. [CrossRef]

nanomaterials

MDPI

Article

Sunlight-Induced Photocatalytic Removal of Paracetamol Using Au-TiO_2 Nanoparticles

Lamine Aoudjit [1,†], Joana M. Queirós [2,3,4,*,†], A. S. Castro [2,3,4,5], Djamila Zioui [1], Noelia González-Ballesteros [6], S. Lanceros-Mendez [2,7,8,*] and Pedro M. Martins [3,4]

1 Unité de Développement des Equipements Solaires, UDES/Centre de Développement des Energies Renouvelables, CDER, Bou Ismail 42415, Algeria; aoudjit.lamine@udes.dz (L.A.); zioui.djamila@udes.dz (D.Z.)
2 Physics Centre of Minho and Porto Universities (CF-UM-UP) and LaPMET—Laboratory of Physics for Materials and Emergent Technologies, University of Minho, 4710-057 Braga, Portugal; id11653@alunos.uminho.pt
3 Centre of Molecular and Environmental Biology, University of Minho, 4710-057 Braga, Portugal; pamartins@bio.uminho.pt
4 IB-S—Institute for Research and Innovation on Bio-Sustainability, University of Minho, 4710-057 Braga, Portugal
5 Centre of Chemistry, University of Minho, 4710-057 Braga, Portugal
6 Departamento de Química Inorgánica, Universidade de Vigo, 36310 Vigo, Spain; noeliagb@uvigo.gal
7 BCMaterials, Basque Center for Materials, Applications and Nanostructures, UPV/EHU Science Park, 48940 Leioa, Spain
8 Ikerbasque, Basque Foundation for Science, 48009 Bilbao, Spain
* Correspondence: id10031@alunos.uminho.pt (J.M.Q.); lanceros@fisica.uminho.pt (S.L.-M.)
† These authors equally contributed to this work.

Abstract: Using sunlight as the driving force for photocatalytic processes holds great promise for sustainability. As a starting point for developing a material capable of degrading aquatic pollutants using solar energy as a stimulus, this work focuses on synthesizing Au-TiO_2 nanocomposites using the deposition–precipitation method. Characterization of Au-TiO_2 nanoparticles was performed by X-ray diffraction (XRD), Fourier transform infrared spectroscopy (FTIR), and Transmission Electron Microscopy (TEM). A model pollutant, paracetamol, was used to test the synergetic effect of Au (0.05 wt%) nanoparticles (NPs) with TiO_2 on photocatalytic activity. The influence of the parameters pH, loading (0.4, 0.8, and 1 g/L), pollutant concentration (20, 30, 40 ppm), and contact time (30, 60, 90, 120, 150, and 180 min) was studied by exposing the NPs to solar radiation. The photocatalytic degradation was most effective at a contact time of 3 h, an initial concentration of 20 ppm, and a pH of 6.8. Under these conditions, paracetamol in 1 g/L of Au-TiO_2 nanocomposites can be degraded by more than 99.17% under solar irradiation. As a result of the Au-TiO_2 composite's ability to successfully serve as a photocatalyst using sun radiation, water purification processes can be more widespread, cost-effective, and environmentally friendly.

Keywords: Au-TiO_2 catalyst; paracetamol; photocatalytic degradation; solar light; water remediation

1. Introduction

Earth's most valuable and critical natural resource is water, which is subject to high demand and contamination [1]. In recent times, the growth of industrial, agricultural, and residential activities has contributed to releasing numerous hazardous pollutants into wastewater, leading to continuous contamination of available freshwater sources [2]. The increasing sensitivity of analytical methods has revealed the presence of trace compounds such as pharmaceuticals [3], veterinary products [4], toiletries [5], food additives [6], pesticides [7], and other contaminants [8]. These compounds are typically known as Persistent Organic Pollutants (POPs). They are renowned for their toxicity and environmental persistence, which has raised significant concerns due to the potential threat to aquatic ecosystems and human health [9,10].

Pharmaceuticals are a significant source of pollution of water bodies, causing increasing concern due to their extensive use and unregulated discharge [11]. One compound of particular concern is acetaminophen (N-acetyl-p-aminophenol), widely known as paracetamol (PAR), which is a mild analgesic, readily available, and widely used to relieve headaches and other minor pains, among other treatments [11,12]. Municipal wastewater treatment plants (WWTPs) address PAR through traditional methods [13], making it one of the most prevalent compounds in natural and drinking water sources [12]. Conventional methods, such as biological treatment [14], ozonation [15], precipitation [16], reverse osmosis [17], and flotation/coagulation [18], have proven ineffective in removing pharmaceutical molecules owing to their high polarity and solubility in water, along with their molecular weight, structure, functionality, and variations in shape [19].

A recent analysis revealed the presence of PAR in surface water, groundwater, and tap or drinking water in 29 countries worldwide, with average and maximum measured environmental concentrations reaching 0.161 μg/L and 230 μg/L, respectively [20,21]. Considering the ubiquitous presence of PAR in water matrixes, several ecotoxicity studies have been conducted in different types of organisms. Daniel et al. [22], who assessed the acute and chronic effects of paracetamol at ecologically relevant levels on the crustacean *Daphnia magna*, reported no biochemical modifications after chronic exposure; however, high concentration levels of PAR are capable of triggering significant antioxidant responses. In another study, *Oncorhynchus mykiss* (rainbow trout) was shown to be highly reactive to PAR at ecologically relevant levels (12.5–50.0 μg/L), suffering oxidative effects and peroxidative damage [23]. Concerning plant ecotoxicity, a report shows alterations in plant physiology when exposed to PAR (2.5 g/L), such as in its uptake and absorption [24], metabolism and growth [25].

Considering the inefficiency of classical approaches to water remediation and the mentioned ecotoxicity impacts of PAR in the environment, it is paramount to foster the development of new materials and technologies that address this emerging problem. In this context, non-conventional treatments, including adsorption [26], membrane technologies [27], and Advanced Oxidative Processes (AOPs) [28], stand out for their potential to improve the removal of these contaminants from water [19]. Among all these approaches, AOPs, particularly photocatalysis, have attracted widespread interest due to the efficient use of solar energy to degrade organic pollutants [29]. This environmentally friendly process offers several benefits, including cost-effectiveness, use under ambient conditions, and efficient solar energy utilization to decompose organic pollutants [29–34].

The activation of semiconductors by solar light irradiation is a strategy to improve materials' functional response [35]. The semiconductors used as photocatalysts predominantly include titanium dioxide (TiO_2), which is the subject of in-depth research and widely applied in photocatalysis. This popularity is due to its outstanding optical and oxidation properties, superhydrophilicity, chemical stability, and low toxicity [36,37]. However, TiO_2 exhibits certain limitations. These include higher hole/electron recombination rates and reduced efficiency under visible light [38]. To overcome these limitations, TiO_2 is usually combined with other materials that attenuate electron–hole recombination, thus facilitating the transfer of these charges to electron acceptors and donors. Other strategies include introducing morphological modifications like increasing porosity or surface area or doping TiO_2 with different elements [39]. Noble metals such as silver (Ag) [40], platinum (Pt) [41], palladium (Pd) [42], and gold (Au) [43] can improve photocatalytic efficiency under visible light by acting as an electron trap, promoting interfacial charge transfer and therefore slowing down electron–hole pair recombination. Gold nanoparticle functionalization has attracted considerable attention because of its stability and nontoxicity [44–47]. In Au/TiO_2 composites, Au absorbs visible light through localized surface plasmon resonance (LSPR). This phenomenon allows Au nanoparticles to absorb visible light, thereby generating electron–hole pairs that can be injected into the conduction band of TiO_2, thus enhancing its photocatalytic performance [48,49]. Additionally, they have proven effective as photocatalysts in contaminant degradation processes using visible light [50]. This system is, therefore, promising for applications in photocatalytic reactions in water remediation [40,41,51,52].

This research focused on the synthesis and physical–chemical characterization of the Au-TiO_2 and a systematic assessment of their efficiency in the degradation of PAR, aiming to identify the optimal experimental parameters (i.e., PAR initial concentrations, catalyst concentration, pH, and type of radiation) for achieving the most effective photocatalytic activity. Moreover, the functionalization of the material was designed to improve its ability to absorb visible light, thereby enhancing its efficiency under solar radiation. Conducted in Algeria, this study capitalizes on the country's geographical position as one of the regions with the highest solar energy potentials globally, benefiting from abundant solar exposure [53]. By utilizing this natural resource, the approach also emphasizes the reuse of the material over multiple cycles and explores the process's scalability, striving to combine cost-effectiveness with sustainability.

2. Experiment

2.1. Materials and Methods

P25 TiO_2 nanoparticles were provided by Evonik (Essen, Germany), and Gold(III) chloride trihydrate was purchased from Sigma-Aldrich (St. Louis, MI, USA). Sodium hydroxide (98% purity) was purchased from Biochem Chemopharma Company (France). Paracetamol ($C_8H_9NO_2$, $\geq$99%) was purchased from Merck (Darmstadt, Germany), and some of its most relevant characteristics and properties are shown in Table 1. All solutions were prepared with ultrapure water (Millipore Milli-Q—resistivity 18.2 MΩ·cm).

Table 1. Characteristics and properties of paracetamol (PubChem).

Name IUPAC	Chemical Structure	λ_{max} Absorption (nm)	Molar Mass g/mol	UV–Vis Absorption Spectrum	Water Solubility (mg/mL) at 25 °C	Density (g/cm^3)	pk_a
N-(4-hydroxyphenyl)acetamide	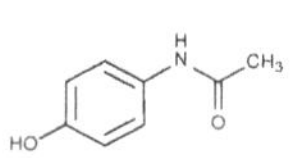	243	151.16	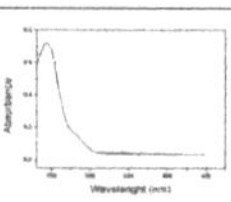	14.00	1.29	9.0–9.5

2.2. Synthesis of Au-TiO_2 Nanocomposite Photocatalysts

The Au-TiO_2 nanocomposites were synthesized by a deposition–precipitation method described by the authors [47,51,52,54,55]. Briefly, 200 mg of TiO_2 nanoparticles was dispersed in 40 mL of ultrapure water (UP) in a sonication bath for 30 min. Afterward, adding 1600 µL of chloroauric acid solution ($HAuCl_4$), an Au loading of 0.05 wt% was obtained.

The solution was stirred at 60 °C for 10 min to ensure homogeneous dispersion of the gold precursor. The pH was adjusted to approximately 9 by adding a sodium hydroxide solution (NaOH) from Sigma Aldrich (St. Louis, MI, USA) and mixing for 10 min. The solution was centrifuged for 20 min before being twice washed with ultrapure water (UP) and subjected to a 5 min ultrasonic process. Finally, the nanocomposite was dried overnight at 80 °C (J.P Selecta from Barcelona- Spain) in an oven and then ground with a mortar and pestle to obtain a fine powder.

2.3. *Au-TiO$_2$ Nanocomposite Characterization*

The morphological characterization of the Au-TiO$_2$ nanocomposites was carried out by transmission electron microscopy (TEM) using a JEOL JEM 1010 microscope (JEOL, Tokyo, Japan) operated at an accelerating voltage of 100 kV. Samples were first dispersed in water and sonicated to ensure uniform dispersion. Subsequently, the material was deposited onto a 400-mesh copper grid coated with Formvar and carbon using the drop-casting method. The crystallographic phases of the Au-TiO$_2$ nanocomposites were evaluated by X-ray diffraction (XRD) using a Bruker D8 Discover diffractometer with incident CuKα radiation (40 kV and 30 mA) from Bruker (Billerica, MA, USA). Fourier-transformed infrared spectroscopy (FTIR) was carried out in the range of 500–4000 cm^{-1}, 64 scans with a resolution of 4 cm^{-1}, using an FTIR Alpha instrument (Bruker Corporation, Billerica, MA, USA) to determine the chemical stability of the nanocomposites. The UV-vis reflectance measurements were performed using a Shimadzu UV 2501-PC (Kyoto, Japan) with and integrating sphere to evaluate the optical properties of both pristine TiO$_2$ and the Au-TiO$_2$ nanocomposite.

2.4. *Photocatalytic Degradation Processes*

The photocatalytic degradation of PAR was conducted in a batch photoreactor, represented by a 250 mL open Erlenmeyer. Before the photocatalytic tests, the Au-TiO$_2$ nanocomposite was kept in the dark for 30 min to ensure the adsorption–desorption equilibrium. The solution was then irradiated with sunlight (northern Algeria—latitude 36°39′; longitude 2°42′ at sea level—map location in Figure S1 in Supplementary Material) and UV radiation under magnetic agitation for three hours, and 3 mL aliquots were withdrawn at 30 min intervals, as illustrated in Figure 1a.

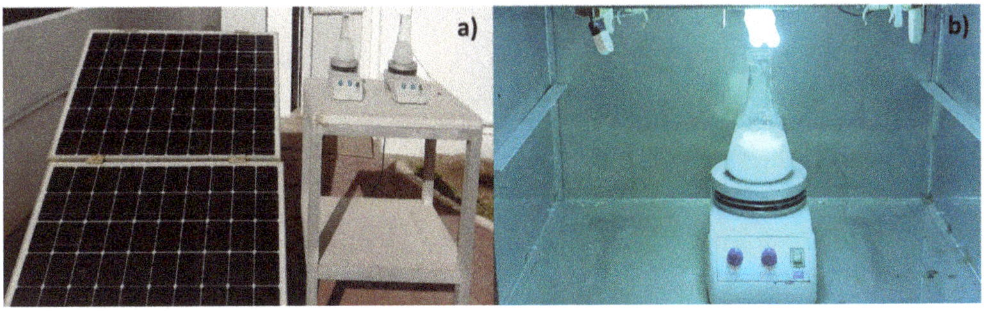

Figure 1. Photographs of the experimental setup to assess the degradation of PAR using Au-TiO$_2$ nanoparticles: (**a**) under solar radiation, where the Erlenmeyer contains the contaminant solution and nanoparticles while being magnetically stirred and powered by a solar panel that collects clean sunlight energy; (**b**) under a UV lamp.

Concurrently, the procedure mentioned above was conducted, subjecting the solution to ultraviolet radiation (UV), which was quantified with a pyranometer (KIPP & ZONZN,

CMP11, Sterling, VA, USA) with a spectral range from 285 to 2800 nm (800 W/m^2). The effect of UV radiation was investigated using a Philips PL-L 24W/10/4P UV lamp with a wavelength of 365 nm (spectra in Supplementary Materials—Figure S3) and an irradiance of 18.6 W/m^2 (Figure 1b). A graph showing the variation of solar intensity over the course of the day is provided in the Supplementary Materials (Figure S2), offering additional context to the experimental conditions.

This photocatalytic study examined several parameters, including the (1) effect of photolysis, adsorption, and photocatalysis on the removal of paracetamol from the solution, the (2) influence of the initial concentration of paracetamol, the (3) effect of the Au-TiO_2 dosage, the (4) pH of the paracetamol solution and the (5) type of radiation. All the parameters used are described in Table 2.

Table 2. Experimental parameters for the photocatalytic degradation of paracetamol, including variations in initial concentration, Au-TiO_2 dosage, pH solution, and types of radiation exposure (natural sunlight and solar ultraviolet radiation).

	(1) Photolysis/ Adsorption/Photocatalysis	(2) Paracetamol Initial Concentration	(3) Au-TiO_2 Dosage	(4) pH Solution	(5) Type of Radiation
Time (min)	120	180	180	180	180
$C_{initial}$[PAR] (mg/L)	20	20, 30, 40	20	20	20
C_{dosage}[Au-TiO_2] (g/L)	1	1	0.3, 0.8, 1	1	1
pH	6.8	6.8	6.6	3, 6.8, 10	6.8

All withdrawn samples were analyzed using a UV–visible spectrophotometer (Shimadzu-1800, Kyoto, Japan), and the peak at 243 nm was used to monitor paracetamol absorbance over the irradiation time. The pollutant concentration was determined using a calibration curve (Figure S5 in the Supplementary Materials), and the degradation percentage was estimated using Equation (1) [56]:

$$Degradation\ (\%) = \frac{C_0 - C_t}{C_0} \times 100 \quad (1)$$

where C_o is the initial pollutant concentration, and C_t is the concentration after a given reaction time t (min).

Still related to the photocatalytic degradation, the nanoparticles' reusability was evaluated based on their effectiveness over three usage cycles. The same catalyst previously employed was resuspended in distilled water after each cycle (180 min), recovered through centrifugation, and dried at room temperature.

2.5. Chemical Oxygen Demand Measurements

Samples were withdrawn at different times during the PAR degradation process to ensure that the mineralization process was monitored. Chemical Oxygen Demand (COD) analysis was performed by boiling an excess of potassium dichromate ($K_2Cr_2O_7$) in an acidic medium in the presence of silver sulphates (Ag_2SO_4) and mercury sulphate ($HgSO_4$) as catalysts. The organic matter partially reduces the dichromate, and the remaining rate, expressed in percentage (%), can be determined by the following equation [29,33,57]:

$$COD(\%) = \frac{COD_0 - COD_t}{COD_0} \times 100 \quad (2)$$

where COD_0 and COD_t are the initial COD and COD at time t, respectively. The COD measurements were performed using a spectrophotometer DR 1900 LANGE HACH (Dusseldorf, Germany).

3. Results and Discussion

3.1. Characterization of Au-TiO_2 Nanoparticles

The TEM technique was used to visualize the size, morphology and distribution of Au nanoparticles on the surface of TiO_2 nanoparticles. The titanium catalysts were functionalized with gold particles obtained at 60 °C with an Au load of 0.05 wt%. Figure 2a shows a homogeneous distribution of Au nanoparticles (round-shaped darker spots) over TiO_2, with a size of 3 nm (measured with Image J software, version 1.54g—200 nanoparticles), in agreement with the study by Martins et al. [47]. In addition, by comparing Figure 2a,b, it is shown that the contact with the paracetamol solution did not cause significant alterations in the morphology of the nanoparticles.

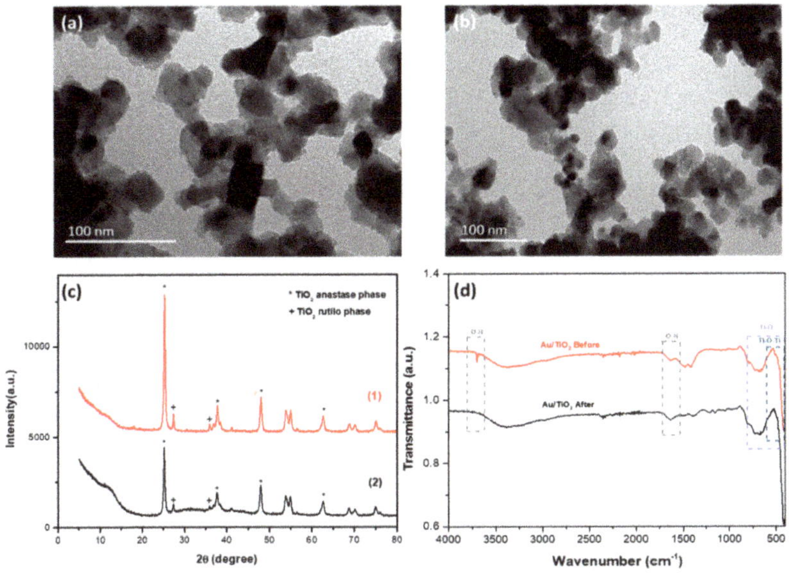

Figure 2. Scanning electron microscopy images of Au-TiO_2 nanoparticles (**a**) before contact and (**b**) after contact with the paracetamol solution; (**c**) XRD pattern of Au-TiO_2 before (1) and after (2) paracetamol degradation; (**d**) FTIR spectra of Au-TiO_2 before and after paracetamol degradation.

X-ray diffraction was carried out to assess the crystal structure of the Au-TiO_2 nanocomposite. The samples in Figure 2c show the typical reflections of anatase (peaks at 25.3, 37.8, 48.0 and 62.6 degrees) and rutile (peaks at 27.49 and 35.99 degrees), in good agreement with the literature [45,51,58,59]. No significant differences existed between the intensities or positions of the diffraction peaks of the two samples before (1) and after (2) contact with the paracetamol solution during the degradation process. In addition, the nanocomposite did not undergo significant changes in its crystalline structure after the radiation and contaminated water exposure, confirming its high stability.

The FTIR spectrum of the Au-TiO_2 nanocomposite shows peaks characteristic of the TiO_2 compound. These include the Ti-O stretching vibrations in the 400–800 cm^{-1} range [60] and the Ti-O-Ti bending modes in the 400–600 cm^{-1} region [61]. In Figure 2d, distinct bands are observed in spectra at 3699 and 1638 cm^{-1}, attributed to the O-H groups' stretching vibrations and the water molecules' deformation mode [62]. After the degradation of paracetamol, a certain amount of surface hydroxyl groups was removed by condensation and elimination of water due to exposure to sunlight during the degradation process of the Au-TiO_2 nanocomposite.

To compare the differences in photocatalyst performance between TiO_2 and Au-TiO_2, their optical properties were investigated using UV-vis diffuse reflectance spectroscopy, as illustrated in Figure S4 of the Supplementary Material. Pristine TiO_2 demonstrated high reflectance (95%) in the visible range (400–700), while the Au-TiO_2 nanocomposite exhibited lower reflectance (64%), with a minimum (44%) at 545 nm, indicating the surface plasmon resonance, typically observed between 520 nm and 560 nm [63,64]. The broader absorption spectrum of the nanocomposite is thus attributed to its characteristic purple/pink coloration [65].

3.2. Photocatalytic Degradation of Paracetamol

The photocatalytic performance of the prepared Au-TiO_2 particles was investigated by monitoring the concentration of paracetamol ($C_{initial}$ = 20 mg/L, pH 6.8, $C_{catalyst}$ = 1 g/L) over 120 min when exposed to sunlight. Control experiments were conducted using TiO_2 nanoparticles to evaluate paracetamol degradation under both solar and UV radiation, and the results are presented in Figure S3 of the Supplementary Materials. Additionally, preliminary studies were conducted to determine paracetamol stability (1), particle interaction in the absence of radiation (2), and Au-TiO_2 photocatalytic performance (3). In the first scenario, a paracetamol solution was subjected to solar radiation (photolysis). In the second, particles were in contact with paracetamol without radiation (adsorption phenomenon). Lastly, in the third scenario, the particles were exposed to energetic radiation while coming into contact with paracetamol (photocatalysis). The outcomes are illustrated in Figure 3a.

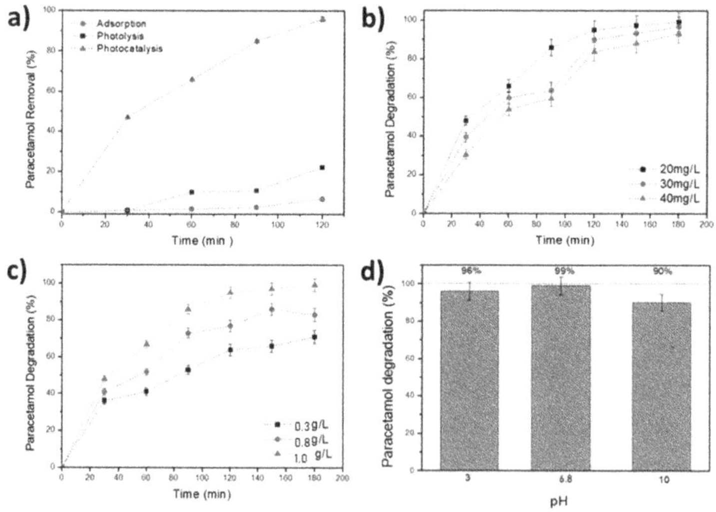

Figure 3. Study of photodegradation of paracetamol using sunlight radiation: (**a**) control process with C_{PAR} = 20 mg/L, $C_{catalyst}$ = 1 g/L and free pH 6.8, for 120 min; (**b**) effects of initial concentration (20, 30 and 40 mg/L); (**c**) different doses of Au-TiO_2 catalyst (0.3, 0.8 and 1.0 g/L); (**d**) impact of pH in solution (pH = 3, pH = 6.8 and pH = 10).

The impact of various parameters on the catalytic process was also evaluated, including initial concentration (20, 30, and 40 mg/L) (Figure 3b), catalyst dosage (0.3, 0.8, and 1.0 g/L) (Figure 3c), pH levels (3, 6.8 and 10) (Figure 3d) and the effect of radiation source: solar radiation located in northern Algeria (latitude 36°39′; longitude 2°42′ at sea level), from June to August, and UV radiation (Figure 4). These experiments were conducted to discern the influence of these factors on the degradation process of paracetamol.

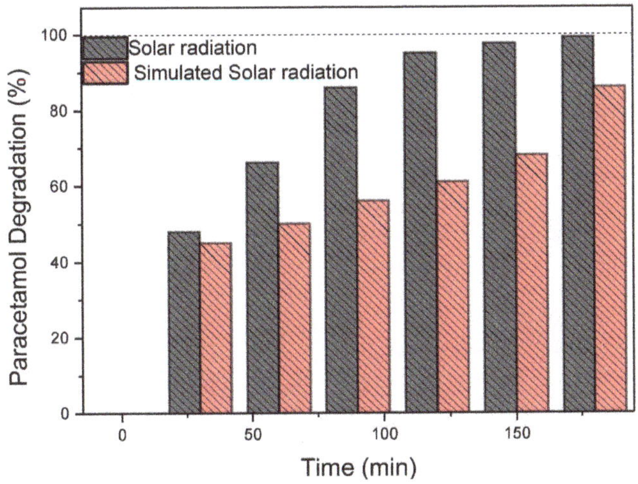

Figure 4. The impact of the irradiation source on the photocatalyst's degradation of PAR in 180 min (C_{PAR} at 20 mg/L, $C_{catalyst}$ at 1 g/L, and a pH of 6.8).

3.2.1. Influence of Parameters on the Photocatalytic Degradation of Paracetamol

The photocatalytic activity of the Au-TiO_2 nanocomposite was evaluated by adsorption, photolysis, and photocatalytic degradation of paracetamol as a function of time. Figure 3a shows that the photocatalysis process allows up to 99% of paracetamol to be degraded after 120 min. At the end of 50 min, around 50% of the contaminant had been degraded.

On the other hand, the degradation rates of the adsorption and photolysis processes were not as efficient as photocatalysis, but adsorption showed a significant value of 22.31%. Photolysis showed a 6.61% rate of paracetamol degradation after 120 min. These results align with other studies [33,36,66] that identify heterogeneous photocatalysis as a method that can effectively oxidize most organic pollutants. Therefore, this study will be based on removing paracetamol by photocatalysis.

The initial pollutant concentration (C_0) effect on the photocatalytic degradation efficiency of paracetamol was examined by varying the concentration within the range of 20 to 40 mg/L at pH (6.8) and a dose of 1 g/L Au-TiO_2 nanoparticles. The variation in the paracetamol concentration was considered regarding the record in surface waters passing through pharmaceutical factories of 50 mg/L [66]. The degradation of paracetamol increases significantly at lower initial concentrations, as shown by the percentage curves in Figure 3b over 180 min. At an initial concentration of 20 mg/L, a minimum concentration of approximately 0.26 mg/L (99.17% degradation yield) was reached, while at 30 and 40 mg/L, the final concentrations were 0.99 mg/L and 2.64 mg/L, corresponding to degradation yields of 98.9% and 93.33%, respectively.

These results indicate a marginal discrepancy in the final PAR concentration across the varying initial concentration levels, suggesting that degradation efficiency slightly decreases as the pollutant concentration increases. Our results corroborate prior research [56,67] focused on removing emerging contaminants. The postulate has been made that high concentrations of pollutants obstruct the formation of free radicals (OH^-) on the photocatalyst's surface by coating active sites with pollutant molecules. This phenomenon hinders the formation of active species responsible for the photocatalytic reactions. In addition, high concentrations prevent light from penetrating, as opposed to low concentrations, which act as a barrier. Hence, the initial pollutant concentration correlates inversely with the degradation efficiency [29,31,36,68].

Various tests with different amounts of Au-TiO_2 catalyst (0.4, 0.8, and 1 g/L) were carried out for photocatalytic studies to better understand the effect of catalyst dosage on paracetamol photocatalysis. Figure 3c illustrates the degradation rate of PAR following a 180 min reaction in the presence of different amounts of catalyst.

The results show that the degradation rate increases as the concentration of the Au-TiO_2 catalyst increases. At 1 g/L Au-TiO_2 dosage, the yield exceeds 99.17%. Thus, the photodegradation yield proportionally increases with the amount of the catalyst. The increased degradation rate is ascribed to increased active sites, which generate free radicals [30,68].

The effect of pH on the kinetics of paracetamol degradation was studied by varying the pH of the solution (3, 6.8, 10) (Figure 3d). The pH of the solutions was maintained at a constant level by adding sodium hydroxide (NaOH) as a basic medium and hydrochloric acid (HCl) as an acidic medium. The experimental outcomes indicate that pH moderately impacts paracetamol degradation, with a neutral pH (6.8) showing the highest degradation rate of 99.17% within 180 min. Moreover, the study demonstrated that PAR continued to degrade regardless of the pH of the solution, exhibiting a degradation efficiency of 90.00% at an alkaline pH of 10 and 96.69% at an acidic pH of 3 within the same time frame. The hydroxyl radical (OH·) is a powerful oxidizing agent of organic pollutants during the photolysis process, with its generation being dependent on pH. The production of OH· is slightly favored by alkaline conditions due to the increased availability of hydroxide ions. However, this effect is counterbalanced by an increase in electrostatic repulsion between the TiO_2 surface and paracetamol molecules as pH rises (pKa = 9.38), which results in a marginal reduction in adsorption and, consequently, in degradation efficiency. In acidic conditions (pH < 6.3), the TiO_2 surface carries a positive charge, which may enhance the adsorption of paracetamol.

As pH increases, however, the negative charge on the TiO_2 surface strengthens, slightly hindering the adsorption of paracetamol molecules. This, combined with a reduction in certain reactive species, such as $O_2^{-}\cdot$, due to the lower rate of oxygen reduction at high pH, may contribute to the observed decline in degradation efficiency. It is important to note that paracetamol degradation still occurs regardless of the pH level of the solution. A pH level of neutrality was selected as the optimal pH for the study at hand [69,70].

3.2.2. Radiation Type

The effect of varying the source UV radiation and natural solar radiation on the degradation of the contaminant with Au-TiO_2 nanoparticles was evaluated. The test was conducted using optimized parameters based on results obtained previously. Thus, the initial concentration of PAR was 20 mg/L, with 1 g/L of Au-TiO_2 NP at a pH of 6.8. Figure 4 shows the results obtained over 180 min under solar irradiation (800 W/m^2), where a degradation efficiency of approximately 99% was observed. Under UV radiation, significant degradation was also achieved, with efficiency reaching 86%. The degradation process is influenced by multiple factors, including radiation intensity and wavelength distribution, which affect the activation of electron–hole pairs and overall photocatalytic activity [36,56,68].

Solar radiation, which encompasses both highly energetic short ultraviolet wavelengths (<300 nm) and visible light, provides sufficient energy to activate TiO_2 and facilitate the generation of hydroxyl radicals, thereby significantly enhancing degradation efficiency [71].

Control experiments using TiO_2 were performed under both solar and UV radiation to assess the impact of Au functionalization (Figure S6). Under solar irradiation, TiO_2 exhibited ~47% degradation within 180 min, whereas Au-TiO_2 achieved 99% degradation. A similar trend was observed under UV light, where TiO_2 led to ~40% degradation, while Au-TiO_2 exhibited a higher efficiency of 82%. The enhanced performance of Au-TiO_2 under solar irradiation can be attributed to the localized surface plasmon resonance (LSPR) effect of gold, which extends light absorption into the visible range and improves charge separation, ultimately boosting photocatalytic efficiency [72,73]. Additionally, gold nanoparticles contribute to reducing electron–hole recombination, further enhancing the photocatalytic efficiency of the system [52].

Based in the literature, the degradation mechanism involves the generation of hydroxyl radicals as primary oxidizing agents, regardless of the radiation source [74,75]. The enhanced efficiency under solar radiation may be attributed to a higher production of these radicals, driven by its broader spectrum and higher intensity. Initially, hydroxylation of the aromatic ring leads to the formation of intermediates such as 4-acetamidocatechol (4-AC) and 4-acetamidoresorcinol (4-AR), along with hydroquinone (HQ). These intermediates undergo successive oxidation and ring cleavage, yielding smaller carboxylic acids before complete mineralization. Under UV radiation, competitive pathways may arise, leading to alternative degradation intermediates and influencing the overall efficiency [76]. This mechanistic insight explains the superior performance of Au-TiO_2 under solar radiation compared to UV exposure.

The results demonstrate the potential of Au-TiO_2 as an efficient photocatalyst, particularly in real-world applications where natural sunlight is the primary energy source. The localized surface plasmon resonance (LSPR) effect of gold allows for better solar spectrum utilization, enhancing efficiency and making the system viable for large-scale environmental applications. The study carried out in Algeria—one of the sunniest countries in the world with an abundance of solar energy resources [53]—took advantage of the plentiful solar energy available, making the process highly profitable. This feature positions the photocatalytic system as an optimal, sustainable solution for addressing local contamination challenges [77,78]. By leveraging sunlight, a cost-free, renewable resource, the approach further enhances economic viability, aligning efficiency with environmental and cost-conscious practices. Furthermore, the intensity of solar radiation during the experimental day, as illustrated in Figure S2, was measured to provide a more accurate reflection of real-world conditions. Solar radiation is dynamic, varying over the course of the day, and this directly influences the photocatalytic performance. The literature contains numerous studies that have explored the use of sunlight as a primary energy source for the photocatalytic degradation of PAR, as seen in Table 3.

Table 3. Comparison of studies on the efficiency of paracetamol degradation using different photocatalysts.

Material	C_{PAR} (mg/L)	Radiation	Time (Minutes)	Degradation (%)	References
Ce/Ag co-doped ZnO	20	Solar light	180	97.00	[51]
TiO_2-Au	20	Solar light	180	99.17	This work
		UV light		86	
TiO_2/activated carbon	50	Solar light	180	70	[79]
TiO_2/cellulosic fiber	40	Solar light	150	83	[80]
Zr-WO_3	20	UV light	210	73	[81]
Ag/ZnO	50	Solar light	240	83	[73]
		UV light		55	

Table 3 presents a selection of studies focusing on the functionalization of semiconductors to enhance their ability to absorb visible radiation to degrade PAR in an aqueous solution. Direct comparisons in photocatalytic experiments are challenging due to the inherent complexity of these systems, which are conducted under varying conditions, including pH, contaminant concentration, nanoparticle dosage, and radiation intensity/wavelength. While this complexity hinders direct result comparisons, the data remain valuable for contextualization. Thus, in one of the studies, Bouarroudj and colleagues [67] achieved a paracetamol degradation rate of 97.00%, marginally lower than the rate obtained in this investigation. In this study, Ce/Ag-co-doped ZnO composites were used and employed for 180 min to degrade an initial pollutant concentration of 20 mg/L.

For instance, the work developed by Ivanova et al. indicates a marked enhancement in the degradation efficiency of ZnO when compared with its functionalization with Ag. This increase was more than twofold, from 40% to 83%. This improvement was evident when both materials were subjected to solar radiation and UV lamps [73]. The works from Table 3 show different approaches to functionalizing semiconductors to absorb a range of radiation, resulting in a degradation time of less than four hours, yielding higher than 55% degradation percentages.

Another relevant study explores the photocatalytic activity of Zr-WO_3 for degradation of 20 mg/L of PAR under UV irradiation (254 nm, 25 W), achieving a degradation efficiency of 73% within 210 min. In comparison to the 86% degradation achieved in our Au-TiO_2 system under UV light in a shorter reaction time, our photocatalyst demonstrated a higher efficiency under similar conditions [81]. Regarding the mentioned works, our material also presents high efficiencies (for similar conditions). However, our work systematically evaluates different parameters, such as the type of radiation and reusability, making it more consistent and robust. Additionally, the materials we employ, namely TiO_2 and Au, represent a sustainable approach regarding aquatic ecosystems' ecotoxicity compared to materials such as ZnO and Ag.

3.3. Chemical Oxygen Demand Analysis

Chemical Oxygen Demand (COD) measures the oxygen required to oxidize organic compounds in water chemically. It assesses the quantity of oxygen required for the oxidation of organic matter present in wastewater and the overall oxygen consumption by said organic matter. This feature is of particular relevance and can be readily quantified, thereby facilitating the characterization of a range of waterbodies, including sewage, industrial waste, and treatment plant effluent [82]. At equivalent paracetamol degradation times (as illustrated in Figure 5a), the oxygen demand was calculated over 180 min (Figure 5b).

Figure 5a shows the absorption intensity of paracetamol throughout the photocatalyst degradation assay. It is possible to observe that the peak at 243 nm diminishes over 180 min of solar radiation exposure, indicative of the fragmentation of the primary PAR molecule. Complementing this, Figure 5b presents COD results, with values declining from 32 mg/L to 0.72 mg/L, confirming the mineralization of degradation by-products. The 99.17% degradation rate corresponds to the reduction of paracetamol. In comparison, the 98.53% mineralization rate reflects the near-complete conversion of intermediates into simpler compounds, demonstrating the effectiveness of the process in eliminating PAR. These results are consistent with those of Bouarroudj et al. [67], wherein the degradation percentage reached 97% within the corresponding time frame.

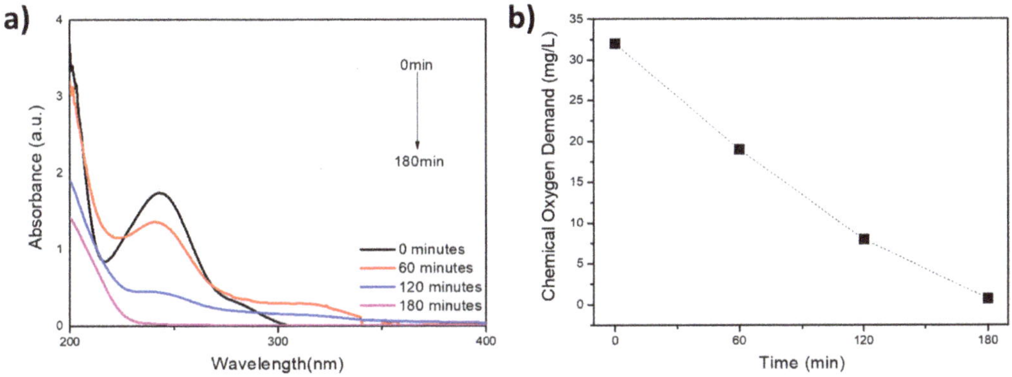

Figure 5. (**a**) UV–Vis spectra of paracetamol degradation under different sun irradiation times at pH = 6.8; (**b**) chemical oxygen demand (COD) at different times.

3.4. Reusability of the Au-TiO_2 Nanocomposite

Reusability constitutes an essential material feature in sustainability and the circular economy. To evaluate the reusability of Au-TiO_2 nanoparticles in the photocatalyst degradation of PAR under solar radiation, the nanoparticles were recovered and reused in two new cycles, as shown in Figure 6.

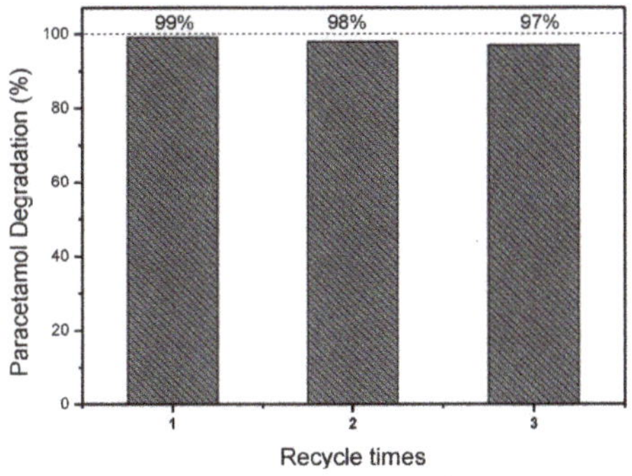

Figure 6. Photodegradation of paracetamol over three catalytic cycles in sunlight (C_{PAR} = 20 mg/L, $C_{catalyst}$ = 1 g/L, and pH 6.8).

The results indicate that Au-TiO_2 continues to exhibit a remarkable photocatalytic performance after three consecutive cycles with an efficiency loss below 2% [43]. This slight decrease in photocatalytic degradation efficiency can be attributed mainly to the residual loss of photocatalysts during the nanoparticle recovery through centrifugation and rinsing steps [83]. These outcomes are similar to reports in studies involving different catalysts. For context, ZnO nanoparticles demonstrated no substantial decline in photocatalytic activity after five cycles [84], and ZnO co-doped with Ce/Ag exhibited a decline from 97% to 95% after three reuse cycles [67], while another study reported a reduction from 98.4% to 94% [85]. In contrast, the Au-TiO_2 photocatalyst demonstrated superior stability, maintaining higher efficiency over several cycles with minimal loss. This reinforces its stability, making it particularly advantageous for environmental remediation applications. Moreover, it indicates that the process can be scaled up, as the catalyst retains high activity even after recovery and reuse, ensuring a cost-effective and environmentally friendly approach for large-scale applications.

The system has demonstrated significant efficiency in PAR removal under the studied conditions, achieving complete degradation, with the process being driven primarily by the photocatalyst under solar radiation. However, to further enhance reusability, reduce operational costs, and minimize potential environmental impacts associated with nanoparticle dispersion, future work will focus on immobilizing these photocatalysts into polymeric substrates. This strategy has been successfully employed in previous studies, improving catalyst recovery and stability while maintaining high photocatalyst efficiency [54,86].

4. Conclusions

The efficient removal of PAR from aquatic environments is a critical challenge, requiring innovative, economically viable techniques and environmentally sustainable solutions. This study successfully developed Au-TiO_2 nanocomposites using the deposition–precipitation technique and demonstrated their exceptional performance in the solar-powered degradation of PAR. A significant strength of this work lies in the comprehensive characterization of the nanocomposite using TEM, XRD, and FTIR, which confirmed their structural stability and chemical integrity, even after exposure to the contaminant and degradation process.

The photocatalytic activity of Au-TiO_2 was optimized by meticulously evaluating key parameters, including the PAR concentration, catalyst dosage, pH value, and application of different radiation sources. Under optimal conditions (20 mg/L PAR, 1 g/L AuTiO_2, pH 6.8, and 180 min of sunlight exposure), the nanocomposite achieves an outstanding degradation efficiency, reducing the PAR initial concentration to 0.25 mg/L (99% efficiency). Photocatalysis played the dominant role compared to adsorption, which removed 4.95 mg/L (22.31%), and photolysis, which contributed to a reduction of 1.46 mg/L (6.61%). Notably, the functionalization of TiO_2 with Au nanoparticles significantly enhanced the photocatalytic performance. Under solar radiation, the nanocomposite achieved 99% degradation, while 86% degradation was observed under UV exposure. This improvement is attributed to the LSPR of Au, which extends light absorption into the visible spectrum. Furthermore, the nanocomposite exhibits excellent stability during reuse, maintaining over 98% of its photocatalytic efficiency after three consecutive cycles, emphasizing its recyclability and cost-effectiveness for long-term applications.

The findings of this study highlight the potential of Au-TiO_2 to act as a resilient, effective, and recoverable photocatalyst, thereby establishing it as a promising contender for addressing pharmaceutical contamination in water bodies. The study also establishes a robust framework for advancing sustainable and scalable water purification technologies, paving the way for effective and environmentally friendly solutions.

Supplementary Materials: The following supporting information can be downloaded at: https://www.mdpi.com/article/10.3390/nano15050358/s1, Figure S1. Geographic location of the experimental site in northern Algeria (latitude 36.39° N, longitude 2.42° E). Figure S2. Temporal variation of solar intensity throughout the experimental day. Figure S3. UV lamp spectra provided by the manufacturer. Figure S4. UV–vis reflectance spectra of pristine TiO_2 and Au/TiO_2 measured between 250 and 700 nm. Figure S5. Calibration curve of Paracetamol. Figure S6. TiO_2 and Au-TiO_2 photocatalytic activities of photodegradation of paracetamol under (a) sunlight irradiation and (b) UV lamp irradiation (C_{PAR} = 20 mg/L, $C_{catalyst}$ = 1 g/L and free pH 6.8, for 180 min).

Author Contributions: Conceptualization, L.A., J.M.Q., A.S.C. and D.Z.; Methodology, L.A. and D.Z.; Software, L.A., J.M.Q., A.S.C., D.Z. and N.G.-B.; Validation, J.M.Q., N.G.-B. and P.M.M.; Formal analysis, L.A., J.M.Q. and A.S.C.; Investigation, L.A., J.M.Q., A.S.C., D.Z. and N.G.-B.; Resources, S.L.-M.; Writing—original draft, L.A. and D.Z.; Writing—review & editing, J.M.Q., A.S.C., N.G.-B., S.L.-M. and P.M.M.; Visualization, J.M.Q., A.S.C. and P.M.M.; Supervision, S.L.-M. and P.M.M.; Project administration, S.L.-M. and P.M.M.; Funding acquisition, S.L.-M. All authors have read and agreed to the published version of the manuscript.

Funding: This work was supported by the Solar Equipment Development Unit (UDES) Algeria. The authors thank the Fundação para a Ciência e Tecnologia (FCT) for financial Support under the framework of Strategic Funding UIDB/04650/2020, UID/FIS/04650/2020 and UID/04050. Joana M. Queirós thanks the FCT for the fellowship 10.54499/2021.08822.BD (https://doi.org/10.54499/2021.08822.BD), A. S. Castro thanks to FCT for the fellwoship 2024.01433.BD and Pedro M. Martins for contract 10.54499/2020.02802.CEECIND/CP1600/CT0017 (https://doi.org/10.54499/2020.02802.CEECIND/CP1600/CT0017). NGB was financed by the Ministry of Universities under application 33.50.460A.752 and by the European Union NextGenerationEU/PRTR through a contract Margarita Salas from the University of Vigo. This study formed part of the Advanced Materials program and was supported by MCIN with funding from European Union NextGenerationEU (PRTR-C17.I1) as well as by IKUR Strategy under the collaboration agreement between Ikerbasque Foundation and Fundación BCMaterials on behalf of the Department of Education of the Basque Government. Funding from the Basque Government Industry Department under the ELKARTEK program is also acknowledged.

Data Availability Statement: The raw data supporting the conclusions of this article will be made available by the authors on request.

Conflicts of Interest: The authors declare no conflict of interest.

References

1. Gopinath, K.P.; Madhav, N.V.; Krishnan, A.; Malolan, R.; Rangarajan, G. Present applications of titanium dioxide for the photocatalytic removal of pollutants from water: A review. *J. Environ. Manag.* **2020**, *270*, 110906. [CrossRef] [PubMed]
2. Bhatnagar, A.; Minocha, A. Conventional and non-conventional adsorbents for removal of pollutants from water–A review. *Indian J. Chem. Technol.* **2006**, *13*, 203–217.
3. Zanni, S.; Cammalleri, V.; D'Agostino, L.; Protano, C.; Vitali, M. Occurrence of pharmaceutical residues in drinking water: A systematic review. *Environ. Sci. Pollut. Res.* **2024**, 1–28. [CrossRef] [PubMed]
4. Delgado, N.; Orozco, J.; Zambrano, S.; Casas-Zapata, J.C.; Marino, D. Veterinary pharmaceutical as emerging contaminants in wastewater and surface water: An overview. *J. Hazard. Mater.* **2023**, *460*, 132431. [CrossRef] [PubMed]
5. Ashraf, I.; Agarwal, A.; Singh, N.B. Removal of pharmaceuticals, cosmetics, and toiletries from water by nanomaterials. In *Nanotechnology to Monitor, Remedy, and Prevent Pollution*; Elsevier: Amsterdam, The Netherlands, 2024; pp. 323–340.
6. Morin-Crini, N.; Lichtfouse, E.; Liu, G.; Balaram, V.; Ribeiro, A.R.L.; Lu, Z.; Stock, F.; Carmona, E.; Teixeira, M.R.; Picos-Corrales, L.A. Worldwide cases of water pollution by emerging contaminants: A review. *Environ. Chem. Lett.* **2022**, *20*, 2311–2338. [CrossRef]

7. de Souza, R.M.; Seibert, D.; Quesada, H.B.; de Jesus Bassetti, F.; Fagundes-Klen, M.R.; Bergamasco, R. Occurrence, impacts and general aspects of pesticides in surface water: A review. *Process Saf. Environ. Prot.* **2020**, *135*, 22–37. [CrossRef]
8. Spaltro, A.; Pila, M.N.; Colasurdo, D.D.; Grau, E.N.; Román, G.; Simonetti, S.; Ruiz, D.L. Removal of paracetamol from aqueous solution by activated carbon and silica. Experimental and computational study. *J. Contam. Hydrol.* **2021**, *236*, 103739. [CrossRef] [PubMed]
9. Benosmane, S.; Bendjelloul, M.; Elandaloussi, E.H.; Touhami, M.; de Ménorval, L.-C. Experimental and modeling study on adsorption of emerging contaminants onto hyper-crosslinked cellulose. *Chem. Pap.* **2021**, *75*, 4021–4034. [CrossRef]
10. Ashraf, M.A. Persistent organic pollutants (POPs): A global issue, a global challenge. *Environ. Sci. Pollut. Res.* **2017**, *24*, 4223–4227. [CrossRef] [PubMed]
11. Parolini, M. Toxicity of the Non-Steroidal Anti-Inflammatory Drugs (NSAIDs) acetylsalicylic acid, paracetamol, diclofenac, ibuprofen and naproxen towards freshwater invertebrates: A review. *Sci. Total Environ.* **2020**, *740*, 140043. [CrossRef] [PubMed]
12. Villaroel, E.; Silva-Agredo, J.; Petrier, C.; Taborda, G.; Torres-Palma, R.A. Ultrasonic degradation of acetaminophen in water: Effect of sonochemical parameters and water matrix. *Ultrason. Sonochem.* **2014**, *21*, 1763–1769. [CrossRef] [PubMed]
13. Fawzi Suleiman Khasawneh, O.; Palaniandy, P. Photocatalytic degradation of pharmaceuticals using TiO_2 based nanocomposite catalyst-review. *Civ. Environ. Eng. Rep.* **2019**, *29*, 1–33.
14. Taoufik, N.; Boumya, W.; Janani, F.; Elhalil, A.; Mahjoubi, F. Removal of emerging pharmaceutical pollutants: A systematic mapping study review. *J. Environ. Chem. Eng.* **2020**, *8*, 104251. [CrossRef]
15. Issaka, E.; Amu-Darko, J.N.-O.; Yakubu, S.; Fapohunda, F.O.; Ali, N.; Bilal, M. Advanced catalytic ozonation for degradation of pharmaceutical pollutants—A review. *Chemosphere* **2022**, *289*, 133208. [CrossRef] [PubMed]
16. Golbaz, S.; Jafari, A.J.; Rafiee, M.; Kalantary, R.R. Separate and simultaneous removal of phenol, chromium, and cyanide from aqueous solution by coagulation/precipitation: Mechanisms and theory. *Chem. Eng. J.* **2014**, *253*, 251–257. [CrossRef]
17. Lee, W.; Goh, P.; Lau, W.; Ismail, A. Removal of pharmaceutical contaminants from aqueous medium: A state-of-the-art review based on paracetamol. *Arab. J. Sci. Eng.* **2020**, *45*, 7109–7135. [CrossRef]
18. Kooijman, G.; De Kreuk, M.; Houtman, C.; Van Lier, J. Perspectives of coagulation/flocculation for the removal of pharmaceuticals from domestic wastewater: A critical view at experimental procedures. *J. Water Process Eng.* **2020**, *34*, 101161. [CrossRef]
19. dos Santos, D.F.; Moreira, W.M.; de Araújo, T.P.; Bergamasco, R.; Ostroski, I.C.; de Barros, M.A.S.D. Non-conventional processes applied for the removal of pharmaceutics compounds in waters: A review. *Process Saf. Environ. Prot.* **2022**, *167*, 527–542. [CrossRef]
20. aus der Beek, T.; Weber, F.A.; Bergmann, A.; Hickmann, S.; Ebert, I.; Hein, A.; Küster, A. Pharmaceuticals in the environment—Global occurrences and perspectives. *Environ. Toxicol. Chem.* **2016**, *35*, 823–835. [CrossRef]
21. Pham, T. Occurrence of Pharmaceutical Residues in Water and Treatment Solutions. Bachelor's Thesis, Metropolia University of Applied Sciences, Vantaa, Finland, 2018.
22. Daniel, D.; Dionísio, R.; de Alkimin, G.D.; Nunes, B. Acute and chronic effects of paracetamol exposure on *Daphnia magna*: How oxidative effects may modulate responses at distinct levels of organization in a model species. *Environ. Sci. Pollut. Res.* **2019**, *26*, 3320–3329. [CrossRef]
23. Ramos, A.; Correia, A.; Antunes, S.; Gonçalves, F.; Nunes, B. Effect of acetaminophen exposure in *Oncorhynchus mykiss* gills and liver: Detoxification mechanisms, oxidative defence system and peroxidative damage. *Environ. Toxicol. Pharmacol.* **2014**, *37*, 1221–1228. [CrossRef] [PubMed]
24. Hammad, H.M.; Zia, F.; Bakhat, H.F.; Fahad, S.; Ashraf, M.R.; Wilkerson, C.J.; Shah, G.M.; Nasim, W.; Khosa, I.; Shahid, M. Uptake and toxicological effects of pharmaceutical active compounds on maize. *Agric. Ecosyst. Environ.* **2018**, *258*, 143–148. [CrossRef]
25. Kummerová, M.; Zezulka, Š.; Babula, P.; Tříska, J. Possible ecological risk of two pharmaceuticals diclofenac and paracetamol demonstrated on a model plant Lemna minor. *J. Hazard. Mater.* **2016**, *302*, 351–361. [CrossRef]
26. Akhtar, J.; Amin, N.A.S.; Shahzad, K. A review on removal of pharmaceuticals from water by adsorption. *Desalination Water Treat.* **2016**, *57*, 12842–12860. [CrossRef]
27. Alfonso-Muniozguren, P.; Serna-Galvis, E.A.; Bussemaker, M.; Torres-Palma, R.A.; Lee, J. A review on pharmaceuticals removal from waters by single and combined biological, membrane filtration and ultrasound systems. *Ultrason. Sonochem.* **2021**, *76*, 105656. [CrossRef]
28. Moreira, F.C.; Soler, J.; Alpendurada, M.; Boaventura, R.A.; Brillas, E.; Vilar, V.J. Tertiary treatment of a municipal wastewater toward pharmaceuticals removal by chemical and electrochemical advanced oxidation processes. *Water Res.* **2016**, *105*, 251–263. [CrossRef]
29. Anucha, C.B.; Altin, I.; Bacaksiz, E.; Stathopoulos, V.N. Titanium dioxide (TiO_2)-based photocatalyst materials activity enhancement for contaminants of emerging concern (CECs) degradation: In the light of modification strategies. *Chem. Eng. J. Adv.* **2022**, *10*, 100262. [CrossRef]

30. Murgolo, S.; De Ceglie, C.; Di Iaconi, C.; Mascolo, G. Novel TiO2-based catalysts employed in photocatalysis and photoelectrocatalysis for effective degradation of pharmaceuticals (PhACs) in water: A short review. *Curr. Opin. Green Sustain. Chem.* **2021**, *30*, 100473. [CrossRef]
31. Aoudjit, L.; Martins, P.; Madjene, F.; Petrovykh, D.; Lanceros-Mendez, S. Photocatalytic reusable membranes for the effective degradation of tartrazine with a solar photoreactor. *J. Hazard. Mater.* **2018**, *344*, 408–416. [CrossRef] [PubMed]
32. Ghribi, F.; Sehailia, M.; Aoudjit, L.; Touahra, F.; Zioui, D.; Boumechhour, A.; Halliche, D.; Bachari, K.; Benmaamar, Z. Solar-light promoted photodegradation of metronidazole over ZnO-$ZnAl_2O_4$ heterojunction derived from 2D-layered double hydroxide structure. *J. Photochem. Photobiol. A Chem.* **2020**, *397*, 112510. [CrossRef]
33. Bouarroudj, T.; Aoudjit, L.; Djahida, L.; Zaidi, B.; Ouraghi, M.; Zioui, D.; Mahidine, S.; Shekhar, C.; Bachari, K. Photodegradation of tartrazine dye favored by natural sunlight on pure and (Ce, Ag) co-doped ZnO catalysts. *Water Sci. Technol.* **2021**, *83*, 2118–2134. [CrossRef] [PubMed]
34. Zioui, D.; Aoudjit, L.; Touahra, F.; Bachari, K. Preparation and characterization of TiO_2-chitosan composite films and application for tartrazine dye degradation. *Cellul. Chem. Technol* **2022**, *56*, 1101–1107. [CrossRef]
35. Gao, C.; Low, J.; Long, R.; Kong, T.; Zhu, J.; Xiong, Y. Heterogeneous single-atom photocatalysts: Fundamentals and applications. *Chem. Rev.* **2020**, *120*, 12175–12216. [CrossRef] [PubMed]
36. Aoudjit, L.; Zioui, D.; Touahra, F.; Mahidine, S.; Bachari, K. Photocatalytic degradation of tartrazine dyes using TiO_2–Chitosan beads under sun light irradiation. *Russ. J. Phys. Chem. A* **2021**, *95*, 1069–1076. [CrossRef]
37. Martins, P.; Miranda, R.; Marques, J.; Tavares, C.J.; Botelho, G.; Lanceros-Mendez, S. Comparative efficiency of TiO_2 nanoparticles in suspension vs. immobilization into P (VDF–TrFE) porous membranes. *RSC Adv.* **2016**, *6*, 12708–12716. [CrossRef]
38. Murcia Mesa, J.J.; Guarín Romero, J.R.; Cely Macías, Á.C.; Rojas Sarmiento, H.A.; Cubillos Lobo, J.A.; Hidalgo López, M.d.C.; Navío Santos, J.A. Methylene blue degradation over M-TiO_2 photocatalysts (M= Au or Pt). *Cienc. Desarro.* **2017**, *8*, 109–117. [CrossRef]
39. Anpo, M.; Kamat, P.V. *Environmentally Benign Photocatalysts: Applications of Titanium Oxide-Based Materials*; Springer Science & Business Media: Berlin/Heidelberg, Germany, 2010.
40. Wang, W.; Zhang, J.; Chen, F.; He, D.; Anpo, M. Preparation and photocatalytic properties of Fe^{3+}-doped Ag@ TiO_2 core–shell nanoparticles. *J. Colloid Interface Sci.* **2008**, *323*, 182–186. [CrossRef]
41. Zeng, Y.; Wu, W.; Lee, S.; Gao, J. Photocatalytic performance of plasma sprayed Pt-modified TiO_2 coatings under visible light irradiation. *Catal. Commun.* **2007**, *8*, 906–912. [CrossRef]
42. Cybula, A.; Nowaczyk, G.; Jarek, M.; Zaleska, A. Preparation and Characterization of Au/Pd Modified-TiO_2 Photocatalysts for Phenol and Toluene Degradation under Visible Light—The Effect of Calcination Temperature. *J. Nanomater.* **2014**, *2014*, 918607. [CrossRef]
43. Li, X.; Li, F. Study of Au/Au^{3+}-TiO_2 photocatalysts toward visible photooxidation for water and wastewater treatment. *Environ. Sci. Technol.* **2001**, *35*, 2381–2387. [CrossRef]
44. Ayati, A.; Ahmadpour, A.; Bamoharram, F.F.; Tanhaei, B.; Mänttäri, M.; Sillanpää, M. A review on catalytic applications of Au/TiO_2 nanoparticles in the removal of water pollutant. *Chemosphere* **2014**, *107*, 163–174. [CrossRef] [PubMed]
45. Zheng, F.; Martins, P.M.; Queirós, J.M.; Tavares, C.J.; Vilas-Vilela, J.L.; Lanceros-Méndez, S.; Reguera, J. Size effect in hybrid TiO_2: Au nanostars for photocatalytic water remediation applications. *Int. J. Mol. Sci.* **2022**, *23*, 13741. [CrossRef] [PubMed]
46. Primo, A.; Corma, A.; García, H. Titania supported gold nanoparticles as photocatalyst. *Phys. Chem. Chem. Phys.* **2011**, *13*, 886–910. [CrossRef] [PubMed]
47. Martins, P.; Kappert, S.; Nga Le, H.; Sebastian, V.; Kühn, K.; Alves, M.; Pereira, L.; Cuniberti, G.; Melle-Franco, M.; Lanceros-Méndez, S. Enhanced photocatalytic activity of Au/TiO_2 nanoparticles against ciprofloxacin. *Catalysts* **2020**, *10*, 234. [CrossRef]
48. Mondal, S.; Reyes, M.E.D.A.; Pal, U. Plasmon induced enhanced photocatalytic activity of gold loaded hydroxyapatite nanoparticles for methylene blue degradation under visible light. *RSC Adv.* **2017**, *7*, 8633–8645. [CrossRef]
49. Du, L.; Furube, A.; Hara, K.; Katoh, R.; Tachiya, M. Ultrafast plasmon induced electron injection mechanism in gold–TiO_2 nanoparticle system. *J. Photochem. Photobiol. C Photochem. Rev.* **2013**, *15*, 21–30. [CrossRef]
50. Huang, J.; Lin, M.; Cai, Y.; Luo, Y.-H.; Zhou, D.; Rittmann, B.E. Complete dehalogenation of chloramphenicol by bimetallic alloy Pd-Au nanoparticles in a H2-Based membrane Catalyst-Film reactor. *Chem. Eng. J.* **2024**, *497*, 154758. [CrossRef]
51. González-Ballesteros, N.; Martins, P.M.; Tavares, C.J.; Lanceros-Méndez, S. Quercetin-mediated green synthesis of Au/TiO_2 nanocomposites for the photocatalytic degradation of antibiotic ciprofloxacin. *J. Ind. Eng. Chem.* **2024**, *143*, 526–537. [CrossRef]
52. Espada, I.C.G.; González-Ballesteros, N.; Tavares, C.J.; Lanceros-Méndez, S.; Martins, P.M. Towards green visible range active photocatalytic Au/TiO_2 nanocomposites through rutin-based synthesis and their application in the degradation of ciprofloxacin. *RSC Sustain.* **2024**, *2*, 3090–3099. [CrossRef]

53. Stambouli, A.B.; Koinuma, H. A primary study on a long-term vision and strategy for the realisation and the development of the Sahara Solar Breeder project in Algeria. *Renew. Sustain. Energy Rev.* **2012**, *16*, 591–598. [CrossRef]
54. Zheng, F.; Martins, P.M.; Queirós, J.M.; Tavares, C.J.; Vilas-Vilela, J.L.; Lanceros-Méndez, S.; Reguera, J. Hybrid TiO_2:Au nanostars based polymeric membranes for photocatalytic degradation of ciprofloxacin in water samples. *Chemosphere* **2023**, *313*, 137630. [CrossRef] [PubMed]
55. Zheng, F.; Queirós, J.M.; Martins, P.M.; de Luis, R.F.; Fidalgo-Marijuan, A.; Vilas-Vilela, J.L.; Lanceros-Méndez, S.; Reguera, J. Au-sensitised TiO_2 and ZnO nanoparticles for broadband pharmaceuticals photocatalytic degradation in water remediation. *Colloids Surf. A Physicochem. Eng. Asp.* **2023**, *671*, 131594. [CrossRef]
56. Aoudjit, L.; Salazar, H.; Zioui, D.; Sebti, A.; Martins, P.M.; Lanceros-Mendez, S. Reusable Ag@ TiO_2-based photocatalytic nanocomposite membranes for solar degradation of contaminants of emerging concern. *Polymers* **2021**, *13*, 3718. [CrossRef] [PubMed]
57. Lv, Z.; Shan, X.; Xiao, X.; Cai, R.; Zhang, Y.; Jiao, N. Excessive greenhouse gas emissions from wastewater treatment plants by using the chemical oxygen demand standard. *Sci. China Earth Sci.* **2022**, *65*, 87–95. [CrossRef]
58. Fonseca-Cervantes, O.R.; Pérez-Larios, A.; Romero Arellano, V.H.; Sulbaran-Rangel, B.; Guzmán González, C.A. Effects in band gap for photocatalysis in TiO_2 support by adding gold and ruthenium. *Processes* **2020**, *8*, 1032. [CrossRef]
59. Martins, P.; Santos, B.; Salazar, H.; Carabineiro, S.A.; Botelho, G.; Tavares, C.J.; Lanceros-Mendez, S. Multifunctional hybrid membranes for photocatalytic and adsorptive removal of water contaminants of emerging concern. *Chemosphere* **2022**, *293*, 133548. [CrossRef]
60. de Almeida, M.F.; Bellato, C.R.; Miranda, L.D.; Milagres, J.L. Preparation of calcined hydrotalcite/TiO_2-Ag composite and enhanced photocatalytic properties. *Ceram. Int.* **2017**, *43*, 1843–1852. [CrossRef]
61. Li, Z.; Hou, B.; Xu, Y.; Wu, D.; Sun, Y.; Hu, W.; Deng, F. Comparative study of sol–gel-hydrothermal and sol–gel synthesis of titania–silica composite nanoparticles. *J. Solid State Chem.* **2005**, *178*, 1395–1405. [CrossRef]
62. Trujillano, R.; Rives, V.; García, I. Photocatalytic degradation of paracetamol in aqueous medium using TiO_2 prepared by the sol–gel method. *Molecules* **2022**, *27*, 2904. [CrossRef] [PubMed]
63. Tu, C.-Y.; Wu, J.M. Localized surface plasmon resonance coupling with piezophototronic effect for enhancing hydrogen evolution reaction with Au@MoS2 nanoflowers. *Nano Energy* **2021**, *87*, 106131. [CrossRef]
64. Guo, Z.; Dai, F.; Yin, H.; Zhang, M.; Xing, J.; Wang, L. The dual role of Au nanoparticles in the surface plasmon resonance enhanced photocatalyst Au/g-C3N4. *Colloid Interface Sci. Commun.* **2022**, *48*, 100615. [CrossRef]
65. Chen, N.; Xia, J.; Li, L.; Lv, Q.; Zhao, K.; Ahmad, M.; Xiao, Z.; Wang, S.; Ye, F.; Zhang, Q. Comprehensive enhancement of photocatalytic H_2O_2 generation and antibacterial efficacy on carbon nitride through a straightforward polydopamine coating strategy. *Surf. Interfaces* **2025**, *56*, 105566. [CrossRef]
66. Sheikhmohammadi, A.; Asgari, E.; Alinejad, N.; Hashemzadeh, B. Photocatalytic oxidation of ciprofloxacin by UV/α-Fe_2O_3/sulfite: Mechanism, kinetic, degradation pathway. *Int. J. Environ. Health Res.* **2023**, *33*, 192–205. [CrossRef]
67. Bouarroudj, T.; Aoudjit, L.; Nessaibia, I.; Zioui, D.; Messai, Y.; Bendjama, A.; Mezrag, S.; Chabbi, M.; Bachari, K. Enhanced Photocatalytic Activity of Ce and Ag Co-Doped ZnO Nanorods of Paracetamol and Metronidazole Antibiotics Co-Degradation in Wastewater Promoted by Solar Light. *Russ. J. Phys. Chem. A* **2023**, *97*, 1074–1087. [CrossRef]
68. Salima, M.; Youcef, M.; Bouarroudj, T.; Chetoui, A.; Belkhettab, I.; Bezzi, H.; Aoudjit, L.; Zioui, D.; Ziouche, A.; eddine Mekki, D. Sunlight-assisted photocatalytic degradation of tartrazine in the presence of Mg doped ZnS nanocatalysts. *Solid State Sci.* **2023**, *143*, 107260. [CrossRef]
69. Shourong, Z.; Qingguo, H.; Jun, Z.; Bingkun, W. A study on dye photoremoval in TiO_2 suspension solution. *J. Photochem. Photobiol. A Chem.* **1997**, *108*, 235–238. [CrossRef]
70. Yang LiMing, Y.L.; Yu, L.; Ray, M. Degradation of paracetamol in aqueous solutions by TiO_2 photocatalysis. *Water Res.* **2008**, *42*, 3480–3488. [CrossRef]
71. Gurkan, Y.Y.; Kasapbasi, E.; Cinar, Z. Enhanced solar photocatalytic activity of TiO_2 by selenium (IV) ion-doping: Characterization and DFT modeling of the surface. *Chem. Eng. J.* **2013**, *214*, 34–44. [CrossRef]
72. Piedra-López, J.; Calzada, L.A.; Guerra-Blanco, P.; Ortíz-Landeros, J.; Elizalde-Martínez, I.; Valenzuela, M.A.; Albiter, E. Enhancing photocatalytic H_2 production and dye degradation: Comparative analysis of gold reduction techniques on Au/TiO_2 nanocomposites. *Catalysis Today* **2024**, *432*, 114610. [CrossRef]
73. Ivanova, D.; Tzvetkov, G.; Kaneva, N. Degradation of Paracetamol in Distilled and Drinking Water via Ag/ZnO Photocatalysis under UV and Natural Sunlight. *Water* **2023**, *15*, 3549. [CrossRef]
74. Dalmázio, I.; Alves, T.; Augusti, R. An appraisal on the degradation of paracetamol by TiO_2/UV system in aqueous medium: Product identification by gas chromatography-mass spectrometry (GC-MS). *J. Braz. Chem. Soc.* **2008**, *19*, 81–88. [CrossRef]

75. Moctezuma, E.; Leyva, E.; Aguilar, C.A.; Luna, R.A.; Montalvo, C. Photocatalytic degradation of paracetamol: Intermediates and total reaction mechanism. *J. Hazard. Mater.* **2012**, *243*, 130–138. [CrossRef] [PubMed]
76. Nasr, O.; Mohamed, O.; Al-Shirbini, A.-S.; Abdel-Wahab, A.-M. Photocatalytic degradation of acetaminophen over Ag, Au and Pt loaded TiO_2 using solar light. *J. Photochem. Photobiol. A Chem.* **2019**, *374*, 185–193. [CrossRef]
77. Kermia, A.E.B.; Fouial-Djebbar, D.; Trari, M. Occurrence, fate and removal efficiencies of pharmaceuticals in wastewater treatment plants (WWTPs) discharging in the coastal environment of Algiers. *Comptes Rendus Chim.* **2016**, *19*, 963–970. [CrossRef]
78. Ouda, M.; Kadadou, D.; Swaidan, B.; Al-Othman, A.; Al-Asheh, S.; Banat, F.; Hasan, S.W. Emerging contaminants in the water bodies of the Middle East and North Africa (MENA): A critical review. *Sci. Total Environ.* **2021**, *754*, 142177. [CrossRef]
79. Alalm, M.G.; Tawfik, A.; Ookawara, S. Enhancement of photocatalytic activity of TiO_2 by immobilization on activated carbon for degradation of pharmaceuticals. *J. Environ. Chem. Eng.* **2016**, *4*, 1929–1937. [CrossRef]
80. Jallouli, N.; Elghniji, K.; Trabelsi, H.; Ksibi, M. Photocatalytic degradation of paracetamol on TiO_2 nanoparticles and TiO_2/cellulosic fiber under UV and sunlight irradiation. *Arab. J. Chem.* **2017**, *10*, S3640–S3645. [CrossRef]
81. Guettaıa, D.; Gaffour, H. Photo-catalytıc degradatıon of paracetamol using a novel photocatalyst Zr–WO_3 doped charcoal. *React. Kinet. Mech. Catal.* **2024**, *138*, 533–550. [CrossRef]
82. Khan, S.; Ali, J. Chemical analysis of air and water. In *Bioassays*; Elsevier: Amsterdam, The Netherlands, 2018; pp. 21–39.
83. Cherif, Y.; Azzi, H.; Sridharan, K.; Ji, S.; Choi, H.; Allan, M.G.; Benaissa, S.; Saidi-Bendahou, K.; Damptey, L.; Ribeiro, C.S. Facile Synthesis of Gram-Scale Mesoporous Ag/TiO_2 Photocatalysts for Pharmaceutical Water Pollutant Removal and Green Hydrogen Generation. *ACS Omega* **2022**, *8*, 1249–1261. [CrossRef]
84. Hendrix, Y.; Rauwel, E.; Nagpal, K.; Haddad, R.; Estephan, E.; Boissière, C.; Rauwel, P. Revealing the Dependency of Dye Adsorption and Photocatalytic Activity of ZnO Nanoparticles on Their Morphology and Defect States. *Nanomaterials* **2023**, *13*, 1998. [CrossRef] [PubMed]
85. Yahi, N.; Aoudjit, L.; Kahina, I.; Baalache, I.; Sellam, D. Photocatalytic Degradation of Safranin O Dye Under Visible Light Using NiO-MgO Catalysts. *Cellul. Chem. Technol.* **2024**, *58*, 133–139. [CrossRef]
86. Costa, C.M.; Cardoso, V.F.; Martins, P.; Correia, D.M.; Gonçalves, R.; Costa, P.; Correia, V.; Ribeiro, C.; Fernandes, M.M.; Martins, P.M. Smart and multifunctional materials based on electroactive poly (vinylidene fluoride): Recent advances and opportunities in sensors, actuators, energy, environmental, and biomedical applications. *Chem. Rev.* **2023**, *123*, 11392–11487. [CrossRef] [PubMed]

nanomaterials

MDPI

Article

Solar Light Elimination of Bacteria, Yeast and Organic Pollutants by Effective Photocatalysts Based on $Ag/Cr-TiO_2$ and $Pd/Cr-TiO_2$

Mónica Hernández-Laverde [1,2], Nicola Morante [3], Blanca Liliana Gutiérrez [2], Julie Joseane Murcia [1], Katia Monzillo [3], Diana Sannino [3,*] and Vincenzo Vaiano [3]

1 Grupo de Catálisis, Escuela de Ciencias Químicas, Universidad Pedagógica y Tecnológica de Colombia (UPTC), Avenida Central del Norte, Tunja 150002, Boyacá, Colombia; monica.hernandez06@uptc.edu.co (M.H.-L.); julie.murcia@uptc.edu.co (J.J.M.)

2 Escuela de Ciencias Básicas Tecnología e Ingeniería, Universidad Nacional Abierta y a Distancia (UNAD), Calle 5 #1-08, Sogamoso 152217, Boyacá, Colombia; blgutierrezp@unadvirtual.edu.co

3 Department of Industrial Engineering, University of Salerno, Via Giovanni Paolo II 132, 84084 Fisciano, SA, Italy; nmorante@unisa.it (N.M.); kmonzillo@unisa.it (K.M.); vvaiano@unisa.it (V.V.)

* Correspondence: dsannino@unisa.it

Abstract: This study focused on searching for more effective nanomaterials for environmental remediation and health protection; thus, coliform bacteria, yeast and the organic food dye sunset yellow were selected as target pollutants to be eliminated under solar light by $Ag/Cr-TiO_2$ and $Pd/Cr-TiO_2$. Firstly, Cr^{3+} was in situ incorporated into the anatase crystalline lattice by the sol-gel method; then, Ag or Pd nanoparticles were deposited on $Cr-TiO_2$ by chemical photoreduction. The scientific challenge addressed by the development of these composites was to analyse the recovery of Cr, to be employed in photocatalyst formulation and the enhancement of the TiO_2 photocatalytic activity by addition of other noble metals. By extensive characterization, it was found that after TiO_2 doping with chromium, the parameters of the crystal lattice slightly increased, due to the incorporation of Cr ions into the lattice. The TiO_2 band gap decreased after Cr addition, but an increase in the optical absorptions towards the visible region after noble metals deposition was also observed, which was dependent of the Ag or Pd loading. Generally, it was observed that the noble metals type is a factor that strongly influenced the effectiveness of the photocatalysts concerning each substrate studied. Thus, by using $Ag(0.1\%)/Cr-TiO_2$, the complete elimination of *E. coli* from samples of water coming from a highly polluted river was achieved. $Pd(0.5\%)/Cr-TiO_2$ showed the highest efficiency in the elimination of *S. cerevisiae* from a lab prepared strain. On the other hand, the $Pd(0.1\%)/Cr-TiO_2$ sample shows the highest dye degradation rate, achieving 92% of TOC removal after 180 min.

Keywords: Cr-Ag; Cr-Pd; TiO_2; *E. coli*; *S. cerevisiae*; sunset yellow

1. Introduction

The increase in pollutant emission and its negative effects on the environment are important concerns worldwide [1]. For years the natural water sources have been contaminated with organic compounds such as dyes, solvents, oils, and fertilizers [2] with inorganic substances such as heavy metals, metallic compounds, inorganic salts [3], and enteric microorganisms [4,5]. These pollutants come from different anthropogenic activities. In order to reduce river pollution, a number of technologies have been studied such as heterogeneous photocatalysis, which has been demonstrated to be effective in the removal of emerging pollutants, heavy metals, organic pollutants, and pathogenic bacteria [6,7].

TiO_2 has been the most widely studied photocatalyst due to its non-toxic nature, high stability, low cost, and abundance [8,9]. However, this semiconductor presents some disadvantages when applied at a large scale, such as: (i) the wide band gap [10], (ii) the high rate of photo-induced electron-hole pair recombination [11], and (iii) this oxide only absorbs light in the ultraviolet region of the electromagnetic spectrum, which represents approximately 5% of the entire solar spectrum [12].

There are many strategies to overcome the limitations of TiO_2 such as: the application of metallic and non-metallic dopants [13–15], coupling or heterojunctions [16], and the deposition of noble metals on the titania surface such as Au [17], Ag [18], Pd [19], and Pt [20,21], which are promising pathways for the production of more efficient photocatalytic materials.

TiO_2 doping involves the integration of transition metals into the oxide structure; this procedure replaces a certain amount of Ti atoms with doping metal ions, thus inserting new energy levels into the band structure [22]. Some researchers have suggested that the redshift is necessary to enhance optical absorption, and it could be achieved by increasing the concentration of impurity ions or oxygen defects in the TiO_2 structure [23,24]. Thus, the main objective of doping with Cr is to decrease the extensive band gap and to modify the electronic structure of this semiconductor, which is focused on enhancing the TiO_2 absorption of the visible region of the electromagnetic spectrum.

Additionally, the metal co-catalysis aims to decrease the electron-hole recombination and therefore enhance the photocatalytic performance of TiO_2 [25]. In general, chemical photoreduction is employed to include plasmonic noble metals on the TiO_2 surface, thus, the metal nanoparticles addition leads to the formation of the Schottky junction [26]. The metals can also enhance the visible excitation of surface electrons by surface plasmon resonance (SPR) effects [27].

It is worth mentioning that the criteria to choose Cr as a dopant for TiO_2, lies in the fact that this metal is commonly employed in different industries as a wood protector, in the tanning of fibers, and in paintings; thus, this metal is leached in these industrial effluents. As an alternative to valorize and recycle these effluents, this metal is employed in the present study to enhance TiO_2 light absorption.

In this study, we also selected Pd for properties such as its high activity when it interacts with the oxide's surface [28]. The small size of highly dispersed Pd particles increases the activity concerning strong metal-support interactions [29], as well as the metal favoring the removal of organic pollutants and its antimicrobial properties, as reported in several studies [30,31]. In the case of Ag, the effect that this noble metal has on the optical and electrical properties of TiO_2 nanoparticles [32], its high intrinsic antimicrobial activity, and the capacity of silver ions to cause the denaturation of proteins present in bacterial cell walls and to delay bacterial growth, make it an interesting choice for the design of new photocatalysts for environmental applications [33].

Based on all the aforementioned considerations, achieving a high-performance photocatalyst with an appropriate bandgap and high electron-hole pair separation efficiency is the main target of this work. Based on a previous individuation of a suitable formulation of visible light activity, this study was focused on the design of two series of photocatalysts based on Cr-doped TiO_2. The first one was modified by Ag photodeposition and the second series of materials was obtained by Pd addition. The effect of the loading of these metals on the photocatalytic activity of the obtained materials was also evaluated by addition of Ag or Pd contents from 0.1 to 0.5 wt.%. The photocatalytic performance of the materials prepared was evaluated in the degradation of the coloring food azo dye Sunset yellow (SY FCF), and in the elimination of enteric bacteria and yeast.

2. Materials and Methods

2.1. Chemicals

Titanium tetraisopropoxide ($C_{12}H_{28}O_4Ti$ > 97% (*w*/*w*)), Sigma Aldrich, St. Louis, MO, USA), chromium (III) nitrate nonahydrate ($Cr(NO_3)_3{\cdot}9H_2O \geq 99\%$), Sunset Yellow FCF ($C_{16}H_{10}N_2Na_2O_7S_2$ 90%, Sigma Aldrich) (SY), Tetraamminepalladium (II) nitrate solution ($Pd(NH_3)_4(NO_3)_2$, 10 wt.% in H_2O, 99.99%, Sigma Aldrich), Silver Nitrate ($AgNO_3$, 99%, Sigma Aldrich), hydrochloric acid solution (HCl, 37% (*w*/*w*) in H_2O, Carlo Erba, Milan,

Italy), sodium hydroxide solution (NaOH, 0.5 M in H_2O, Sigma Aldrich), and distilled water (Carlo Erba) were purchased and used as received. The molecular structure of SY is composed of benzene and naphtalene rings connected by an azo-linkage and substituted with sulphonate groups, OH, and CH_3.

2.2. *Materials Preparation*

2.2.1. Preparation of Cr-TiO_2

The preparation method of Cr-TiO_2 was reported in earlier works [33,34]; thus, in a general procedure, the amount of $Cr(NO_3)_3 \cdot 9H_2O$ necessary to obtain a nominal Cr content of 0.7 wt.% was first dissolved in 50 mL of distilled water and this solution was mixed under magnetic stirring for 5 min. Then, 12.5 mL of $C_{12}H_{28}O_4Ti$ was dropped, under stirring at room temperature for 10 min. The suspension obtained was centrifuged and washed with distilled water three times; and finally, the powder obtained was calcined at 450 °C for 30 min, with a heating ramp of 20 °C min^{-1}.

2.2.2. Preparation of Pd/Cr-TiO_2 and Ag/Cr-TiO_2

6.25 mL of isopropanol was employed as a sacrificial agent, which was mixed with distilled water and the required amount of the metal precursor to achieve 0.1 to 0.5 wt.% of Ag or Pd with respect to the Cr-TiO_2 content [34,35]. The resulting suspension was sonicated for 10 min in an ultrasonic bath at 20 °C. Then, the suspension was stirred for 10 min in the dark under N_2 atmosphere. The suspension was irradiated for 2 h under continuous stirring and N_2 flux by two UV-A lamps (240 W m^{-2}). Finally, the suspension was centrifuged, and the resulting solids were dried at 90 °C for 8 h and labelled as Pd/Cr-TiO_2 and Ag/Cr-TiO_2 [34,35].

2.2.3. Physicochemical Analysis of the Photocatalytic Materials

The photocatalytic materials prepared as previously described were characterized by different techniques and a detailed description of the experimental and equipment conditions are described below:

X-ray fluorescence (XRF) was employed to determine the chemical composition of the photocatalysts, using Panalytical Minipal PW4025 equipment (Malvern Panalytical, Westminster, London, UK). The parameters used in these analyses were: He flux, 20 KV and 180 s.

X-ray diffraction (XRD) analysis was performed with an Xpert pro Panalytical diffractometer, using the Cu Kα radiation (35 mA and 40 KV) and with a scanning angle of 2θ from 10 to 90° with 0.05° passes. The crystallite anatase sizes ($D_{Anatase}$) were estimated by the Scherrer equation.

The specific surface area (S_{BET}) of the prepared nanomaterials was obtained using the Costech Sorptometer 1042 analyser by the volumetric N_2 adsorption at −196° C. Before the measurements, a degassing pretreatment of the photocatalytic particles was carried out at 150° C for 30 min in He flux.

The light absorption characteristics of the photocatalysts were studied by Diffuse reflectance UV-Vis spectrophotometry (UV-Vis DRS) in a Thermo scientific spectrometer model Evolution 300 (Thermo Fisher Scientific, Waltham, MA, USA) equipped with a diffuse reflectance $BaSO_4$ covered accessory (RSA-PE-20, Labsphere Inc., North Sutton, NH, USA) and applying a reference certificated blank (SRS-99-010, Labsphere Inc., North Sutton, NH, USA). Band-gap values were calculated from Kubelka–Munk functions by the Tauc plot for indirect semiconductors [36].

Fourier Transform Infrared Spectroscopy (FTIR) analyses were carried out by using an ATR cell in a Thermo Scientific Nicolet TM iS50 FT-IR spectrometer (Thermo Fisher Scientific), samples of the photocatalysts were evaluated from 4000 to 1000 cm^{-1} with a resolution of 2 cm^{-1}.

Electrochemical impedance spectroscopy (EIS) analysis of the photocatalysts was carried out using an Origalys potentiostat (ElectroChem OrigaLys, Lyon, France) and a conventional three-electrode cell equipped with an Ag/AgCl/3 M KCl reference electrode, a graphite rod was used as the counter electrode, and the working electrode. A film of the photocatalytic material was placed on the working electrode via drop-casting, wherein 20 mg of photocatalyst was mixed with 200 μL of Nafion solution (5 wt.%). The mixture coated a titanium sheet (1.5 × 1.5 cm) and was dried at 100 °C. EIS spectra were recorded from 0.1 to 100 KHz with a current amplitude of 10 mV and using a 0.5 M Na_2SO_4 aqueous solution as the electrolyte [37]. Measurements were performed at room temperature at open-circuit potential.

2.3. *Photocatalytic Activity Tests*

2.3.1. Sunset Yellow FCF Photocatalytic Removal

The laboratory photoreactor used was a Pyrex cylindrical vessel (I.D. = 10 cm; height = 6.0 cm). A magnetic stirrer was employed to keep the photocatalyst suspended in the solution. To prevent the reaction temperature from exceeding 35 °C, a cooling fan was positioned near the reaction system. Oxygen saturation was assured by an air flow of fine bubbling.

The dye solution volume was 100 mL with an initial fixed concentration of SY ranging between 5 and 25 ppm. The spontaneous pH of the solutions containing the dye was 5.5. In some tests, the initial pH was adjusted to 4 and 11, by adding fixed amounts of HCl or NaOH concentrated aqueous solutions, respectively.

Before irradiation, the photoreactor was kept in the dark for 2 h to achieve adsorption-desorption equilibrium of SY on the catalyst surface.

The reactor was then irradiated using two solar light lamps (Sun Glo, 8 W; wavelength range 300–900 nm; intensity 2 W m^{-2}) positioned 15 cm above the reactor's upper surface (Figure 1).

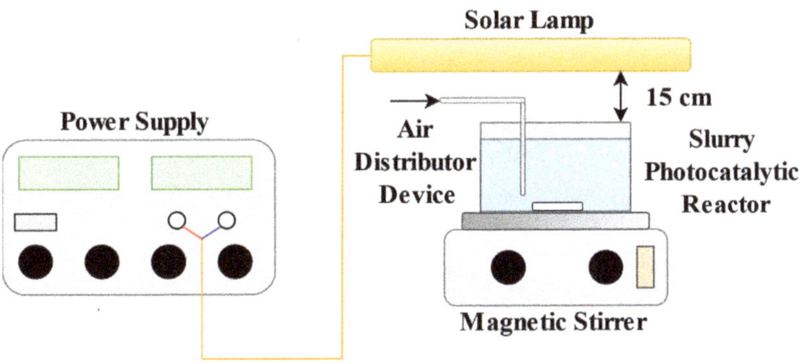

Figure 1. Schematic picture of the experimental setup for Sunset yellow photodegradation.

The photoreactor was covered with reflective aluminum foil to ensure only the upper surface was exposed to irradiation.

At regular intervals, approximately 3 mL of the suspension was withdrawn and centrifuged to remove catalyst particles. The aqueous solution was analysed with a UV-Vis spectrophotometer (Thermo Scientific Evolution 201) to monitor the reaction progress. Specifically, the discoloration of the dye was tracked by measuring the maximum absorbance at 485 nm [38]. The mineralization of the pollutant was assessed by measuring the total organic carbon (TOC) content in the solutions over the irradiation period. TOC was determined from CO_2 produced by high-temperature (680 °C) catalytic combustion [39].

A kinetic analysis of the photocatalytic discoloration of SY was performed using the Langmuir–Hinshelwood model, which is typically used to describe the kinetics of photocatalytic processes [40]. The degradation rate (r(t)) is expressed as follows:

$$r(t) = \frac{dc(t)}{dt} = \frac{k_r\, K_{ad}\, c}{1 + K_{ad}\, c} \quad (1)$$

where k_r, K_{ad}, and $c(t)$ represent the intrinsic kinetic constant, adsorption equilibrium constant, and dye concentration at a given irradiation time, respectively. Assuming weak adsorption and low compound concentrations, the above equation simplifies to a first-order kinetics expression with an apparent degradation kinetic constant (k):

$$\ln\left(\frac{c_0}{c}\right) = k_r\, K_{ad}\, c = k\, t \quad (2)$$

The value of the apparent discoloration kinetic constant can be determined from the slope of the straight line resulting by plotting ln(c_0/c) Vs time t. The TOC removal (mineralization) and SY discoloration efficiency at a given irradiation time were evaluated using the following relationship:

$$\text{TOC removal efficiency } (t) = \left(1 - \frac{TOC(t)}{TOC_0}\right) 100 \quad (3)$$

$$\text{SY discoloration efficiency } (t) = \left(1 - \frac{c(t)}{c_0}\right) 100 \quad (4)$$

where TOC(t) is the total organic carbon at a given irradiation time (mg L^{-1}), TOC_0 is the initial total organic carbon (mg L^{-1}), and c_0 is the initial SY concentration (mg L^{-1}).

2.3.2. Photocatalytic Treatment of Polluted River Water Samples

In these tests, water samples taken from a Colombian river called the Jordan River (geographic coordinates 5.553981, −73.350224) were studied. This river is highly polluted by industrial and domestic wastewater. The samples of the river were collected following the instructions given in the Standard Methods for the Examination of Water and Wastewater [41].

The river water samples were analysed by different physicochemical methods for the determination of chlorides, nitrates, pH, and conductivity in a Spectroquant® Move 100 instrument (Merck, Darmstadt, Germany). To ensure the reproducibility of the results, each test was carried out twice.

The microbiological analysis was performed by the membrane filtration method Merck SM 9222B and by ISO 9308 method part 1 [42].

For the photocatalytic activity test on the river water, a discontinuous batch reactor was employed and the reaction parameters were: (i) 250 mL of the water sample, (ii) continuous stirring, (iii) photocatalyst dosage: 1 g L^{-1}, (iv) light source: an Osram Ultra-Vitalux lamp (300 W) (Munich, Germany) with sun-like radiation spectrum in the UVA and UVB, (v) light intensity: 30 W m^{-2}, (vi) oxygen flow: 0.84 STP L h^{-1}, and (vii) total treatment time: 4 h.

After water treatment, the photocatalyst was recovered by filtration, and the treated water was analysed by the physicochemical and microbiological methods previously described in this section. To ensure the reproducibility of the results, all tests were carried out twice, with a standard deviation of 0.05. The arithmetic averages correspond to values reported in each test.

2.3.3. Photocatalytic Elimination of Yeast

For this test, a suspension of active dry yeast (*Saccharomyces cerevisiae*) from Levapan® (Bogotá, Colombia), was prepared with 3.8×10^5 number yeast/mL.

The photocatalytic activity test was conducted using the prepared yeast suspension under the conditions described in the previous Section 2.3.2. After 4 h of treatment, the photocatalyst was recovered by filtration, and 1 mL of the treated sample was taken for microbiological analysis. This analysis was carried out by the plate count method by surface seeding and following the ISO 21527 method for the enumeration of yeasts and moulds—Part 1 [43]. In order to ensure the reproducibility of the results, all the analyses were performed twice.

3. Results and Discussion

3.1. Physicochemical Analysis of Photocatalytic Materials

XRF: By using XRF, it was possible to confirm the effective incorporation of Cr, Ag, or Pd in the photocatalytic materials prepared.

The Cr content in the Cr-TiO_2 material was 0.6 wt.%, which is slightly lower than the nominal content (i.e., 0.7 wt.%). On the contrary, in the case of Ag and Pd, the content of these metals in the materials obtained was slightly higher than the nominal content, thus demonstrating the effectiveness of the photodeposition method. However, the results obtained can be in the error margin of the analysis technique.

XRD: X-ray diffraction patterns of all photocatalysts are plotted in Figure 2. As it can be seen in this figure, the samples present diffraction patterns assigned to the Anatase phase of TiO_2 (ICDD #21-1272) with the signal of highest intensity located at 25.6° 2θ, which corresponds to the (101) plane. Peaks at 38.1°, 48.1°, 55.0°, 62.6°, 70.2°, and 75.2° correspond to the (004), (200), (211), (204), (220) and (215) planes, respectively [44]. It was also possible to observe a typical signal associated with the brookite phase located at 2θ = 31.0° [45].

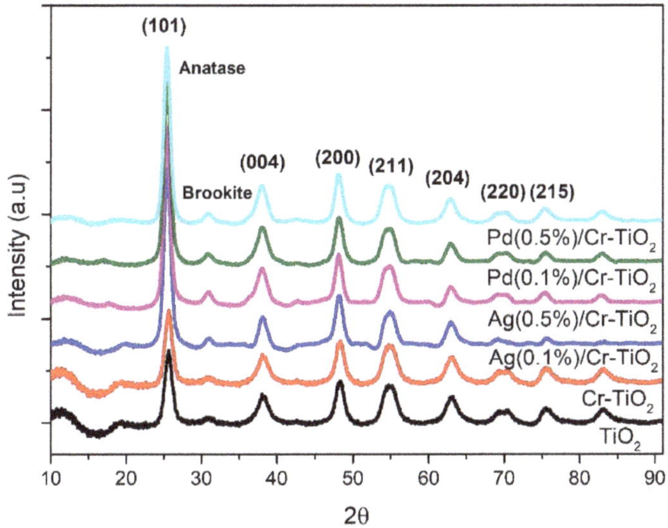

Figure 2. XRD patterns of all photocatalytic powders.

On the other hand, the calculated anatase lattice parameters for planes (101) and (004) are presented in Table 1. The results obtained indicated that the doped samples present a slight increase in parameters "a", "b", and "c" due to the expansion of the unit cell volume compared to the bare TiO_2, which is probably due to the lattice modification in the anatase structure arrangement by the incorporation of Cr ions into the Titania lattice. The Cr^{3+} ions can enter the lattice sites of Ti^{4+} via substitutional doping [22,46], where the expansion of the unit cell volume could be possible due to the ionic radii, which is very similar for Cr and the host titanium ions, i.e., 0.69 Å and 0.68 Å, respectively. Any significant modification of the TiO_2 crystalline structure was observed after the addition of the metals, and similar results have been previously reported by different authors [47].

UV-Vis DRS: The UV-Vis DR spectra of the synthesized photocatalyst are plotted in Figure 3. As is observed in Figure 3a, the undoped TiO_2 exhibits absorption only at wavelengths lower than 400 nm; after its doping with Cr, the absorption in the visible region of the electromagnetic spectrum significantly increased. The red shift in the reflectance spectra of the doped TiO_2 samples can be due to: (i) the inclusion of Cr within the network of the TiO_2 and (ii) the change of colour of the white undoped TiO_2 to light yellow in the Cr-TiO_2 powders.

Table 1. Main photocatalysts characterization results.

Photocatalysts	$D_{Anatase}$ (nm)	S_{BET} (m^2g^{-1})	Lattice Parameters (A°)		Band Gap (eV)
			a = b	c	
TiO_2	7.71	107	3.69	9.33	3.22
Cr-TiO_2	7.59	113	3.73	9.41	2.15
Ag (0.1%)/Cr-TiO_2	7.68	91	3.75	9.44	2.16
Ag (0.5%)/Cr-TiO_2	8.14	92	3.75	9.44	2.12
Pd (0.1%)/Cr-TiO_2	7.66	95	3.75	9.44	2.00
Pd (0.5%)/Cr-TiO_2	7.74	103	3.75	9.44	2.03

As observed in Figure 3a, all the Cr-TiO_2 materials show two bands: the first one, with the highest intensity, is located between 400 and 600 nm, which is due to the $^4A_2$ (F) to $^4T_1$ (F) transition. The second band is located between 625 and 800 nm and can be due to the spin-forbidden $^4A_2$ (F) to $^4T_1$ (F) transition. In the octahedral crystalline field, Cr^{+3} ions can exhibit this type of *d-d* electronic transition [22].

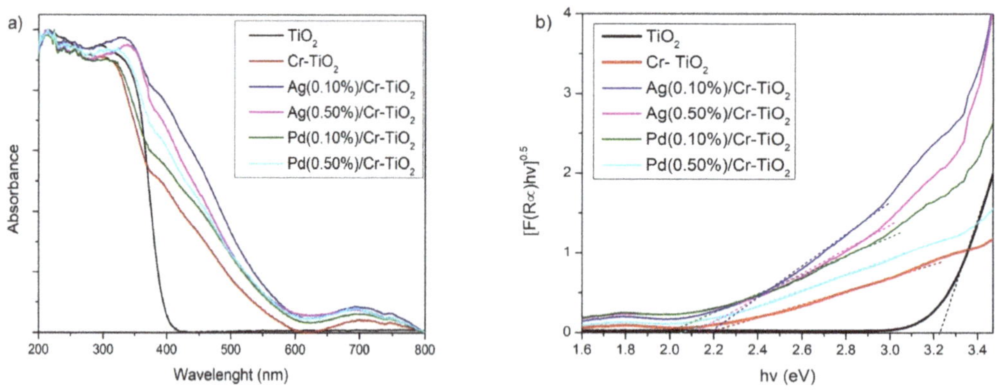

Figure 3. (**a**) UV-Vis DR spectra and (**b**) Band gap calculation by Kubelka–Munk functions and Tauc plots of all photocatalysts analysed.

On the other hand, with the incorporation of Ag or Pd nanoparticles, the visible absorption of the Cr-TiO_2 grows even more. It is also observed in Figure 3a that the metal content has a strong influence on the width and intensity of the signals, which is mainly due to the metal nanoparticle size and distribution on the TiO_2 surface. In general, an adsorption edge shifting toward the visible region of Titania after the deposition of Ag nanoparticles was observed, similar to that observed by other authors [48]. It was also evidenced that the silver materials exhibit higher visible absorption than that observed in the materials containing Pd, which is probably due to the more intense grey colour of the silver-based materials. Ag(0.10%)/$Cr\text{-}TiO_2$ showed the highest absorption in the visible region thus indicating a potential better photocatalytic performance in this material.

The materials modified with Ag show absorption bands of charged and metallic clusters, namely $Ag_n{}^{\delta+}$ and $Ag_m{}^0$ where n and m are low numbers, respectively; these absorptions can be observed at 220 and 365 nm in Figure 3a [49]. Ag(0.1%)/$Cr\text{-}TiO_2$ and Ag(0.5%)/$Cr\text{-}TiO_2$ present higher absorption at 450 nm than the $Cr\text{-}TiO_2$ material, which can be due to the surface plasmon resonance (SPR) characteristic of Ag [50,51].

The addition of Pd on the surface of $Cr\text{-}TiO_2$ also increased the visible light absorption which is due to the SPR of the Pd nanoparticles [52,53]. It is also evident that the Pd(0.5%)/$Cr\text{-}TiO_2$ material shows higher absorption than Pd(0.1%)/$Cr\text{-}TiO_2$, which was due to the higher content of Pd on the surface of the material.

Figure 3b includes the Tauc plots of the materials analysed, employed to calculate the band gap (eV) value, and the results obtained were included in Table 1. As it can be observed, the band gap value of Titania decreased after Cr doping from 3.22 to 2.15 eV, then, after the addition of Ag or Pd, the band gap of the doped material ($Cr\text{-}TiO_2$) slightly decreased, thus showing the effect of the metal nanoparticles on the optical properties of TiO_2.

The energy band structure of the catalysts is a crucial factor influencing the degradation efficiency in photocatalytic reaction systems, as it governs the redox potential of photogenerated charge carriers, which drive the catalytic reactions. In particular, the Mulliken relationship was employed to determine the edge position of the valence band (E_{VB}) for TiO_2, and $Cr\text{-}TiO_2$ at pH equal to 5.5 [54,55]:

$$E_{VB} = \chi - E_c + \frac{1}{2}E_{bg} \quad (5)$$

where χ is the electronegativity of the semiconductor and is the geometric mean of the electronegativity of the constituent atoms (the electronegativity of an atom is the arithmetic mean of the atomic electron affinity and the first ionization energy), E_c is the energy of free electrons on the hydrogen scale (ca. 4.5 eV), and E_{bg} is the band gap energy of TiO_2 or $Cr\text{-}TiO_2$ (eV) obtained from UV-Vis DRS analysis.

Therefore, to estimate the absolute electronegativity of the photocatalyst, we calculated the geometric mean of the electronegativities of the atoms constituting the photocatalyst:

$$\chi_{TiO_2} = \left(\chi_{Ti}\chi_O^2\right)^{\frac{1}{(1+2)}} \quad (6)$$

$$\chi_{Cr-TiO_2} = \left(\chi_{Cr}^{0.0188}\chi_{Ti}\chi_O^{1.9718}\right)^{\frac{1}{(0.0188+1+1.9718)}} \quad (7)$$

$$\chi_i = \frac{1}{2}\left(E_{EA_i} + E_{FI_i}\right) \quad (8)$$

where χ_i is the absolute electronegativity of the i-th element, E_{EA_i} is the electro-affinity of the i-th element, and E_{FI_i} is the energy of the first ionization of the i-th element. In detail, to evaluate the absolute electronegativity of $Cr\text{-}TiO_2$, the nominal Cr/Ti molar ratio for the photocatalyst (0.0188) was assumed to write the minimum formula that characterizes the doped semiconductor ($TiCr_{0.0188}O_{1.9718}$).

Finally, to determine the position of the conduction bands (E_{CB}) of the synthesized samples, the following equation was used [56,57]:

$$E_{CB} = E_{VB} - E_{bg} \quad (9)$$

In Figure 4, the calculated band edge positions adopted from the Mulliken electronegativity approach for TiO_2 and $Cr\text{-}TiO_2$ photocatalysts are reported.

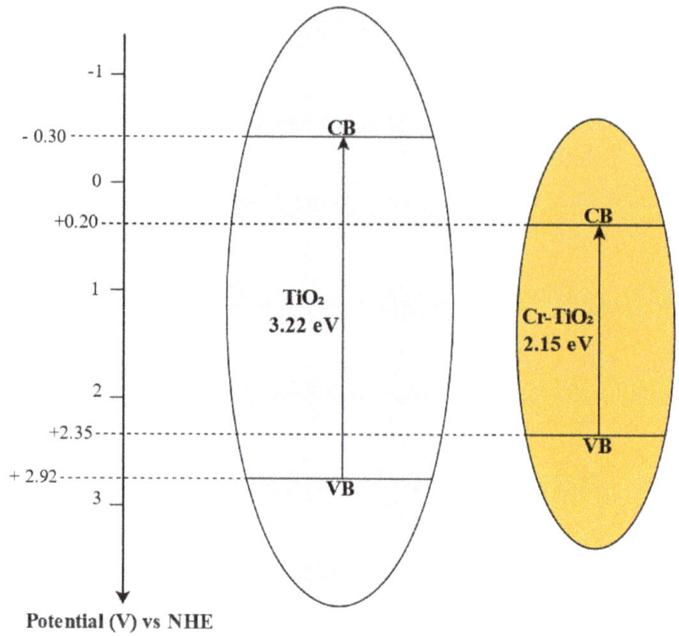

Figure 4. The calculated band edge positions adopted from the Mulliken electronegativity approach for (**left**) TiO_2 and (**right**) Cr-TiO_2 samples.

FTIR: Figure 5 includes the FTIR spectra of the analysed photocatalysts; in this figure is possible to identify for all the samples the presence of isolated OH^- groups by bands located over 3600 cm^{-1} [58].

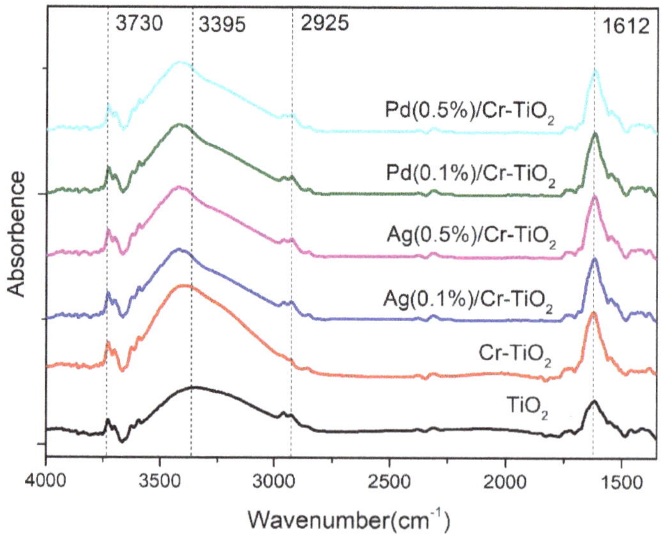

Figure 5. FTIR spectra of the photocatalysts analysed.

The band centered at 3395 cm^{-1} in all the spectra is assigned to terminal Ti-OH species and the signal at 2925 cm^{-1} is assigned to adsorbed water Ti-OH_2 [59]. The signal located at

1612 cm^{-1} in all catalysts is attributed to the water bending mode (υ2) while the broadband between 3600–3000 cm^{-1} corresponds to antisymmetric (υ3) and symmetric (υ1) stretching vibration modes of the water [60]. It is interesting to note that the intensity of the band assigned to the terminal Ti-OH species in the TiO_2 spectrum slightly increased after Cr addition, which can indicate that this metal acts as a dopant agent in the Titania structure. Then, after the addition of Ag or Pd, the width and intensity of this band slightly decreased, which indicates that the nanoparticles of these metals are covering the Titania surface, thus leading to decreased surface hydroxylation, as has been also observed in previous works for metals such as Au or Pt [20,21].

EIS: Figure 6 shows the EIS Nyquist plot of the photocatalysts analysed, this evidence the graphical representation of the real part of impedance (Zr) versus the imaginary part of impedance (Zi). Firstly, Figure 6a includes the Nyquist diagrams, and then Figure 6b shows the zoom of these diagrams at low frequencies (i.e., from 0 to 0.6). As is observed in Figure 6b, the bare TiO_2 presented a line at high frequencies, which indicates that this material presents resistance to electronic transfer. This resistance slightly decreased after Cr doping, thus indicating that the modification of Titania with this metal improved the charge transfer.

On the other hand, the modification of Cr-TiO_2 photocatalysts by chemical photoreduction of Pd or Ag led to even more improvement in the charge transfer. Thus, in the materials modified with Pd, a curved trend was evidenced at low frequencies; the charge transfer slightly decreased in the Pd (0.1)/Cr-TiO_2 photocatalyst.

In the case of the materials modified by Ag addition, a curvature located at low frequencies which is greater than that observed in the Pd modified materials is evidenced in Figure 6b. This observation suggests the formation of a semicircle with a small radius, thus indicating low resistance to the charge transfer at the interface. Taking into account that the Ag(0.1%)/Cr-TiO_2 photocatalyst presents the lowest radius arc, it is possible to assume that this material can have improved photocatalytic activity.

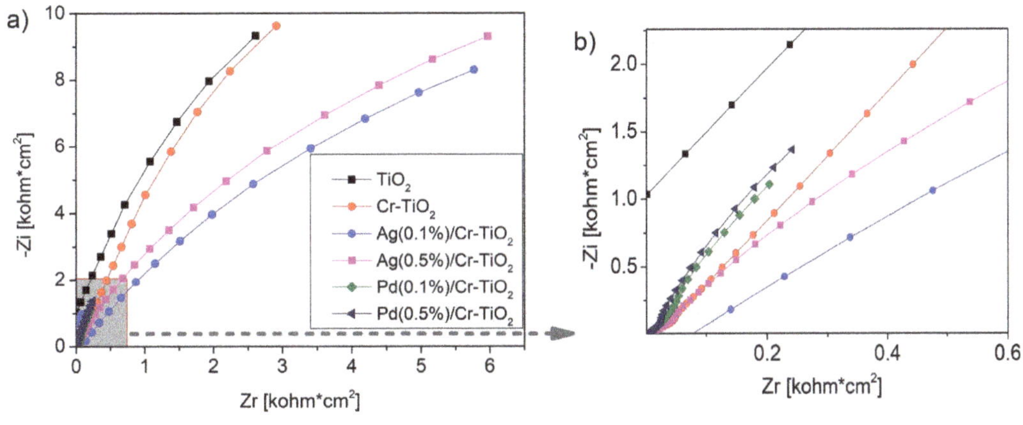

Figure 6. (**a**) EIS Nyquist plot of all photocatalysts and (**b**) Zoom for the EIS Nyquist plot at low frequencies.

3.2. Photocatalytic Test Results

3.2.1. SY Degradation Results

Preliminary Tests: Photolysis and Process Conducted Using TiO_2 or Cr-TiO_2

A preliminary test was conducted to demonstrate that SY is mineralized only due to the photodegradation activity of the photocatalysts of our reaction system. For this reason, the photolysis test was carried out in the absence of photocatalytic powders (photolysis), using a solution of distilled water with an initial SY concentration of 10 ppm, at spontaneous pH, while irradiated by the two solar lamps.

On the other hand, the photocatalytic tests were carried out in the same operating conditions but with a photocatalyst dosage equal to 3 g L^{-1}, using TiO_2 and Cr-TiO_2 nanopowders. The results of the tests are shown in Figure 7, from which it can be seen that photolysis alone has a negligible effect on the removal of the dye.

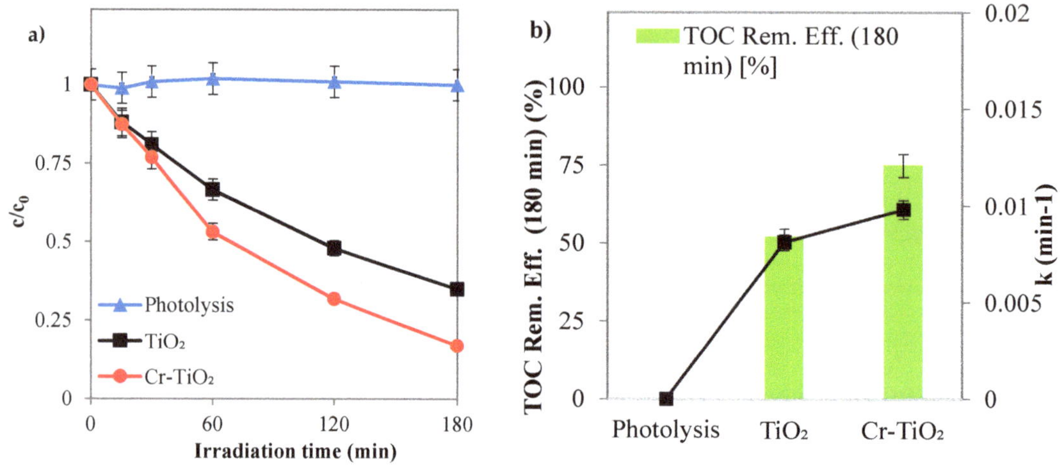

Figure 7. (**a**) SY discoloration and (**b**) mineralization under solar light registered for TiO_2 and Cr-TiO_2 compared with SY photolysis (photocatalyst dosage = 3 g L^{-1}; c_0 = 10 ppm; pH = 5.5).

Cr-TiO_2 recorded the highest values in SY discoloration and mineralization efficiencies, 87% (Figure 7a) and 75% (Figure 7b), respectively, confirming the influence of Cr metal ion doping in TiO_2 on the intrinsic properties of the semiconductor by reducing charge carrier recombination and improving its photo-response in the solar field [61,62].

Influence of Pd and Ag Loading

Experimental tests were conducted to evaluate the photocatalytic activity of Pd- and Ag-based samples with varying active phase loads photodeposited on the photocatalyst; specifically, the palladium and silver loads were adjusted within the range of 0.1% to 0.5% by weight. These tests were conducted under the same operating conditions as the activity tests carried out with TiO_2 and Cr-TiO_2.

The results, illustrated in Figure 8, indicate that the optimal photocatalyst is Cr-TiO_2 decorated with a palladium load of 0.1% by weight. This specific loading significantly enhances the discoloration and mineralization rates of the azo dye compared to unmodified Cr-TiO_2 (Figure 9a,b). Indeed, it achieved the highest values for the apparent discoloration kinetic constant and TOC removal efficiency after 180 min of solar irradiation, with values of 0.06 min^{-1} and 92%, respectively. For this reason, subsequent tests to identify the optimal operating conditions for the mineralization of SY were conducted using the Pd(0.1%)/Cr-TiO_2 sample.

The best performance observed with this photocatalyst can be associated with its physicochemical properties, thus, as observed by UV-Vis DRS, this material presents higher absorption in the UV-visible regions of the electromagnetic spectrum than the Cr-TiO_2 sample, which improves the radiation available for the SY photodegradation. The Pd(0.1%)/Cr-TiO_2 sample also presents lower resistance to the charge transfer at its interface, compared with the Cr-TiO_2 material, thus indicating improved photogenerated charges during the photocatalytic process.

On the other hand, it is important to note that the increase of the Pd content in the materials had a detrimental effect on the photocatalytic efficiency, thus, as is observed in Figure 8, the dye degradation represented by the TOC removal significantly decreases from

81 to 37% in the material containing 0.5 wt.% of Pd. In the case of the Ag/Cr-TiO_2 material, the effectiveness in the TOC removal efficiency significantly decreased compared to pristine Cr-TiO_2, which can be associated with the Ag particle size. These nanoparticles can be more homogeneously distributed on the Cr-TiO_2 surface than the Pd particles, which can obstruct the light interaction with the TiO_2 surface, and act as recombination centers.

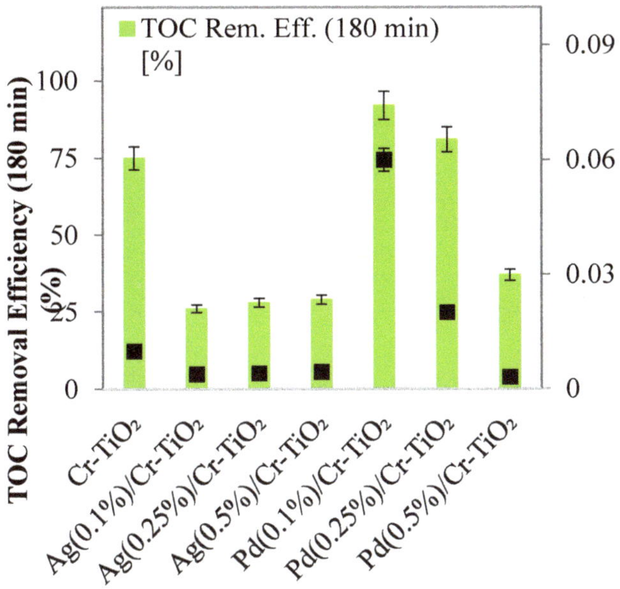

Figure 8. Mineralization efficiency after 180 min of solar irradiation and apparent kinetic discoloration constant values obtained using the different samples (photocatalyst dosage = 3 g L^{-1}; c_0 = 10 ppm; pH = 5.5).

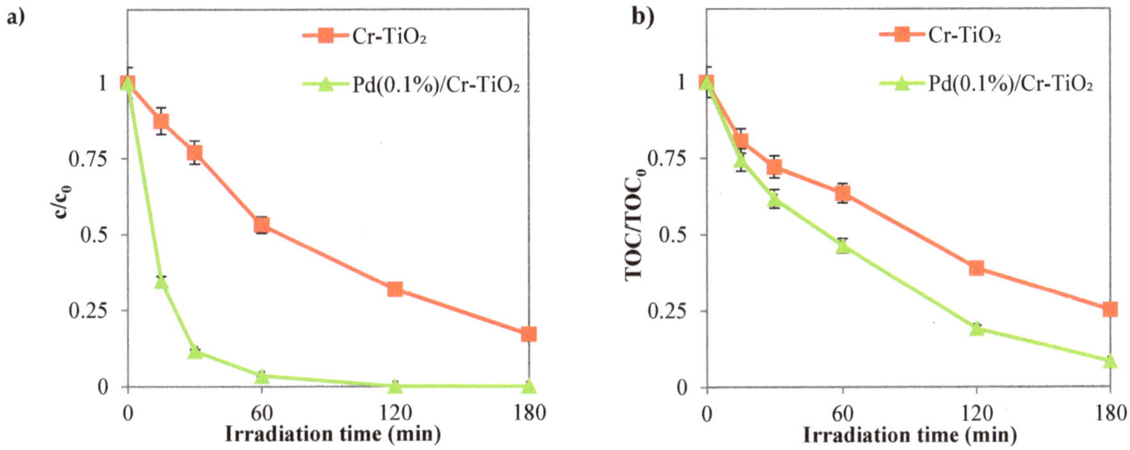

Figure 9. (**a**) SY discoloration and (**b**) mineralization under solar light registered with Cr-TiO_2 and Pd(0.1%)/Cr-TiO_2 (photocatalyst dosage = 3 g L^{-1}; c_0 = 10 ppm; pH = 5.5).

Influence of Photocatalyst Dosage

Since Pd(0.1%)/Cr-TiO_2 showed the best photocatalytic performance in SY discoloration and mineralization, subsequent experimental tests were conducted with this sample. In particular, the effect of photocatalyst dosage on the SY degradation was examined and the optimal value was determined. An adequate dosage of photocatalyst can improve the photocatalytic performance and lower the energy cost of the treatment. Since the number of active sites and photo-adsorption capability of the photocatalyst have a significant impact on the photodegradation efficiency, it is a crucial parameter for photocatalytic processes [63]. From Figure 10 it can be seen that the optimal dosage of photocatalyst is equal to 3 g L^{-1}, allowing the maximization of both the discoloration (Figure 10a) and the mineralization of SY, with the highest value of apparent discoloration kinetic constant of 0.06 min^{-1} (Figure 10b). Indeed, a sufficient increase in photocatalyst dosage raised the amount of photons absorbed, leading to an improvement in photodegradation rates [63,64]. Instead, when the loading was increased from 3 to 4.5 g L^{-1}, photocatalytic degradation and mineralization of SY decreased. Indeed, the overdosage of the photocatalyst enhanced the opacity of the solution, decreasing the penetration of the photon flux in the reactor and, consequently, the rate of photocatalytic degradation, even in the presence of a significant number of active sites [63,65].

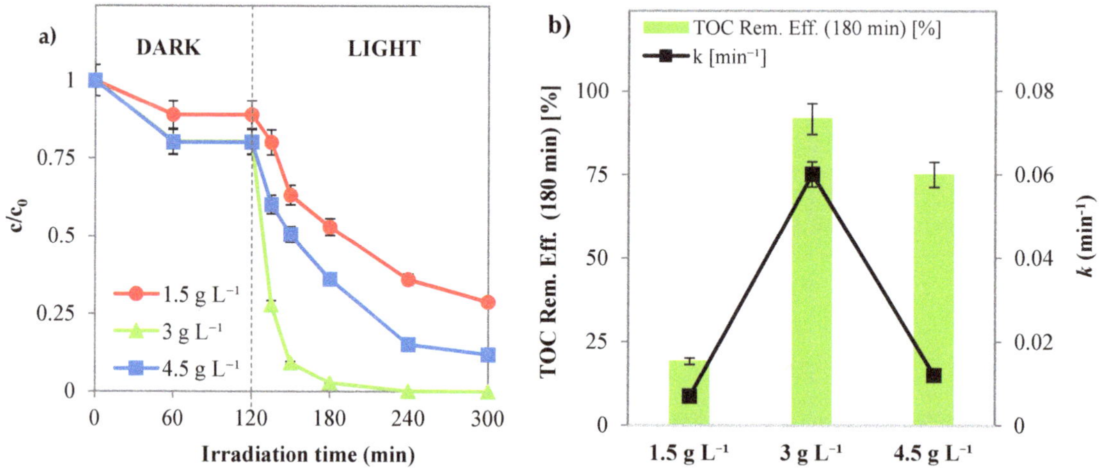

Figure 10. (**a**) SY discoloration and (**b**) mineralization under solar light registered with Pd(0.1%)/Cr-TiO_2 during the experimental tests by varying the photocatalyst dosage (c_0 = 10 ppm, pH = 5.5).

Influence of Initial Dye Concentration

Subsequently, the photocatalytic process at various initial SY concentrations was further evaluated using the optimal photocatalyst dosage (3 g L^{-1}). From Figure 11a,b, the data indicates that 10 ppm is the optimal initial concentration for the photocatalytic reaction. Indeed, with an initial SY concentration of 10 ppm, the discoloration efficiency of 97.3% occurred after 60 min of solar irradiation, while the SY discoloration efficiency at initial concentrations of 5 and 25 ppm after 180 min of irradiation time was 91.7% and 61%, respectively (Figure 11a).

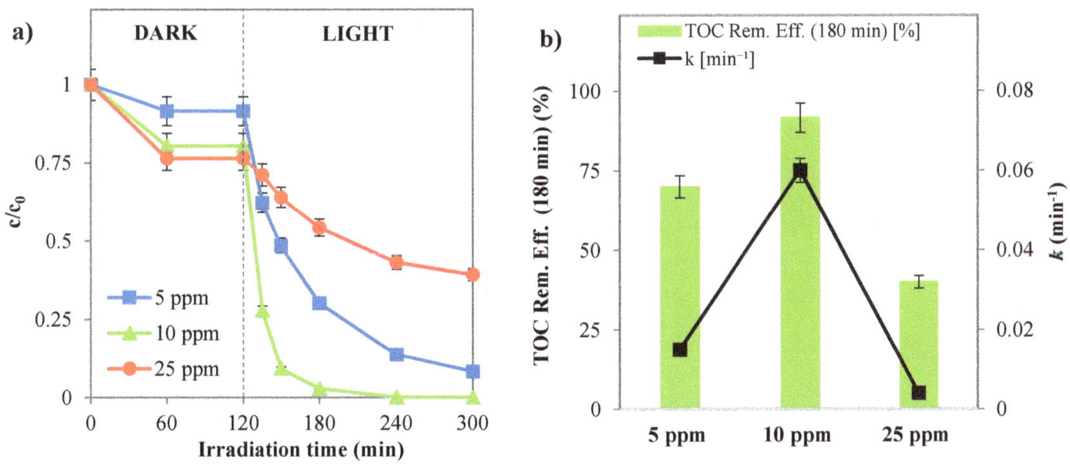

Figure 11. (**a**) SY discoloration and (**b**) mineralization under solar light registered with Pd(0.1%)/Cr-TiO_2 during the experimental tests by varying the initial SY concentration (photocatalyst dosage = 3 g L^{-1}, pH = 5.5).

Furthermore, the results for mineralization also confirmed that 10 ppm of SY as the initial concentration is the optimal one, recording 91.6% TOC removal efficiency after 180 min of irradiation time, against 70% and 40% at 5 and 25 ppm, respectively (Figure 11b). The registered data illustrate that the pollutant discoloration and mineralization increase when the starting concentration passes from 5 to 10 ppm, but 25 ppm is recorded as the lowest photocatalytic performance. Indeed, because the adsorbed SY molecules occupy the active sites of the photocatalyst particles, an excessive increase in dye concentration inhibits the production of reactive oxygen species (ROS) such as hydroxyl radicals, superoxides, and positive holes that interact with the dye molecules. Moreover, the intensity of the incident light on the photocatalyst surface is reduced by an excessive increase in dye concentration in the solution because more photons from the solar irradiations will be absorbed by the SY molecules [34,63].

Influence of Initial pH

The influence of the initial pH of the aqueous SY solution on photocatalytic activity was subsequently analysed. Various tests were conducted using the Pd(0.1%)/Cr-TiO_2 sample under previously identified optimal operating conditions. The initial pH of the solution was varied by adding a fixed amount of aqueous HCl or NaOH solution. A key factor in this analysis is the point of zero charge (PZC), which corresponds to the condition where no net charge is present on the photocatalytic surface. In fact, the PZC significantly affects the adsorption of solution species and the photocatalytic activity of materials [66]. The PZC of the optimal synthesized sample (Pd(0.1%)/Cr-TiO_2) was determined using the mass titration method [67], registering a value of 4.8. At pH values lower than the PZC of the photocatalyst, the surface becomes positively charged due to the protonation of surface hydroxyl groups (Equation (10)). On the other hand, as the pH increases, deprotonation gradually becomes dominant (Equation (11)). When the pH exceeds the PZC, the surface of the photocatalyst becomes negatively charged [68].

$$\mathrm{pH} < \mathrm{PZC}, \mathrm{Ti}-\mathrm{OH}+\mathrm{H}^{+} \longrightarrow \mathrm{TiOH}_2^{+} \quad (10)$$

$$\mathrm{pH} > \mathrm{PZC}, \mathrm{Ti}-\mathrm{OH}+\mathrm{OH}^{-} \longrightarrow \mathrm{TiO}^{-}+\mathrm{H_2O} \quad (11)$$

SY is typically an anionic dye. When the pH is lower than the PZC of the photocatalyst, the sulfonate groups in SY (D–SO_3Na) remain deprotonated, resulting in a negative charge [69]. Under acidic conditions, electrostatic attraction facilitates the adsorption of SY onto the positively charged photocatalyst surface. However, when the pH is higher than the PZC, the deprotonation of surface hydroxyl groups reduces the electrostatic attraction, leading to decreased adsorption of SY. Figure 12a illustrates that the high adsorption of the dye on the photocatalyst material causes a rapid decrease in dye concentration in the acidic pH range. These results indicate that the initial pH of the solution influences the kinetics of the photocatalytic process by affecting the amount of dye adsorbed on the photocatalyst. The optimal initial pH for achieving the highest dye discoloration and mineralization is 4 (Figure 12a,b). Under these conditions, the reaction system achieved complete discoloration after 60 min and mineralization of the azo dye within 180 min of solar irradiation.

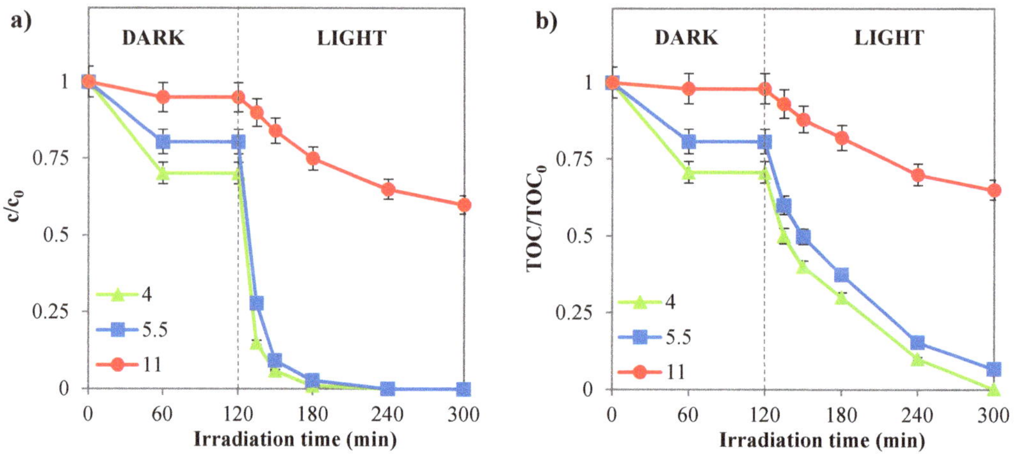

Figure 12. (**a**) SY discoloration and (**b**) mineralization observed with Pd(0.1%)/Cr-TiO_2 during the experimental tests by varying the initial pH (photocatalyst dosage = 3 g L^{-1}; c_0 = 10 ppm).

Electric Energy Consumption Evaluation

Finally, the electric energy consumption associated with the photodegradation of 90% of SY in 1 m^3 of water contaminated with the azo dye was assessed under optimal operating conditions. This analysis utilized the correlation proposed by Bolton et al. [70], expressed as:

$$E_{E/O} = \frac{P\, t_{90\%}\, 1000}{V\, 60 \ln\left(\frac{c(t_0)}{c(t)}\right)} \tag{12}$$

where $E_{E/O}$ is the electric energy consumption (kWh), P is the nominal power of the light source (kW), $t_{90\%}$ is the irradiation time to achieve 90% removal of SY (min), V is the volume of treated solution (L), $c(t_0)$ is the concentration of SY (ppm) at the start of irradiation, and $c(t)$ is the concentration of SY (ppm) at generic time t.

Next, the following best-performing literature works on the degradation of SY have been identified:

i. W1 [71]: The photocatalytic reaction was conducted in a reactor under visible light irradiation using a 200 W tungsten bulb from Osram. Optimal performance was achieved with the addition of 800 mg of $CuCr_2O_4$ photocatalytic nanoparticles to 0.05 L of SY aqueous solution.
ii. W2 [72]: The photocatalytic performance of a Pd-$BiFeO_3$ composite was evaluated for the degradation of SY under visible light irradiation. The experiments were conducted in a photoreactor illuminated by a 105 W visible lamp. Each test analysed

the photodegradation of 100 mL of dye aqueous solution, starting with an initial SY concentration of 10 ppm.

iii. W3 [73]: The photocatalytic activity of the TiO_2/CAC sample was evaluated in a reactor under UV-A light using four parallel medium-pressure mercury lamps (8 W). Specifically, 50 mL of SY solution with an appropriate amount of catalyst was tested in each experiment.

iv. W4 [72]: A slurry reactor containing a suspension of $HoAG_5$/g-C_3N_4 nanocomposite was used, irradiated by a 350 W xenon-mercury lamp. The volume of the SY aqueous solution was 100 mL.

v. W5 [74]: The degradation experiment with the best performance was conducted by treating 200 mL of SY aqueous solution with an initial concentration of 10 ppm. The BiOBr dosage was 1.31 g L^{-1}, and the system was irradiated using a 55 W light bulb as the visible light source.

vi. W6 [75]: The mineralization of SY using a ZnS-TiO_2 photocatalyst was evaluated under UV-A light, provided by four parallel medium-pressure mercury lamps (8 W each). Optimal performance was observed with a 50 mL SY solution (~450 ppm) when the photocatalyst dosage was set to 5 g L^{-1}.

vii. W7 [76]: The photocatalytic activity of selenium nanoparticles (Se-NPLs) was evaluated in a slurry reactor under UV-C light, utilizing an 8 W Herolab lamp. Each batch experiment involved using a 100 mL beaker containing 40 mg of Se-NPLs and 100 mL of a 5 mg L^{-1} aqueous SY solution.

From the data of the seven works in the literature, the apparent discoloration kinetic constant of the dye was identified. Additionally, the electricity consumption required for a 90% reduction of the pollutant in 1 m^3 of solution was calculated using Equation (12) and compared to the consumption required by the reaction system of the present work under the following conditions: photocatalyst dosage = 3 g L^{-1}, c_0 = 10 ppm, and initial pH = 4. The results are presented in Table 2.

Table 2. Comparison of the energy cost for the degradation of 90% of the Sunset Yellow FCF in 1 m^3 of solution for our reaction system at the optimal operative conditions and the other works reported in the literature.

Photocatalyst	Type of Light	*P* (kW)	*k* (min^{-1})	$t_{90\%}$ (min)	*V* (L)	$E_{E/O}$ (kWh m^{-3})	System
$CuCr_2O_4$	Visible Light	0.20	0.014	164.5	0.05	4761.9	[70]
Pd-$BiFeO_3$	Visible Light	0.11	0.017	136.2	0.1	1035.5	[71]
TiO_2/CAC	UV-A	0.03	0.039	59.0	0.05	273.5	[75]
HoAG5/g-C_3N_4	Visible Light	0.35	0.037	61.6	0.1	1559.7	[73]
BiOBr	Visible Light	0.06	0.007	343.7	0.2	684.1	[74]
ZnS-TiO_2	UV-A	0.03	0.040	57.6	0.05	266.7	[75]
Se-NPLs	UV-C	0.01	0.002	1151.3	0.1	666.7	[76]
Pd(0.1%)/Cr-TiO_2	Solar Light	0.02	0.070	32.9	0.1	38.1	Our system

The experimental results demonstrated that using small quantities of palladium significantly enhances the photodegradation process of azo dyes under solar irradiation. Our photocatalytic system exhibits the highest energy efficiency in SY removal, with the lowest energy consumption for 90% pollutant removal at 38.1 kWh m^{-3}.

Analysis of the Possible Reaction Mechanism

The photocatalytic activity tests were subsequently conducted to investigate the reaction mechanism for the mineralization of the azo dye, using a fixed initial concentration of specific scavenger probe molecules. These experiments were performed under optimal conditions determined from previous tests, with an initial SY concentration of 10 ppm and a Pd(0.1%)/Cr-TiO_2 dosage of 3 g L^{-1}.

The scavenger probe molecules employed included isopropanol (IPA, 10 mM) to avoid the presence of hydroxyl radicals ($OH^{\bullet}$) [77–79], benzoquinone (BQ, 1 µM) for superoxide radicals ($O_2^{-\bullet}$) [80–82] and methanol (MetOH, 10 mM) for positive holes (h^+) [83,84].

The results shown in Figure 13 depict a significant reduction in photocatalytic activity when the formation of hydroxyl and superoxide radicals is inhibited. Specifically, in the presence of isopropanol, which scavenges hydroxyl radicals, the SY apparent degradation kinetic constant decreased by 95% compared to the control test conducted under identical conditions without the scavenger. These experimental data underline that the degradation of the azo dye using the Pd(0.1%)/Cr-TiO_2 primarily occurs through hydroxyl radicals which are generated by both positive holes and superoxide ions. However, the Pd(0) photodeposited on the surface of the photocatalytic nanoparticles acts as an electron collector allowing the formation of H_2O_2, increasing the hydroxyl radicals for the oxidation of pollutant [85,86].

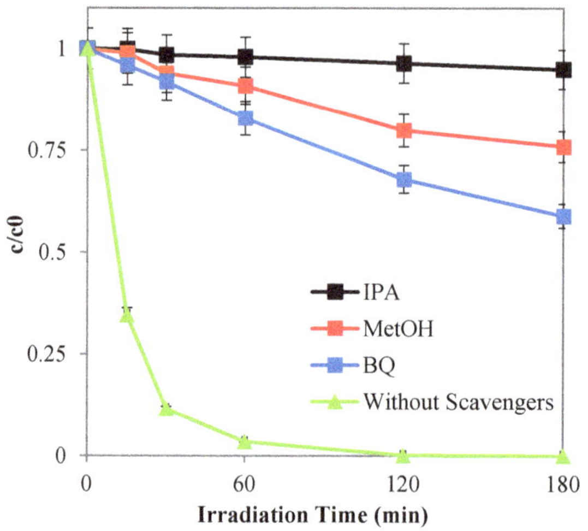

Figure 13. SY discoloration observed with Pd(0.1%)/Cr-TiO_2 during the experimental tests with the presence of IPA, MetOH, and BQ, and without the scavengers.

The following reactions describe a potential mechanism for the mineralization of SY:

$$\mathrm{Pd(0.1\%)/Cr-TiO_2} + h\nu \rightarrow h^+ + e^- \quad (13)$$

$$O_2 + e^- \rightarrow O_2^{-\bullet} \quad (14)$$

$$OH^- + h^+ \rightarrow OH^{\bullet} \quad (15)$$

$$O_2^{-\bullet} + H^+ \rightarrow HO_2^{\bullet} \quad (16)$$

$$HO_2^{\bullet} + HO_2^{\bullet} \rightarrow H_2O_2 + O_2 \quad (17)$$

$$H_2O_2 + e^- \rightarrow OH^{\bullet} + OH^- \quad (18)$$

$$SY + OH^{\bullet} \rightarrow \text{intermediates} \rightarrow CO_2 + H_2O \quad (19)$$

$$e^- + h^+ \rightarrow E + N \quad (20)$$

where N is a neutral center, and E is the energy generated by the recombination of the photoexcited pairs (light $h\nu\prime \leq h\nu$, or heat).

3.2.2. Photocatalytic Treatment of Polluted River Water

The effectiveness of the photocatalysts prepared with 0.1 and 0.5 wt.% of Ag or Pd were also evaluated in the treatment of water samples from a polluted river. Different physicochemical parameters were evaluated before and after photocatalytic treatment on the water river samples and the main results are presented in Table 3.

Table 3. Water river Quality Control Parameters analysed before and after photocatalytic treatment.

Treatment	pH	Chlorides (mg/L)	Nitrates (mg/L)	Conductivity (μS/cm)
Starting river water sample	6.78	118	1.3	119.2
Photolysis	6.60	116	1.2	118.8
TiO_2	7.03	27	0.6	24.1
Cr-TiO_2	6.91	24	1.1	79.3
Ag (0.1%)/Cr-TiO_2	6.61	16	1.1	87.0
Ag (0.5%)/Cr-TiO_2	6.85	20	0.9	82.1
Pd (0.1%)/Cr-TiO_2	6.92	22	1.1	16.2
Pd (0.5%)/Cr-TiO_2	6.79	<3	0.3	12.5

As observed in Table 3, the pH value slightly increased after the photocatalytic treatments, which is due to a mild alkalinity of the reaction medium by the hydroxyl groups generated during the photocatalytic process [87]. Other physicochemical parameters such as chlorides, nitrates, and conductivity also decrease after treatments, which is mainly due to the adsorption of ionic species such as NO^{3-}, Cl^-, and CO_3^{2-} on the surface of the photocatalytic materials, thus reducing the concentration of pollutants in the fluid phase [88]. The Pd(0.5%)/Cr-TiO_2 exhibited the best performance in surface ion adsorption, which can be related to its highest surface area compared with the materials modified with metals.

The content of enteric bacteria (i.e., total coliforms, *E. coli*, and other Enterobacteriaceae) in the river water sample was also analysed before and after the photocatalytic treatment, and the results obtained are represented in Figure 14. As observed in Figure 14a, and as expected, the UV-Vis light presents a certain bactericidal effect, thus leading to a decrease in the bacteria content in the photolysis test which was carried out in the presence of radiation and without photocatalytic powders in the reaction medium. Then, as observed in Figure 14b, the bacteria removal significantly increased by close to 99% in the tests performed in the presence of the photocatalytic materials.

It is also observed in Figure 14a that the efficiency of TiO_2 in the *E. coli* and other Enterobacteria removal slightly improved by Cr doping, thus showing the improved optical properties of this material (increased light absorption in the visible region and decreased band gap) as it was determined by UV-Vis DRS analysis previously described.

When Ag or Pd was added to Cr-TiO_2, the bacteria removal efficiency was improved, which is mainly due to the effect of the metallic particles acting as electron traps, thus decreasing the photogenerated charges recombination.

It is interesting to note that the substrate to be treated is an important factor influencing the photocatalytic performance in the materials evaluated; thus, different from the results observed in the SY degradation (Section 3.2.1). In the water river treatment, the best performance was observed with the silver modified materials. Thus, the total elimination of *E. coli* was only achieved by using Ag(0.1%)/Cr-TiO_2 in the photocatalytic treatment. With this material, the highest elimination of total coliforms and other Enterobacteriaceae was also obtained. The results obtained can be explained by taking into account the well-known bactericidal effect of the silver nanoparticles compared with the Pd; likewise, as observed by EIS analysis, the Ag(0.1%)/Cr-TiO_2 presented the lowest charge transfer resistance, thus demonstrating the highest effectiveness of this material in the river water sample treatment. It was also observed that an increase in the loading of Ag or Pd from 0.1% to 0.5% had a detrimental effect on the biocidal efficiency of the Cr-TiO_2 photocatalyst, which can be due to the obstruction of active sites on the surface, which could affect the amount of reactive oxygen species (ROS) in the reaction medium.

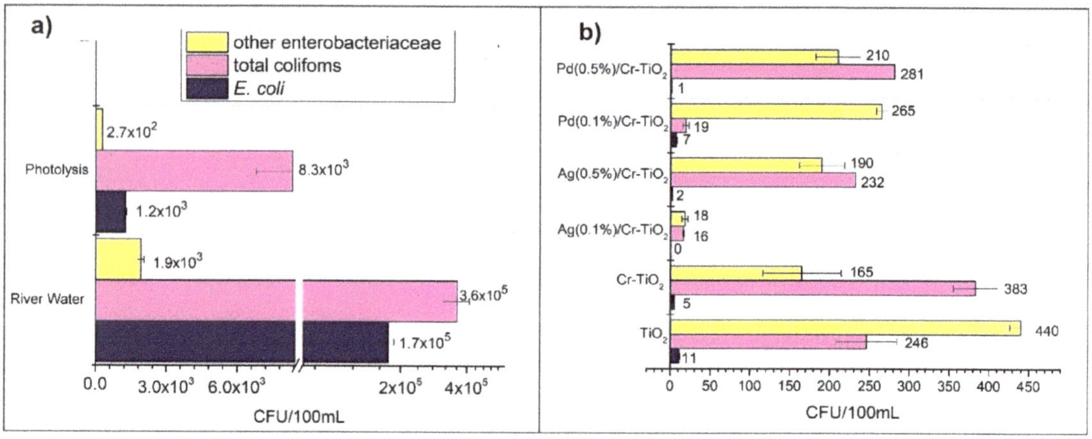

Figure 14. Population of bacteria in the water river sample, (**a**) before and after photolysis, (**b**) after the different photocatalytic treatments.

On the other hand, in order to ensure the real effectiveness in the *E. coli* inactivation, regrowth tests were also performed by storing the treated water at room temperature for 24 h. After this time passed, the bacteria content was analysed, and no *E. coli* regrowth was observed, thus indicating the effectiveness of the treatment over time. However, it is important to consider that total coliforms and other enterobacteria still remained after treatment.

3.2.3. Photocatalytic Test of Yeast Elimination

It is also interesting to consider that different microorganisms can be more or less sensitive to the photocatalytic treatment. For that reason, we decided to compare two completely different microorganism species, the first one the enteric bacteria previously reported, and in this section the yeast photocatalytic inactivation. *S. cerevisiae* presents different properties in the cell wall compared to gram-negative bacteria studied in Section 3.2.2 and the results obtained in its elimination are presented in Figure 15.

As observed in this figure and similar to that observed in the bacteria elimination, the UV-Vis light also has a negative effect on the yeast cell, thus leading to a decrease in the cell content after the photolysis test (Figure 14a). On the other hand, the yeast elimination significantly increased after photocatalytic treatment (Figure 14b), and the effectiveness tendency in the analysed materials exhibits the following order: Pd(0.5%)/Cr-TiO_2 > Ag (0.1%)/Cr-TiO_2 > Cr-TiO_2 > Ag (0.5%)/Cr-TiO_2 > Pd (0.1%)/Cr-TiO_2 > TiO_2.

In general, it was observed that the yeast presented more resistance to be eliminated by photocatalysis compared with *E. coli*, which can be due to the structural differences presented in the cell walls of these microorganisms. Thus, the yeast cell wall is thicker compared with bacteria; likewise, yeast is composed of dry cell mass (15–25%) [89], β-glucan (29% to 64%), mannan (31%), protein (13%), lipids (9%), and chitin (1–2%) [90,91]. Indeed, this highlights the significant protection provided by the yeast cell wall to the internal organelles of the cell during the photocatalytic treatment.

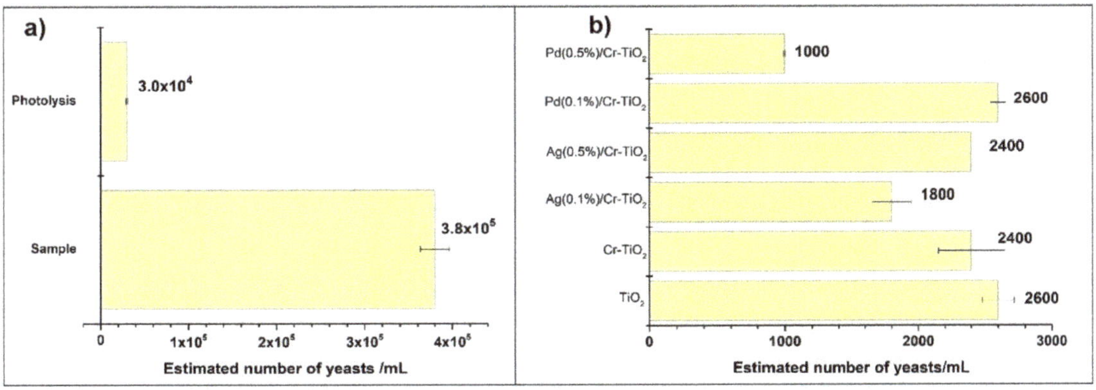

Figure 15. Yeast content (**a**) Sample and after photolysis, (**b**) after all photocatalytic treatments.

The damage caused to the yeast cell wall by ROS leads to cell inactivation, as evidenced by the decrease in the estimated number of yeast cells/mL after photocatalytic treatment. However, there is a possibility that the by-products of cell wall rupture, such as β-glucan and mannan, create a chemical environment unfavorable for the generation of ROS. Some studies have shown that the hydroxyl groups of β-glucan and mannan combine to inhibit the generation of hydroxyl radicals [92,93].

This could explain why the photocatalytic materials were less effective in yeast elimination. It was also important to highlight that yeast can generate mechanisms of resistance to oxidative stress, which prevents complete elimination from occurring.

In this case, it was important to highlight the biocidal effect of Pd NPs, as the most effective material was Pd(0.5%)/Cr-TiO_2. The mechanism of action can be explained from two perspectives: (i) the small size of Pd NPs allows them to contact the yeast cells, and the metal ions tend to penetrate inside the cell, causing oxidative stress that leads to cell death [94], and (ii) previously, authors have reported the ability of Pd to generate reactive oxygen species (ROS) [95]. This could promote the rupture of the yeast cell wall by ROS generated from the metal and ROS from TiO_2.

3.3. Discussion on Photocatalytic Results

The results in the elimination of the different pollutants point out that a high efficiency with the application of solar light could be obtained. In the case of organic anionic food dye SY, the boost to its decolorization and mineralization is due to the selection of Pd as a co-catalyst with respect to silver on Cr-TiO_2. To explain the higher activity of Pd, it could be considered that a higher work function of this metal increases the height of the Schottky barrier with respect to silver, leading to a more favorable electron-hole separation. As demonstrated by the scavenger tests, the most active ROS species is the OH radical, generously generated by the interaction of the photogenerated holes with OH- adsorbed groups, and interaction with the adsorbed water leads to almost total mineralization of SY FCF. However, an increase in the Pd content induces a slight increase in the band gap values, resulting in a reduction of photoactivity. The optimal load of Pd is concluded to be 0.1 wt.%. Concerning the microorganism, Ag performs better than Pd owing to its capacity to also generate ROS in the dark and to its high toxicity to the bacteria, again with an optimal low load of 0.1 wt.% on Cr-TiO_2, although Pd(0.1%)/Cr-TiO_2 had comparable results in the removal of *E. coli* and total coliform and an Enterobacteriaceae conversion higher than 90% in the water river sample. With regards to yeast removal, the increase of Pd presence on Cr-TiO_2 induces the highest value of 96.67%, but all the photocatalysts show high ability in their elimination, with performances higher than 90%.

On the basis of these considerations, a suitable balance among the limiting of the costs of the photocatalyst and the high photoactivities showed allows Pd(01%)/Cr-TiO_2 to be considered as the optimal composition for a strong reduction of the four polluting agents under solar light.

4. Conclusions

The Cr-TiO_2 photocatalysts were successfully synthesized by the sol-gel method, and some physicochemical properties of this material were improved by chemical photoreduction of Ag or Pd nanoparticles. Thus, the heterojuncted nanomaterials presented higher absorption in the visible region of the electromagnetic spectrum and higher charge transference compared with the bare TiO_2.

The functionalizing photocatalysts with noble metals are highly effective for maximizing the efficiency of photocatalytic degradation of SY FCF azo dye and other micropollutants. Among the different compositions, Pd(0.1%)/Cr-TiO_2, with the lowest E_{BG}, emerged as a powerful nanomaterial able to give a very high and desired mineralization of the organic dye.

Indeed, in the study of photocatalytic activity for bacteria removal, it was determined that Ag(0.1%)/Cr-TiO_2 exhibited the highest efficiency. Thus, the minimum amount of Ag nanoparticles is sufficient to achieve complete elimination of *E. coli*.

Regarding the elimination of *S. cerevisiae*, the material that showed the best performance was Pd(0.5%)/Cr-TiO_2. In this case, the Pd nanoparticles play a crucial role in their affinity with the cell and the biocidal effect of the metal. The higher amount of 0.5% showed better results compared to the same material in lower amounts. Similarly, the larger surface area of this material could enhance the surface adsorption of ions such as nitrate, chloride, and carbonate present in river water.

Despite these differences, Pd-based nanomaterial demonstrated activities higher than 90% with respect to the removal of all tested pollutants. The lower amount of Pd is sufficient to guarantee an efficient and low-cost photocatalyst for future solar applications.

This study demonstrated that the formulation of a solar photocatalyst for the removal of different natural pollutants can be finely tuned to get more advantageous results when they prevail within wastewater.

Author Contributions: Conceptualization, J.J.M. and D.S.; Formal analysis, N.M., J.J.M., K.M., D.S. and V.V.; Funding acquisition, M.H.-L., J.J.M., D.S. and V.V.; Investigation, N.M. and B.L.G.; Methodology, M.H.-L., N.M. and B.L.G.; Resources, M.H.-L., J.J.M. and V.V.; Software, N.M.; Supervision, J.J.M. and D.S.; Writing—original draft, M.H.-L., N.M. and K.M.; Writing—review & editing, J.J.M. and D.S. All authors have read and agreed to the published version of the manuscript.

Funding: This work was funded by the Ministerio de Ciencia, Tecnología e Innovación—Minciencias and Ministerio de Salud y Protección Social, Project 110991891727, and Universidad Pedagógica y Tecnológica de Colombia (UPTC) Project SGI 3393.

Data Availability Statement: Data are contained within the article.

Acknowledgments: M. Hernández-Laverde would like to thank Ministerio de Ciencia Tecnología e Innovación de Colombia (Minciencias) and Gobernación de Boyacá for the concession of a researcher grant (OCAD-Fondo Nacional de Financiamiento para la Ciencia, la Tecnología e Innovación (FCTel) del Sistema general de regalías, Becas de Excelencia Doctoral del Bicentenario).

Conflicts of Interest: The authors declare no conflict of interest.

References

1. Verma, P.; Ratan, J.K. Assessment of the negative effects of various inorganic water pollutants on the biosphere—An overview. In *Inorganic Pollutants in Water*; Elsevier: Amsterdam, The Netherlands, 2020; pp. 73–96. [CrossRef]
2. Hernández, F.; Ibáñez, M.; Portolés, T.; Cervera, M.I.; Sancho, J.V.; López, F.J. Advancing towards universal screening for organic pollutants in waters. *J. Hazard. Mater.* **2015**, *282*, 86–95. [CrossRef] [PubMed]

3. Lu, Y.; Ma, T.; Lan, Q.; Liu, B.; Liang, X. Single entity collision for inorganic water pollutants measurements: Insights and prospects. *Water Res.* **2023**, *248*, 120874. [CrossRef] [PubMed]
4. Bitton, G. *Wastewater Microbiology*, 3rd ed.; John Wiley & Sons: Hoboken, NJ, USA, 2005.
5. Gholipour, S.; Nikaeen, M.; Rabbani, D.; Mohammadi, F.; Manesh, R.M.; Besharatipour, N.; Bina, B. Occurrence of enteric and non-enteric microorganisms in coastal waters impacted by anthropogenic activities: A multi-route QMRA for swimmers. *Mar. Pollut. Bull.* **2023**, *188*, 114716. [CrossRef]
6. Zhu, D.; Zhou, Q. Action and mechanism of semiconductor photocatalysis on degradation of organic pollutants in water treatment: A review. *Environ. Nanotechnol. Monit. Manag.* **2019**, *12*, 100255. [CrossRef]
7. Rahman, A.; Khan, M.M. Chalcogenides as photocatalysts. *New J. Chem.* **2021**, *45*, 19622–19635. [CrossRef]
8. Hu, Y.H. A highly efficient photocatalyst—Hydrogenated black TiO_2 for the photocatalytic splitting of water. *Angew. Chem. Int. Ed.* **2012**, *51*, 12410–12412. [CrossRef]
9. Li, W.; Deng, Y.; Wu, Z.; Qian, X.; Yang, J.; Wang, Y.; Gu, D.; Zhang, F.; Tu, B.; Zhao, D. Hydrothermal etching assisted crystallization: A facile route to functional yolk-shell titanate microspheres with ultrathin nanosheets-assembled double shells. *J. Am. Chem. Soc.* **2011**, *133*, 15830–15833. [CrossRef]
10. Li, Z.; Ding, D.; Liu, Q.; Ning, C. Hydrogen sensing with Ni-doped TiO_2 nanotubes. *Sensors* **2013**, *13*, 8393–8402. [CrossRef]
11. Shehzad, N.; Tahir, M.; Johari, K.; Murugesan, T.; Hussain, M. A critical review on TiO_2 based photocatalytic CO_2 reduction system: Strategies to improve efficiency. *J. CO2 Util.* **2018**, *26*, 98–122. [CrossRef]
12. Xu, W.; Xie, Z.; Han, W.; Zhang, K.; Guo, D.; Chang, K. Rational design of interfacial energy level matching for $CuGaS_2$ based photocatalysts over hydrogen evolution reaction. *Int. J. Hydrogen Energy* **2022**, *47*, 11853–11862. [CrossRef]
13. Asahi, R.; Morikawa, T.; Ohwaki, T.; Aoki, K.; Taga, Y. Visible-light photocatalysis in nitrogen-doped titanium oxides. *Science* **2001**, *293*, 269–271. [CrossRef] [PubMed]
14. Li, D.; Haneda, H.; Hishita, S.; Ohashi, N. Visible-light-driven N–F–codoped TiO_2 photocatalysts. 2. Optical characterization, photocatalysis, and potential application to air purification. *Chem. Mater.* **2005**, *17*, 2596–2602. [CrossRef]
15. Yu, J.; Yip, H.; Wong, P.K.; Zhao, J. Efficient visible-light-induced photocatalytic disinfection on sulfur-doped nanocrystalline titania. *Environ. Sci. Technol.* **2005**, *39*, 1175–1179. [CrossRef] [PubMed]
16. Khan, A.; Altaf, M.; Shahid, M.; Zeyad, M.T. In situ solid-state fabrication of Z-Scheme $BiVO_4/g\text{-}C_3N_4$ heterojunction photocatalyst with highly efficient-light visible activity and their antibacterial properties against bacterial pathogens. *J. Mol. Struct.* **2024**, *1300*, 137222. [CrossRef]
17. Zhou, H.; Li, B.; Liu, J.; Yang, W.; Wang, W.; Hu, X.; Wang, S. Construction of novel tandem reaction system coupling H_2O_2 production with in situ bleaching over Au/TiO_2 photocatalyst with different metal-support interactions. *Process Saf. Environ. Prot.* **2024**, *183*, 355–364. [CrossRef]
18. Saroha, J.; Rani, E.; Devi, M.; Pathi, P.; Kumar, M.; Sharma, S.N. Plasmon-assisted photocatalysis of organic pollutants by $Au/Ag–TiO_2$ nanocomposites: A comparative study. *Mater. Today Sustain.* **2023**, *23*, 100466. [CrossRef]
19. Lan, Y.; Wang, X.; Chen, Y.; Xu, J.; Zhang, Y. Simultaneous removal of bisphenol A and Cr (VI) by Pd–TiO_2: Significantly improve the photocatalytic activity through synergistic effect. *Solid State Sci.* **2023**, *146*, 107346. [CrossRef]
20. Murcia, J.J.; Hidalgo, M.C.; Navío, J.A.; Araña, J.; Doña-Rodríguez, J.M. Correlation study between photo-degradation and surface adsorption properties of phenol and methyl orange on TiO_2 Vs platinum-supported TiO_2. *Appl. Catal. B Environ.* **2014**, *150–151*, 107–115. [CrossRef]
21. Murcia, J.J.; Hidalgo, M.C.; Navío, J.A.; Araña, J.; Doña-Rodríguez, J.M. Study of the phenol photocatalytic degradation over TiO_2 modified by sulfation, fluorination, and platinum nanoparticles photodeposition. *Appl. Catal. B Environ.* **2015**, *179*, 305–312. [CrossRef]
22. Ruzimuradov, O.; Musaev, K.; Mamatkulov, S.; Butanov, K.; Gonzalo-Juan, I.; Khoroshko, L.; Riedel, R. Structural and optical properties of sol-gel synthesized TiO_2 nanocrystals: Effect of Ni and Cr (co) doping. *Opt. Mater.* **2023**, *143*, 114203. [CrossRef]
23. Jiang, X.; Zhang, Y.; Jiang, J.; Rong, Y.; Wang, Y.; Wu, Y.; Pan, C. Characterization of oxygen vacancy associates within hydrogenated TiO_2: A positron annihilation study. *J. Phys. Chem. C* **2012**, *116*, 22619–22624. [CrossRef]
24. Naldoni, A.; Allieta, M.; Santangelo, S.; Marelli, M.; Fabbri, F.; Cappelli, S.; Dal Santo, V. Effect of nature and location of defects on bandgap narrowing in black TiO_2 nanoparticles. *J. Am. Chem. Soc.* **2012**, *134*, 7600–7603. [CrossRef] [PubMed]
25. Song, K.; Han, X.; Shao, G. Electronic properties of rutile TiO_2 doped with 4*d* transition metals: First-principles study. *J. Alloys Compd.* **2013**, *551*, 118–124. [CrossRef]
26. Kočí, K.; Matějů, K.; Obalová, L.; Krejčíková, S.; Lacný, Z.; Plachá, D.; Šolcová, O. Effect of silver doping on the TiO_2 for photocatalytic reduction of CO_2. *Appl. Catal. B Environ.* **2010**, *96*, 239–244. [CrossRef]
27. Linic, S.; Christopher, P.; Ingram, D.B. Plasmonic-metal nanostructures for efficient conversion of solar to chemical energy. *Nat. Mater.* **2011**, *10*, 911–921. [CrossRef]
28. Banerjee, A.N.; Hamnabard, N.; Joo, S.W. A comparative study of the effect of Pd-doping on the structural, optical, and photocatalytic properties of sol–gel derived anatase TiO_2 nanoparticles. *Ceram. Int.* **2016**, *42*, 12010–12026. [CrossRef]
29. Nguyen, C.H.; Fu, C.-C.; Juang, R.-S. Degradation of methylene blue and methyl orange by palladium-doped TiO_2 photocatalysis for water reuse: Efficiency and degradation pathways. *J. Clean. Prod.* **2018**, *202*, 413–427. [CrossRef]
30. Rossi, L.; Villabrille, P.I.; Morales-Torres, S.; Rosso, J.A. Palladium-modified TiO_2 Photocatalysts: Synthesis, Characterization, and Environmental Application. *Mater. Chem. Phys.* **2023**, *302*, 127740. [CrossRef]

31. Chlumsky, O.; Purkrtova, S.; Michova, H.; Sykorova, H.; Slepicka, P.; Fajstavr, D.; Ulbrich, P.; Viktorova, J.; Demnerova, K. Antimicrobial Properties of Palladium and Platinum Nanoparticles: A New Tool for Combating Food-Borne Pathogens. *Int. J. Mol. Sci.* **2021**, *22*, 7892. [CrossRef]
32. Tekin, D.; Tekin, T.; Kiziltas, H. Photocatalytic degradation kinetics of Orange G dye over ZnO and Ag/ ZnO thin film catalysts. *Sci. Rep.* **2019**, *9*, 17544. [CrossRef]
33. Augustine, R.; Kalarikkal, N.; Thomas, S. Electrospun PCL membranes incorporated with biosynthesized silver nanoparticles as antibacterial wound dressings. *Appl. Nanosci.* **2016**, *6*, 337–344. [CrossRef]
34. Mancuso, A.; Morante, N.; De Carluccio, M.; Sacco, O.; Rizzo, L.; Fontana, M.; Esposito, S.; Vaiano, V.; Sannino, D. Solar driven photocatalysis using iron and chromium doped TiO_2 coupled to moving bed biofilm process for olive mill wastewater treatment. *Chem. Eng. J.* **2022**, *450 Pt 2*, 138107. [CrossRef]
35. Gil, M.A.; Murcia, J.J.; Hernández-Laverde, M.; Morante, N.; Sannino, D.; Vaiano, V. Ag/Cr-TiO_2 and Pd/Cr-TiO_2 for Organic Dyes Elimination and Treatment of Polluted River Water in Presence of Visible Light. *Nanomaterials* **2023**, *13*, 2341. [CrossRef]
36. Landi, S., Jr.; Segundo, I.R.; Freitas, E.; Vasilevskiy, M.; Carneiro, J.; Tavares, C.J. Use and misuse of the Kubelka-Munk function to obtain the band gap energy from diffuse reflectance measurements. *Solid State Commun.* **2022**, *341*, 114573. [CrossRef]
37. Castañeda, C.; Santos, D.; Hernández, J.S.; Álvarez, A.; Rojas, H.; Gómez, R.; Rajabi, F.; Martínez, J.J.; Luque, R. Efficient NiO/F–TiO_2 nanocomposites for 4-chlorophenol photodegradation. *Chemosphere* **2023**, *315*, 137606. [CrossRef]
38. Shokri, S.; Shariatifar, N.; Molaee-Aghaee, E.; Jahed Khaniki, G.; Sadighara, P.; Faramarzi, M.A. Modeling sunset yellow removal from fruit juice samples by a novel chitosan-nickel ferrite nano sorbent. *Sci. Rep.* **2024**, *14*, 208. [CrossRef]
39. Vaiano, V.; Sacco, O.; Sannino, D.; Ciambelli, P. Photocatalytic removal of spiramycin from wastewater under visible light with N-doped TiO_2 photocatalysts. *Chem. Eng. J.* **2015**, *261*, 3–8. [CrossRef]
40. Deepika, S.; Harishkumar, R.; Dinesh, M.; Abarna, R.; Anbalagan, M.; Roopan, S.M.; Selvaraj, C.I. Photocatalytic degradation of synthetic food dye, sunset yellow FCF (FD&C yellow no. 6) by Ailanthus excelsa Roxb. possessing antioxidant and cytotoxic activity. *J. Photochem. Photobiol. B Biol.* **2017**, *177*, 44–55. [CrossRef]
41. Rice, E.W.; Bridgewater, L. *American Public Health Association. Standard Methods for the Examination of Water and Wastewater*; American Public Health Association: Washington, DC, USA, 2012; Volume 10.
42. International Organization for Standardization. *Water Quality-Enumeration of Escherichia coli and Coliform Bacteria*; ISO: Geneva, Switzerland, 2014.
43. International Organization for Standardization. *Microbiology of Food and Animal Feeding Stuffs—Horizontal Method for the Enumeration of Yeasts and Moulds—Part 1*; ISO: Geneva, Switzerland, 2008.
44. Chanda, A.; Rout, K.; Vasundhara, M.; Joshi, S.R.; Singh, J. Structural and magnetic study of undoped and cobalt doped TiO_2 nanoparticles. *RSC Adv.* **2018**, *8*, 10939–10947. [CrossRef]
45. Lee, J.H.; Yang, Y.S. Synthesis of TiO_2 nanoparticles with pure brookite at low temperature by hydrolysis of $TiCl_4$ using HNO_3 solution. *J. Mater. Sci.* **2006**, *41*, 557–559. [CrossRef]
46. Dubey, R.; Singh, S. Investigation of structural and optical properties of pure and chromium doped TiO_2 nanoparticles prepared by solvothermal method. *Results Phys.* **2017**, *7*, 1283–1288. [CrossRef]
47. El-Rady, A.A.A.; El-Sadek, A.M.S.; El-Sayed, B.M.M.; Assaf, F.H. Characterization and Photocatalytic Efficiency of Palladium Doped-TiO_2 Nanoparticles. *Adv. Nanoparticles* **2013**, *2*, 372–377. [CrossRef]
48. Ali, T.; Ahmed, A.; Alam, U.; Uddin, I.; Tripathi, P.; Muneer, M. Enhanced photocatalytic and antibacterial activities of Ag-doped TiO_2 nanoparticles under visible light. *Mater. Chem. Phys.* **2018**, *212*, 325–335. [CrossRef]
49. Azambre, B.; Chebbi, M.; Ibrahim, N. Structure—Activity Relationships between the State of Silver on Different Supports and Their I_2 and CH_3I Adsorption Properties. *Nanomaterials* **2021**, *11*, 1300. [CrossRef] [PubMed]
50. Singaravelan, R.; Bangaru Sudarsan Alwar, S. Electrochemical synthesis, characterisation and phytogenic properties of silver nanoparticles. *Appl. Nanosci.* **2015**, *5*, 983–991. [CrossRef]
51. Morante, N.; Folliero, V.; Dell'Annunziata, F.; Capuano, N.; Mancuso, A.; Monzillo, K.; Galdiero, M.; Sannino, D.; Franci, G. Characterization and Photocatalytic and Antibacterial Properties of Ag- and TiO_x-Based (x = 2, 3) Composite Nanomaterials under UV Irradiation. *Materials* **2024**, *17*, 2178. [CrossRef]
52. Nguyen, H.P.; Pham, M.T.; Nguyen, H.Q.; Cao, T.M.; Van Pham, V. Boosting visible-light-driven photocatalysis of nitrogen oxide degradation by Mott–Schottky Pd/TiO_2 heterojunctions. *Sep. Purif. Technol.* **2024**, *354*, 129012. [CrossRef]
53. Merenda, A.; Weber, M.; Bechelany, M.; Allioux, F.M.; Hyde, L.; Kong, L.; Dumee, L.F. Fabrication of Pd-TiO_2 nanotube photoactive junctions via Atomic Layer Deposition for persistent pesticide pollutants degradation. *Appl. Surf. Sci.* **2019**, *483*, 219–230. [CrossRef]
54. Walczak, M.M.; Dryer, D.A.; Jacobson, D.D.; Foss, M.G.; Flynn, N.T. pH Dependent Redox Couple: An Illustration of the Nernst Equation. *J. Chem. Educ.* **1997**, *74*, 1195. [CrossRef]
55. Jayapandi, S.; Backialakshmi, P.; Soundarrajan, P.; Senthil Pandian, M.; Ramasamy, P.; Suresh Kumar, S.; Gopinathan, C. Construction of p–n junction type Ag_2O/SnO_2 heterostructure photocatalyst for enhanced organic dye degradation under direct sunlight irradiation: Experimental and theoretical investigations. *J. Mater. Res.* **2023**, *38*, 753–766. [CrossRef]
56. Ye, F.; Qian, J.; Xia, J.; Li, L.; Wang, S.; Zeng, Z.; Mao, J.; Ahamad, M.; Xiao, Z.; Zhang, Q. Efficient photoelectrocatalytic degradation of pollutants over hydrophobic carbon felt loaded with Fe-doped porous carbon nitride via direct activation of molecular oxygen. *Environ. Res.* **2024**, *249*, 118497. [CrossRef]

57. Wang, P.; Fan, S.; Li, X.; Duan, J.; Zhang, D. Modulating the molecular structure of graphitic carbon nitride for identifying the impact of the piezoelectric effect on photocatalytic H_2O_2 production. *ACS Catal.* **2023**, *13*, 9515–9523. [CrossRef]
58. Carrizosa, I.; Munuera, G.; Castanar, S. Study of the interaction of aliphatic alcohols with TiO_2: III. Formation of alkyl-titanium species during methanol decomposition. *J. Catal.* **1977**, *49*, 265–277. [CrossRef]
59. Sasi, S.; Chandran, A.; Sugunan, S.K.; Krishna, A.C.; Nair, P.R.; Peter, A.; Shaji, A.N.; Subramanian, K.R.; Pai, N.; Mathew, S. Flexible Nano-TiO_2 Sheets Exhibiting Excellent Photocatalytic and Photovoltaic Properties by Controlled Silane Functionalization—Exploring the New Prospects of Wastewater Treatment and Flexible DSSCs. *ACS Omega* **2022**, *7*, 25094–25109. [CrossRef]
60. El-Alami, W.; Garzón Sousa, D.; Díaz González, J.M.; Fernández Rodríguez, C.; González Díaz, O.; Doña Rodríguez, J.M.; El Azzouzi, M.; Araña, J. TiO_2 and F-TiO_2 photocatalytic deactivation in gas phase. *Chem. Phys. Lett.* **2017**, *684*, 164–170. [CrossRef]
61. Khakpash, N.; Simchi, A.; Jafari, T. Adsorption and solar light activity of transition-metal doped TiO_2 nanoparticles as semiconductor photocatalyst. *J. Mater. Sci. Mater. Electron.* **2012**, *23*, 659–667. [CrossRef]
62. Fang, W.; Yan, J.; Wei, Z.; Liu, J.; Guo, W.; Jiang, Z.; Shangguan, W. Account of doping photocatalyst for water splitting. *Chin. J. Catal.* **2024**, *60*, 1–24. [CrossRef]
63. Rajamanickam, D.; Shanthi, M. Photocatalytic degradation of an azo dye Sunset Yellow under UV-A light using TiO_2/CAC composite catalysts. *Spectrochim. Acta Part A Mol. Biomol. Spectrosc.* **2014**, *128*, 100–108. [CrossRef]
64. Stella, R.J.; Sreevani, I.; Gurugubelli, T.R.; Ravikumar, R.V.S.S.N.; Koutavarapu, R. Enhanced Solar Light-Driven Photocatalytic Degradation of Tetracycline Using Fe^{3+}-Doped CdO/ZnS Nanocomposite: Mechanistic Insights and Performance Evaluation. *Catalysts* **2023**, *13*, 1312. [CrossRef]
65. Hariani, P.L.; Said, M.; Salni; Aprianti, N.; Naibaho, Y.A.L.R. High Efficient Photocatalytic Degradation of Methyl Orange Dye in an Aqueous Solution by $CoFe_2O_4$-SiO_2-TiO_2 Magnetic Catalyst. *J. Ecol. Eng.* **2021**, *23*, 118–128. [CrossRef]
66. Haleem, A.; Shafiq, A.; Chen, S.-Q.; Nazar, M.A. Comprehensive Review on Adsorption, Photocatalytic and Chemical Degradation of Dyes and Nitro-Compounds over Different Kinds of Porous and Composite Materials. *Molecules* **2023**, *28*, 1081. [CrossRef] [PubMed]
67. Reymond, J.P.; Kolenda, F. Estimation of the point of zero charge of simple and mixed oxides by mass titration. *Powder Technol.* **1999**, *103*, 30–36. [CrossRef]
68. Zyoud, A.; Zyoud, A.H.; Zyoud, S.H.; Nassar, H.; Zyoud, S.H.; Qamhieh, N.; Hajamohideen, A.; Hilal, H.S. Photocatalytic degradation of aqueous methylene blue using ca-alginate supported ZnO nanoparticles: Point of zero charge role in adsorption and photodegradation. *Environ. Sci. Pollut. Res. Int.* **2023**, *30*, 68435–68449. [CrossRef]
69. de Sá, F.P.; Cunha, B.N.; Nunes, L.M. Effect of pH on the adsorption of Sunset Yellow FCF food dye into a layered double hydroxide (CaAl-LDH-NO_3). *Chem. Eng. J.* **2013**, *215*, 122–127. [CrossRef]
70. Bolton, J.R.; Bircher, K.G.; Tumas, W.; Tolman, C.A. Figures-of-merit for the technical development and application of advanced oxidation technologies for both electric-and solar-driven systems (IUPAC Technical Report). *Pure Appl. Chem.* **2001**, *73*, 627–637. [CrossRef]
71. Benrighi, Y.; Nasrallah, N.; Chaabane, T.; Belkacemi, H.; Bourkeb, K.W.; Kenfoud, H.; Baaloudj, O. Characterization and application of the spinel $CuCr_2O_4$ synthesized by sol–gel method for sunset yellow photodegradation. *J. Sol-Gel Sci. Technol.* **2022**, *101*, 390–400. [CrossRef]
72. Jaffari, Z.H.; Lam, S.M.; Ng, D.Q.; Sin, J.C. Visible light photocatalytic degradation of sunset yellow dye and industrial textile wastewater using Pd-$BiFeO_3$ composite. *IOP Conf. Ser. Earth Environ. Sci.* **2021**, *835*, 012001. [CrossRef]
73. Pavasaryte, L.; Balu, S.; Yang, T.C.K. Synthesis of sol–gel derived holmium aluminium garnet on exfoliated g-C_3N_4: A novel visible light driven Z-scheme photocatalyst for the degradation of sunset yellow FCF. *J. Mater. Sci. Mater. Electron.* **2019**, *30*, 20132–20143. [CrossRef]
74. Sin, J.C.; Lim, C.A.; Lam, S.M.; Zeng, H.H. Response surface approach for visible-light-driven photodegradation of sunset yellow over flower-like BiOBr hierarchical structures. *AIP Conf. Proc.* **2022**, *2454*, 050020. [CrossRef]
75. Rajamanickam, D.; Shanthi, M. Photocatalytic mineralization of a water pollutant, Sunset Yellow dye by an advanced oxidation process using a modified catalyst. *Toxicol. Environ. Chem.* **2013**, *95*, 1484–1498. [CrossRef]
76. Barani, A.; Alizadeh, S.R.; Ebrahimzadeh, M.A. A Comprehensive Review on Catalytic Activities of Green-Synthesized Selenium Nanoparticles on Dye Removal for Wastewater Treatment. *Water* **2023**, *15*, 3295. [CrossRef]
77. Navidpour, A.; Ahmed, M.; Zhou, J. Photocatalytic Degradation of Pharmaceutical Residues from Water and Sewage Effluent Using Different TiO_2 Nanomaterials. *Nanomaterials* **2024**, *14*, 135. [CrossRef] [PubMed]
78. Hernández-Uresti, D.B.; Ruíz-Gómez, M.A.; Vázquez, A.; Obregón, S. Enhanced electrophoretic deposition of hydrophobic $BiVO_4$ photocatalytic coatings using Cu^{2+} ions as stabilizing agent. *Mater. Today Chem.* **2024**, *38*, 102065. [CrossRef]
79. Tammaro, O.; Morante, N.; Marocco, A.; Fontana, M.; Castellino, M.; Barrera, G.; Allia, P.; Tiberto, P.; Arletti, R.; Fantini, R.; et al. The beneficial role of nano-sized Fe_3O_4 entrapped in ultra-stable Y zeolite for the complete mineralization of phenol by heterogeneous photo-Fenton under solar light. *Chemosphere* **2023**, *345*, 140400. [CrossRef] [PubMed]
80. Mancuso, A.; Sacco, O.; Vaiano, V.; Sannino, D.; Pragliola, S.; Venditto, V.; Morante, N. Visible light active Fe-Pr co-doped TiO_2 for water pollutants degradation. *Catal. Today* **2021**, *380*, 93–104. [CrossRef]
81. Puga, F.; Navío, J.A.; Hidalgo, M.C. A critical view about use of scavengers for reactive species in heterogeneous photocatalysis. *Appl. Catal. A Gen.* **2024**, *685*, 119879. [CrossRef]

82. Martín-Gómez, J.; Pérez-Losada, M.; López-Tenllado, F.; Hidalgo-Carrillo, J.; Herrera-Beurnio, C.; Estévez, R.; Marinas, A.; Urbano, J. Insight into the reaction mechanism of photocatalytic production of solketal. *Catal. Today* **2024**, *429*, 114506. [CrossRef]
83. Mureithi, A.W.; Sun, Y.; Mani, T.; Howell1, A.R.; Zhao, J. Impact of hole scavengers on photocatalytic reduction of nitrobenzene using cadmium sulfide quantum dots. *Cell Rep. Phys. Sci.* **2022**, *3*, 100889. [CrossRef]
84. Augustin, A.; Ganguly, P.; Shenoy, S.; Chuaicham, C.; Pillai, S.; Sasaki, K.; Lee, A.; Sekar, K. Impact of Hole Scavengers on Efficient Photocatalytic Hydrogen Production. *Adv. Susiainable Syst.* **2024**, 2400321. [CrossRef]
85. Fuku, K.; Takioka, R.; Iwamura, K.; Todoroki, M.; Sayama, K.; Ikenaga, N. Photocatalytic H_2O_2 production from O_2 under visible light irradiation over phosphate ion-coated Pd nanoparticles-supported $BiVO_4$. *Appl. Catal. B Environ.* **2020**, *272*, 119003. [CrossRef]
86. Lv, Z.; Song, W.; Pu, Y.; Su, M.; Zheng, H.; Wang, D.; Gao, J.; He, J.; Ouyang, L. Directly Synthesis of H_2O_2 on Pd-In/TiO_2 Catalyst and In-Situ Activation for Efficient Degradation of Tetracycline. *SSRN* **2024**. [CrossRef]
87. Lee, K.M.; Abd Hamid, S.B.; Lai, C.W. Mechanism and kinetics study for photocatalytic oxidation degradation: A case study for phenoxyacetic acid organic pollutant. *J. Nanomater.* **2015**, *2015*, 940857. [CrossRef]
88. Valentine Rupa, A.; Vaithiyanathan, R.; Sivakumar, T. Noble metal modified titania catalysts in the degradation of Reactive Black 5: A kinetic approach. *Water Sci. Technol.* **2011**, *64*, 1040–1045. [CrossRef] [PubMed]
89. Thabet, S.; Simonet, F.; Lemaire, M.; Guillard, C.; Cotton, P. Impact of photocatalysis on fungal cells: Depiction of cellular and molecular effects on *Saccharomyces cerevisiae*. *Appl. Environ. Microbiol.* **2014**, *80*, 7527–7535. [CrossRef]
90. Walker, G.M. *Yeast Physiology and Biotechnology*; John Wiley & Sons: Hoboken, NJ, USA, 1998.
91. Bacon, J.S.; Farmer, V.C.; Jones, D.; Taylor, I.F. The glucan components of the cell wallof baker's yeast (*Saccharomyces cerevisiae*) considered in relation to its ultrastructure. *Biochem. J.* **1969**, *114*, 557–567. [CrossRef] [PubMed]
92. Fleet, G.H.; Manners, D.J. Isolation and composition of an alkali-soluble glucan from the cell walls of Saccharomyces cerevisiae. *J. Gen. Microbiol.* **1976**, *94*, 180–192. [CrossRef] [PubMed]
93. Tang, Q.L.; Huang, G.L.; Zhao, F.Y.; Zhou, L.; Huang, S.; Li, H. The antioxidant activities of six (1→3)-β-d-glucan derivatives prepared from yeast cell wall. *Int. J. Biol. Macromol.* **2017**, *98*, 216–221. [CrossRef]
94. Liu, Y.; Huang, G.L. The derivatization and antioxidant activities of yeast mannan. *Int. J. Biol. Macromol.* **2018**, *107*, 755–761. [CrossRef]
95. Khan, M.; Wu, Z.; Mao, S.; Shah, S.N.A.; Lin, J.M. Controlled grafted poly (quaternized-4-vinylpyridine-co-acrylic acid) brushes attract bacteria for effective antimicrobial surfaces. *J. Mater. Chem.* **2018**, *6*, 3782–3791. [CrossRef]

nanomaterials

MDPI

Article

Solvothermally Grown Oriented WO_3 Nanoflakes for the Photocatalytic Degradation of Pharmaceuticals in a Flow Reactor

Mirco Cescon [1], Claudia Stevanin [2], Matteo Ardit [3,4], Michele Orlandi [5], Annalisa Martucci [4], Tatiana Chenet [2], Luisa Pasti [2,*], Stefano Caramori [1,6,*] and Vito Cristino [1]

1 Department of Chemical, Pharmaceutical and Agricultural Sciences, University of Ferrara, Via L. Borsari 46, 44121 Ferrara, Italy; mirco.cescon@unife.it (M.C.); vito.cristino@unife.it (V.C.)
2 Department of Environmental and Prevention Sciences, University of Ferrara, Via L. Borsari 46, 44121 Ferrara, Italy; claudia.stevanin@unife.it (C.S.); tatiana.chenet@unife.it (T.C.)
3 Department of Geosciences, University of Padova, Via Gradenigo 6, 35131 Padova, Italy; matteo.ardit@unipd.it
4 Department of Physics and Earth Sciences, University of Ferrara, Via Saragat 1, 44121 Ferrara, Italy; annalisa.martucci@unife.it
5 Department of Physics, University of Trento, Via Sommarive 14, 38123 Trento, Italy; michele.orlandi@unitn.it
6 National Interuniversity Consortium of Materials Science and Technology (INSTM), University of Ferrara Research Unit, 44121 Ferrara, Italy
* Correspondence: luisa.pasti@unife.it (L.P.); stefano.caramori@unife.it (S.C.); Tel.: +39-0532455346 (L.P.); +39-0532455474 (S.C.)

Abstract: Contamination by pharmaceuticals adversely affects the quality of natural water, causing environmental and health concerns. In this study, target drugs (oxazepam, OZ, 17-α-ethinylestradiol, EE2, and drospirenone, DRO), which have been extensively detected in the effluents of WWTPs over the past decades, were selected. We report here a new photoactive system, operating under visible light, capable of degrading EE2, OZ and DRO in water. The photocatalytic system comprised glass spheres coated with nanostructured, solvothermally treated WO_3 that improves the ease of handling of the photocatalyst and allows for the implementation of a continuous flow process. The photocatalytic system based on solvothermal WO_3 shows much better results in terms of photocurrent generation and photocatalyst stability with respect to state-of-the-art WO_3 nanoparticles. Results herein obtained demonstrate that the proposed flow system is a promising prototype for enhanced contaminant degradation exploiting advanced oxidation processes.

Keywords: pharmaceuticals; photocatalytic prototype system; solvothermally grown WO_3; water remediation

1. Introduction

In recent years, a significant influx of new contaminants has adversely affected the quality of both surface and groundwater on a global scale. The increase in human life expectancy, coupled with population growth, has led to an escalation in the use of pharmaceuticals [1,2]. Consequently, this has enhanced the presence of drugs in aquatic environments, resulting in a comprehensive list that undergoes continuous updates. The growing reports of the detection of these contaminants in surface waters worldwide is a cause for concern. Their presence in the aquatic environment can induce alterations in the physiological conditions and behavior of aquatic organisms. In particular, among bioactive environmental contaminants, steroid hormones are considered of high priority due to their biological activity, widespread use, and potential to interfere in endocrine functions (e.g., reproductive dysfunction) after exposure [3,4]. Drospirenone (DRO) (Figure 1), a synthetic progestin used in oral contraceptives, is frequently combined with 17-α-ethinylestradiol

(EE2) (Figure 1) [5]. It is one of the most prescribed synthetic progestins in numerous European countries and it is extensively used, especially in the USA. Due to their extensive use, both molecules have been detected in wastewaters and surface waters [6]. It has been demonstrated that fish species, such as adults and embryos of zebrafish (Danio rerio), exposed to DRO showed alteration in the gene transcription involved in hormone homeostasis, leading to reproductive changes and significant alterations of circadian rhythm [6,7]. Similarly, in surface waters, the synthetic estrogen EE2 can be found as a ubiquitous contaminant; indeed, EE2 at environmentally relevant concentration levels presents adverse effects on fish, including decreased egg fertilization, reproduction disruption, feminization, induction of vitellogenin, and increased mortality [8]. Due to its potential adverse effect on biota and human health, a similar estrogen (17β-estradiol) has been listed in Japanese Drinking Water Quality Standard [9]. In fact, EE2 considered as a priority micropollutant, as suggested by the European Union, might be regulated in the Drinking Water Quality Standards of European countries in the future [8].

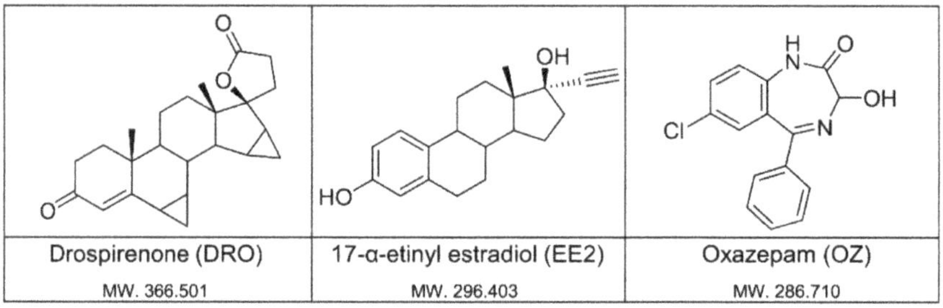

Figure 1. Molecular structure of drospirenone (DRO), 17-α-ethynilestradiol (EE2) and oxazepam (OZ). The molecules are all represented in their neutral form.

Recent research reports that municipal wastewater treatment plants (WWTPs) are the main sources of EE2 and DRO [6,8] in effluents, and consequently, in surface waters, highlighting the importance of enhancing pharmaceutical removal performance of municipal WWTPs. Concentrations of DRO and EE2 in surface waters evaluated in many countries were found up to 4.3 ng L^{-1} and 21.5 ng L^{-1}, respectively [9–11]. For what concerns psychoactive substances, oxazepam (OZ) (Figure 1) belongs to a family of pharmaceuticals called the benzodiazepines. In European river water, oxazepam is the benzodiazepine that is detected with the highest frequency (85%), with a maximum concentration of 61 ng L^{-1} [12]. Close to the outlet of WWTPs, OZ concentrations can reach several hundreds of ng L^{-1} to low ng L^{-1} concentrations, depending on the size of the receiving river and the season [13,14]. Psychoactive substances can raise behavioral alterations that may lead to profound, nonlinear, and perhaps unpredictable ecological effects [15]. One of the main sources of the mentioned contaminants are the effluents of WWTPs; therefore, it is important to develop innovative technologies to efficiently remove those contaminants from water in order to avoid the adverse effects on water and prevent irreversible damage to the aquatic ecosystems. Nowadays, different approaches for the removal of emergent contaminants (ECs) are employed, such as the use of adsorbent materials and advanced oxidation processes (AOPs) [14]. AOPs can be used as a pre-treatment [16,17] process to improve wastewater quality and remove organic contaminants by oxidation through reactions with hydroxyl radicals (OHs) which, being a strong oxidant characterized by an oxidation potential equal to E° = +2.1 V vs. NHE, allows for the oxidation of a variety of organic substances, leading to their mineralization [18–20]. To reduce the energetic request for implementing redox remediation processes, photocatalytic materials show some promise.

The ability of some semiconductors to adsorb UV–vis photons for the production of highly energetic electron/hole pairs makes these materials capable of promoting thermodynamically demanding oxidation reactions. In particular, some metal oxide semiconductors like TiO_2, ZnO and WO_3 display such a deep valence band, whose edge bears the prevailing contributions by π orbitals of the oxygen atoms, that holes therein generated are able to produce hydroxyl radicals following monoelectronic water oxidation. The wide band gap of many metal oxide semiconductors can limit the portion of absorbed solar light; for example, TiO_2 and ZnO have a sizable photochemical activity only upon UV excitation. A good compromise between the need for generating powerfully oxidizing holes and the harvesting of solar light is incorporated in WO_3, being able to absorb photons up to 460 nm (band gap of ca. 2.7 eV). The ability of WO_3 to generate ·OH through water oxidation under visible illumination has been previously reported [21]. A variety of techniques can be used to synthesize efficient WO_3 structures aimed at increasing the photon-to-electron conversion by emphasizing the surface (contact) area of the photocatalyst, thus reducing electron/hole recombinative losses; pulsed laser deposition [22], electrochemical anodization [23–26], sputtering [27], evaporation [28], sol-gel [29–31], electrophoresis [32], hydrothermal and solvothermal [33–35] are some of the techniques described in the scientific literature.

Besides achieving a high photoconversion efficiency, the choice of the preparative technique to afford a practical photocatalytic system for water decontamination should consider other aspects as well. The low cost and abundancy of reagents and of preparative equipment, the potential for scaling-up the synthesis, the possibility to support the catalytic material to facilitate the handling of the photocatalyst, and the ability to implement a flow rather than a "batch" process are some of the issues to consider for the successful implementation of a photocatalytic route-to-water decontamination. Large-scale wastewater treatment plants require a continuous operation unit able to process large volumes of liquid, which is generally difficult and economically disadvantageous with batch reactors. For these reasons, in this study, we tested the capabilities of solvothermal WO_3 (WO_3/st), prepared using a scalable solvothermal route and stabilized with a TiO_x adhesion layer, in the oxidative degradation of selected emerging chemical contaminants using both a batch method and a flow system based on a reactor packed with WO_3/st-modified glass beads.

2. Materials and Methods

2.1. Chemicals for Analytical and Photochemical Experiments

Drospirenone (DRO) was purchased from European Pharmacopoeia Reference Standard (purity 100%), 17-α-ethnylestradiol (EE2) from Dr. Ehrenstorfer GmbH (Augsburg, Germany) (purity > 96.3%), while oxazepam (OZ) was obtained from F.I.S. Fabbrica Italiana Sintetici S.p.a. (Vicenza, Italy) (purity 100%); Na_2SO_4 from VWR International S.r.l. (Milan, Italy) (purity > 99%), acetonitrile (ACN) and phosphoric acid were purchased from Merck KGaA (Darmstadt, Germany) (HPLC grade, purity > 99%), H_2WO_4 and oxalic acid dihydrate were purchased from Alfa Aesar (Haverhill, MA, USA) (purity 99.5–102.5%), hydrogen peroxide (purity 35%) and α-phenyl *N*-tert-butyl nitrone were purchased from Merck KGaA, lithium perchlorate was purchased from Acros Organics (Geel, Belgium) (purity 99%), hydrocloric acid was purchased from Carlo Erba Reagents S.r.l. (Milan, Italy) (37%). All the solutions were prepared using ultrapure waters (Milli-Q® systems, Merck KGaA, Darmstadt, Germany).

2.2. Synthesis of WO_3/st Films and Powder by Solvothermal Method

WO_3/st nanocrystalline film was grown either on commercial glass spheres (GS), with a diameter of 4 mm, microscope glass slides (MGSs) or a fluorinated tin oxide glass (FTO glass, TEC 7 Ω supported on soda lime glass; Pilkington, Lathom, UK) substrate. All substrates were cleaned by ultrasonication for 10 min in 2-propanol prior to use. The preparation of the photocatalytic substrates was accomplished according to three sequential steps.

Step 1: Before the deposition of WO_3/st, the formation of a TiO_x nanometric adhesion layer (<5 nm) was accomplished by the room temperature hydrolysis of $TiCl_4$ on the oxide supports (GS, MGS, FTO). Specifically, the soaking of the supporting surfaces in 0.4 M $TiCl_4$ was carried out for 12 h for GS and MGS and 6 h in the case of FTO. After abundant rinsing of the glass substrates with deionized water, firing for 30 min at 450 °C in air afforded the desired TiO_x adhesion layer which, as we will demonstrate, does not block the collection of the photocarriers generated within WO_3 upon band gap excitation (Figure S1). WO_3 substrates obtained on top of the TiO_x adhesion layer are named WO_3-Ti-substrate (i.e., WO_3-Ti-FTO) and, unless otherwise indicated, all photocatalytic materials herein investigated are of the type WO_3-Ti.

Step 2: In the case of planar substrates, prior to solvothermal deposition, a WO_3 seed layer with a thickness on the order of a few nanometers (typically 10–30 nm) was obtained by spin coating (9 s at 1000 rpm, 20 s at 2000 rpm); the precursor used for this process can be either 0.5 M peroxotungstic acid (PTA) or a colloidal precursor based on previously reported methods [29]. Such a precursor layer is converted into WO_3 by firing at 550 °C in air. The best results were obtained with the peroxotungstic acid precursor. While the presence of the WO_3 seed layer emphasizes the surface coverage and density of the photoactive film, its absence does not preclude the subsequent hydrothermal growth of the porous WO_3/st layer.

Step 3: The solvothermal (st) deposition of WO_3 on top of the glass substrates was treated according to steps 1 and 2 (optional), followed a procedure described by Grimes et al. [35]. In a typical preparation of the synthesis of a dense WO_3 photocatalytic layer based on oriented lamellar arrays or nanoflakes, the precursor solution was prepared by dissolving 1.25 g of H_2WO_4 in 10 mL of ultrapure water and 10 mL of 35% w/w H_2O_2 while heating and stirring bain–marie at 95 °C. In total, 18 mL of this H_2WO_4 solution (0.25 M), plus 1.2 g of oxalic acid and 3 mL of HCl (6 M), were mixed into 60 mL of acetonitrile and 15 mL of ultrapure water. Glass substrates, either flat slides or beads, were immersed in such a solution and heated at 180 °C in a Teflon-lined steel autoclave for a duration variable from 1 h to 30 min for glass spheres and 2 h for FTO. Besides the coated materials, WO_3/st powder was also recovered from the bottom of the Teflon reactor, washed with distilled water, and dried in an oven at 60 °C for a few hours. All WO_3/st-coated substrates were fired at 500 °C in air for 2 h to obtain the sintering and crystallization of WO_3/st. Similar thermal treatment was used with the material in the form of a powder.

2.3. Structural Characterization Techniques

2.3.1. X-ray Powder Diffraction

Data from samples grown on flat surfaces were collected at room-temperature conditions (RT). Data collection was performed on a diffractometer (D8 Advance Davinci, Bruker Corporation, Billerica, MA, USA), working in Bragg–Brentano geometry, and equipped with a Cu-anode X-ray tube, Ni-filter to suppress the Cu Kβ component, and a LynxEye XE silicon strip detector (angular range of the detector window size = 2.585° 2θ) set to discriminate Cu Kα1,2 radiation.

Samples were loaded in a poly-(methyl methacrylate) specimen holder characterized by a cavity with adjustable depth to properly fit the microscope slide thickness, and scanned in a continuous mode from 5 to 90° 2θ, with step size of 0.02° 2θ and a counting time of 3 s per step. A knife perpendicular to the sample holder placed at a sub-millimetric distance from the sample was used to reduce the air-induced scattering.

Rietveld refinement. Collected XRPD patterns were modelled by means of the fundamental-parameter Rietveld approach (TOPAS v.5.0, Bruker Corporation, Billerica, MA, USA). Identified phases were modelled by carrying out a multiphase refinement in which only the scale factors and unit-cell parameters were varied. Instrumental parameters (e.g., goniometer radius, slit sizes, geometrical parameters of the X-ray tube, etc.) were used to calculate the instrumental contribution to the peak profiles, and specimen-related crystallite size information for each phase was extracted from the observed profile. An instrumental zero-error was fixed at the value, determined using the Si Standard Reference Material (640e, National Institute of Standards and Technology, Gaithersburg, MD, USA) standard, and the refinement included a specimen displacement correction and a Chebyshev polynomial to model the background.

2.3.2. Scanning Electron Microscopy

The morphological characterization of WO_3 films, on FTO substrate, was performed using a scanning electron microscopy (SEM, JSM-7001F FEG-SEM, JOEL Ltd., Tokyo, Japan) equipped with an energy-dispersive X-ray spectroscopy detector (EDXs, Oxford INCE PentaFETx3, Oxford Instruments Plc., Abingdon, UK). A 90° stub was used to define the thickness and the different layers of the electrode.

2.3.3. Photoelectrochemical Characterization

Photoelectrochemical studies were carried out by exploiting the WO_3 films grown on FTO. The typical photoelectrochemical cell was composed of an FTO/WO_3 working electrode, a platinum wire counter electrode, and a saturated calomel electrode (SCE) reference electrode. Current/potential (J/V) curves were collected under simulated solar light using a solar simulator (ABET Technologies Inc., Milford, CT, USA) equipped with a AM1.5G filter, with a radiant power of 100 mW/cm^2 and a 380 nm UV cut-off filter. The comparative measurements of WO_3/st and WO_3/coll were performed in an aqueous solution containing 0.7 M Na_2SO_4 as electrolyte, in a potential range between −0.3 V to 1.5 V versus SCE with a scan rate of 0.02 V/s. To reveal the effect of hole scavenging at the WO_3/st interface by the selected contaminants, J/Vs were performed in an acetonitrile (ACN) solution, using 0.1 M $LiClO_4$ as electrolyte, to bypass the low solubility of our target molecules in an aqueous solution. In this case, J/V curves were performed in a potential range between 0.1 and 1.7 V versus SCE with a scan rate of 0.02 V/s. J/V curves under pulsed illumination were obtained with an automatic shutter (SHB1T, Thorlabs, Newton, NJ, USA) at 0.3 Hz.

Cyclic voltammetry was performed in an aqueous solution containing 0.1 M $LiClO_4$ and 0.01 M K_4FeCN_6.

The incident photon-to-current conversion efficiency (IPCE) describes the photocurrent collected per incident photon flux as a function of the illumination wavelength. IPCE data, sampled at 10 nm intervals in the 310–500 nm range, were acquired in 0.7 M Na_2SO_4 using the same three-electrode configuration described above. Limiting photocurrents were collected at 1 V and 1.5 V for WO_3/st and WO_3/coll, respectively. Here, WO_3/coll denotes a mesoporous WO_3 electrode composed of nanoparticles (Figure S2), described in various previous publications [21,23].

The experimental setup used for photo-action spectra consisted of a monochromator (Applied Photophysics Ltd., Leatherhead, UK) equipped with a 175W Xe lamp (LuxteL-Sunoptic Technologies Inc., Jacksonville, FL, USA), various optical elements, a digital multimeter (34401A 6-1/2, Agilent Techologies Inc., Santa Clara, CA, USA), a potentiostat (552, AMEL S.r.l., Milan, Italy), and a calibrated silicon photodiode (OSD 7Q, Centronic Ltd., New Addington, UK).

Absorbed photon-to-current efficiency (APCE) spectra were computed, normalizing the IPCE value for the absorbed photons, after subtracting the reflected or transmitted photons.

$$A(\lambda) = 100 - R\%(\lambda) - T\%(\lambda) \tag{1}$$

APCE is thus dependent on the quantum efficiencies of charge transport and interfacial separation defined as follows:

$$APCE(\lambda) = IPCE(\lambda)/a(\lambda) = \eta_{transport}\,\eta_{interface} \tag{2}$$

considering that the efficiency of charge generation ($\eta e/h$) is unitary in a semiconductor upon band gap excitation.

2.3.4. EPR

EPR spin-trapping experiments were carried out with a Bruker MRD spectrometer equipped with a TE 201 resonator (microwave frequency of 9.4 GHz). The samples were suspensions of WO_3 used for film fabrication in aqueous solutions containing α-phenyl

N-tert-butyl nitrone (pbn, 5×10^{-2} M) as a spin trap. The samples were put into a flat quartz cell and directly irradiated ($\lambda > 400$ nm) in the EPR cavity with a medium-pressure mercury lamp. No EPR signals were obtained in the dark and during the irradiation of the solution in the absence of WO_3.

2.4. Photochemical Degradation Experiments

Batch experiments. Photodegradation tests were performed using the batch method, where a solution with a known concentration of a contaminant is placed in contact with a known quantity of WO_3/st powder (70 mg) as a photocatalytic dispersion. The photodegradation tests were carried out in 20 mL flasks under constant stirring for 300 min at room temperature (Figure S3). The solutions were irradiated using a solar simulator (ABET Technologies Inc., Milford, CT, USA, equipped with a AM1.5G filter, with a radiant power of 100 mW/cm^2 and 380 nm UV cut-off filter), a device that provides illumination approximating natural sunlight. For each solution containing 4 µg/mL of a drug (EE2, DRO and OZ) in a 0.7 mM Na_2SO_4 aqueous solution at pH = 6 ± 0.30, the following four tests were carried out under different conditions: (i) a dark test to evaluate the adsorption of contaminants at the glassware walls; (ii) a dark test, in the presence of powdered WO_3/st, to evaluate the adsorption of drugs at the surface of the solid photocatalyst; (iii) an irradiation test without WO_3/st to verify the direct photochemical degradation of each contaminant; and (iv) irradiation cycles in the presence of the photocatalyst to reveal the photodegradation deriving from the WO_3/st activity. At the end of each experiment, the separation of the supernatant was achieved by centrifugation, and the supernatants containing OZ and DRO were analyzed by HPLC-DAD (see Section 2.6), whereas EE2 solutions were analyzed by HPLC-FLD (see Section 2.7).

Flow system experiments. Moving from batch experiments to the flow system, a 10-fold scale-up in terms of volume was achieved, reaching a total treated water volume of 200 mL. The apparatus was composed of a glass flask, containing the polluted solution, a magnetic stirrer and a glass column containing glass spheres (GSs) coated with WO_3/st (WO_3/st-GS) (Figure 2). The column was completely enveloped by an LED stripe ($395 < \lambda < 400$ nm, Figure 2e) to obtain a uniform illumination of the whole packed tube and increase the probability of photon absorption by the photocatalyst along the whole tube length. The emission range of LEDs was selected in order to maximize the efficiency of photoconversion, peaking, in the WO_3 photocatalyst, at the edge of the visible spectrum (Figure 2d). The average irradiance generated by the LED strip was 49 mW/cm^2.

The water flow was controlled by a peristaltic pump to 7 mL/min. Typical illumination cycles of 480 min under electrolyte recirculation conditions at room temperature were carried out by using an aqueous test solution containing 4 µg/mL of the selected drugs in 0.7 mM Na_2SO_4 at pH = 6.00 ± 0.30 (Five Easy F20, Mettler-Toledo International Inc., Greifensee, Switzerland). Similar to batch tests, various control experiments under different conditions were performed. The flow system connection tubes were selected in consideration of a low surface affinity towards the various contaminants in order to minimize drug losses by adsorption. To keep a tight packing of WO_3/st-GS in the column, PE screw caps and PP frits were placed at the inlet and outlet of the column. The electrolyte was sampled during the degradation experiments to evaluate the kinetics of the process by verifying the concentration of drugs in the solution and the formation of by-products.

2.5. Kinetic Models

Chemical kinetics are a part of science with the theoretical purpose of determining the mechanism of a reaction. Among the different models to study and understand the mechanism, the experimental data were well described by the first order kinetic model (3). First order reactions are those reactions whose rate depends on the concentration of a single reactant raised to an exponent of 1. The mathematical equation is as follows:

$$\ln[C]/[C_0] = -kt \tag{3}$$

where, [C] is the reagent concentration at time t, [C_0] is the initial concentration of the reagent, k is the kinetic constant and t is the time of reaction [36,37].

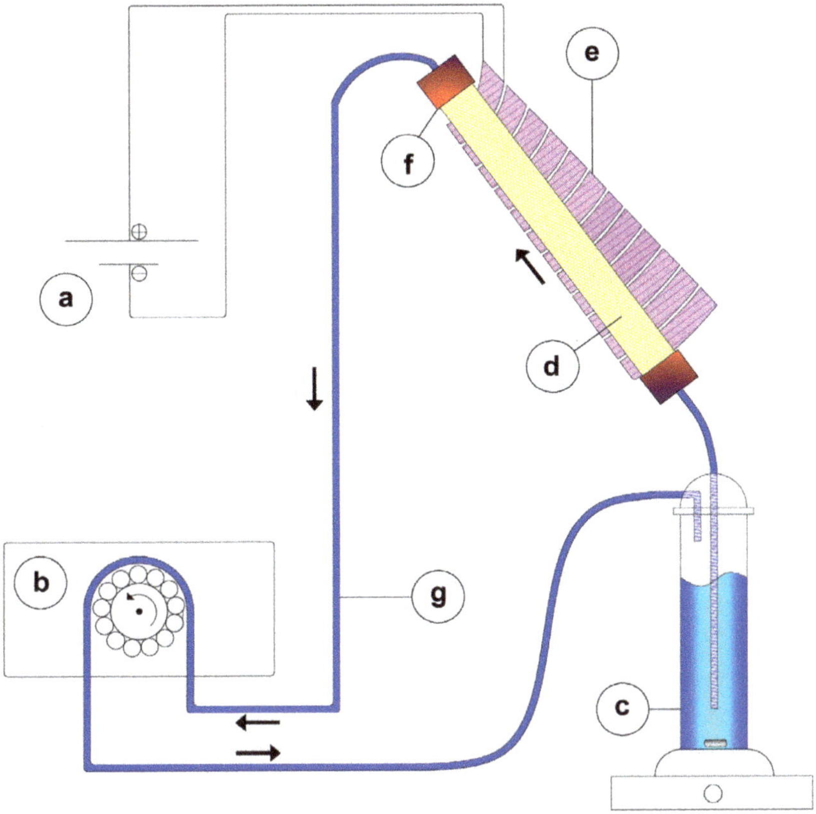

Figure 2. Flow system: (a) 24 V direct-current generator (DC); (b) peristaltic tube pump with two sprung rollers; flow rate set at 7 mL min^{-1}; (c) mixing system on magnetic stirrer; (d) Ø3 mm glass spheres, WO_3-coated (WO_3/st-GS), confined in a glass column with internal Ø19 mm and outer Ø 22 mm; (e) 24 V LED source, λ = 395–400 nm, completely enveloped column; (f) PE screw cap to fix a frit in PP with 2 mm grill in order to lock spheres; and (g) tubing, internal Ø4 mm and outer Ø 7 mm.

2.6. HPLC-DAD Analysis

OZ and DRO removal was evaluated by HPLC-DAD analysis. The apparatus was composed of two pumps (515 HPLC pump, Waters Corp., Milford, MA, USA), Rheodyne injectors (IDEX Corporation, Northbrook, IL, USA) with 20 μm loop and a photodiode array detector (996 DAD, Waters corp., Milford, MA, USA). The column selected was C18 100A 150 × 4.6 mm with a particle size of 5 μm (Kinetex SuXB, Phenomenex, Torrance, CA, USA). The mobile phase was a mixture of acetonitrile (ACN) and ultrapure H_2O, and the analyses were performed under isocratic conditions. The flow rate was 1 mL/min, and the injection volume was 20 μL. For the OZ molecule, the mobile phase mixture was 40% ACN and 60% H_2O, while for the DRO molecule, the eluent was 50% ACN and 50% H_2O. OZ and DRO were detected at 227 nm and 265 nm, respectively.

2.7. HPLC-FLD Analysis

EE2 analysis was performed using an HPLC (1290 Infinity, Agilent Technologies Inc., Santa Clara, CA, USA) equipped with a binary pump, autosampler, thermostatic column compartment, and fluorescence detector (FLD 1290 Infinity II, Agilent Technologies Inc., Santa Clara, CA, USA). The selected column was C18 100A 150 × 4.6 mm with a particle size of 5 μm (Kinetex SuXB, Phenomenex, Torrance, CA, USA). The column compartment temperature was set to 25 °C. The eluent mixture consisted of 70% ACN and 30% ultrapure water, and the analysis was performed under isocratic conditions; the flow rate was 1 mL/min, and the injection volume was 20 μL. The detection conditions were set to an excitation wavelength of 280 nm and an emission wavelength of 310 nm.

2.8. HPLC-MS Analysis

HPLC/MS analyses were carried out by means of Surveyor micro-HPLC hyphenated to a linear trap quadrupole mass spectrometer (LTQ XL, Thermo Fisher Scientific Inc., Waltham, MA, USA). The HPLC was composed of a solvent delivery system, a quaternary pump and an autosampler. The LTQ system was equipped with an electrospray ionization (ESI) ion source. The mobile phase was obtained by mixing ACN/0.1% v/v formic acid with ultrapure water/0.1% v/v formic acid according to the following gradient program: 0 min 10% ACN, 0–10 min 10–90% ACN, and 10–12 min 90% ACN. Then, the column was reconditioned at 10% ACN until reaching 15 min of elution time. The flow rate was 150 μL/min. The column was C-18 100 mm × 2.1 mm × 2.1 μm (Supelco, Bellefonte, PA, USA). The injection volume was 2 μL for all standards and samples. MS experimental conditions were as follows: spray voltage 5 kV, capillary temperature 275 °C, capillary voltage 28 V and tube lens 50 V for positive ESI conditions (OZ and DRO); spray voltage 5 kV, capillary temperature 275 °C, capillary voltage −50 V and tube lens −137 V for negative ESI conditions (EE2).

3. Results and Discussion

3.1. Morphological and Structural Properties of Nanostructured WO_3

The geometrical constraints of the GS make it impossible to perform a reliable structural and morphological characterization of WO_3 nanostructures decorating these curved supports. Nevertheless, the structural properties of WO_3 films obtained on flat surfaces, consisting of both FTO and simple glass, are considered representative of the general properties of the WO_3 coating on the glass beads. In addition, the use of conductive FTO allows us to quantify, via photoelectrochemical measurements, the performance of these WO_3 substrates in generating reactive charge carriers when excited with suitable optical frequencies.

3.1.1. Electron Microscopy

The presence of the TiO_x adhesion layer is not clearly discernible on the FTO surface, indicating that its thickness is below the instrumental detection limit (5 nm in cross section mode on this type of surface) of our SEM. At such low thickness, the crystalline and band structure of TiO_x is probably unable to fully develop, resulting in a defective layer which allows electron percolation to the TCO collector to occur, as shown by cyclic voltammetry with a fast Fe (II)/III) redox probe (Figure S1). Instead, the compact WO_3 seed layer can be clearly seen from Figure 3. The seed layer produced with the PTA precursor conformally covers the FTO substrate, which has an overall rough and irregular appearance due to the presence of large FTO crystals (hundreds of nanometers in size). The seed layer displays a thickness of about 30–40 nm, as shown in Figure 3b, and is quite homogeneous, despite the presence of cracks, in particular in the valleys between adjacent FTO crystallites, wherein the liquid precursor tends to accumulate, giving rise to a locally thicker WO_3 film which fractures upon thermal sintering.

The effect of the TiO_x adhesion layer is macroscopically evident from the mechanical properties of the photoactive films at the end of the solvothermal growth and annealing

(Figure S4a,b). In the WO_3/st-FTO (WO_3/st = solvothermal WO_3) samples (that is, steps 2 and 3 without the TiO_x adhesion layer), uncoated islands appear as a result of the thick WO_3 film detaching from the TCO surface. Moderate magnification optical microscopy shows that the WO_3 film crinkles and peels off the surface because of poor adhesion, or thermal stresses during the annealing, leaving exposed FTO underneath. Actually, the WO_3-FTO shown in Figure S4b is one of the few samples where the solvothermal WO_3 layer did not detach completely from the FTO ohmic contact upon rinsing and annealing. When the same procedure is carried out after the Ti(IV) treatment (step 1), affording the sample WO_3/st-Ti-FTO, one obtains an electrode which remains solidly and uniformly coated after the rinsing/annealing cycle. The hydrophilic TiO_x layer considerably stabilizes the WO_3 adhesion to the electrode, being characterized by the presence of a higher-than-FTO density of surface OH groups. Thus, multiple oxo bridges with WO_3 could be formed, guaranteeing a robust anchor to the FTO of the latter.

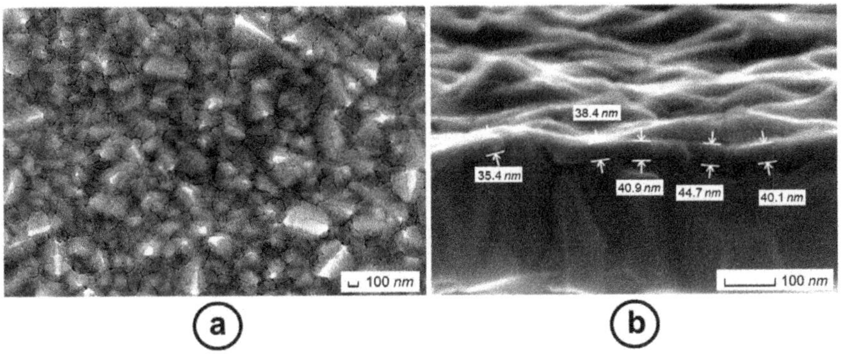

Figure 3. Morphology of the TiO_x-treated WO_3 seed layer on FTO support imaged from the top (**a**) and from the side (**b**), showing the thickness of the deposit.

The typical surface of a WO_3/st-Ti-FTO electrode is characterized by a dense coverage of WO_3 leaves and lamellae, roughly vertically aligned with respect to the surface, and arranged in somewhat organized superstructures which may resemble flowers or flakes. These leaves are relatively thin (<100 nm) (Figure 4b) compared to their length and width, which extend for several microns (Figure 4a) and could, thus, be essentially considered 2D structures. The inspection of the cross section of the electrode reveals that these lamellae extend considerably in length, taking up most of the film thickness (about 8 μm) and confirm their close packing and roughly vertical alignment with respect to the FTO surface. A closer inspection of the cross section reveals a disordered layer, composed mostly of nanoparticles, with the overall thickness of few 100 s of nanometer between the roots of the leaves and the seed layer. This rubble-like layer could be formed either by the partial collapse of the lamellae, or by the precipitation of WO_3 during the initial stage of the solvothermal growth before thermal equilibrium is reached. Nevertheless, it is clear that either in electrodes or decorated glass supports illuminated from the top (electrolyte side), photon absorption will occur mostly within the 2D structures, which will be the primary actors in the charge transfer dynamics involving WO_3/st. In the absence of the TiO_x adhesion layer, the same lamellar morphology of the WO_3 layer of Figure 4 is observed (Figure S5), although a nearly complete detachment of the WO_3 film from the FTO ohmic support is observed. EDS analysis confirms the expected stoichiometry of WO_3, with tungsten and oxygen as the only significant elements in the elemental composition of the nanostructured film, appearing in a 1:3 molar ratio (Figure S6).

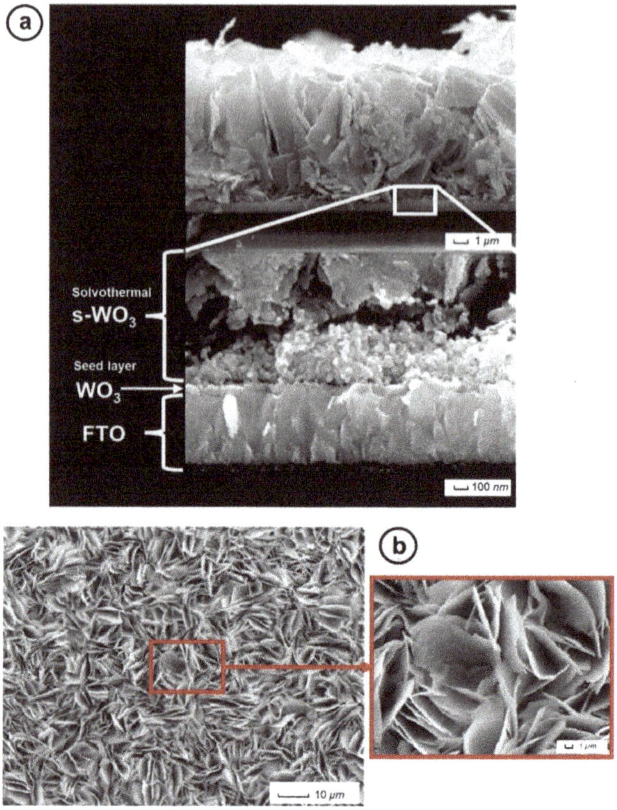

Figure 4. SEM imaging of WO_3-TiO_x-FTO electrode in cross sectional (**a**) and top views (**b**). The magnification of the top view allows us to appreciate the thickness of the WO_3/st leaves, of less than 100 nm, compared to their length and width of several microns.

3.1.2. X-ray Diffraction

X-ray diffraction (XRD) data confirmed that WO_3/st on the Ti-FTO substrate was composed by high-quality monoclinic WO_3, with a P 21/n space group, formed during high-temperature annealing with cell parameters of a = 7.320 Å, b = 7.518 Å, c = 7.670 Å (V(Å3) = 422.1), β = 90.4° (WO_3 database code ICSD 80056) and was marginally different from the data collected on other types of standard WO_3 electrodes produced in our laboratory [37] (our colloidal standard (WO_3/coll) affords a = 7.303 Å, b = 7.536 Å, c = 7.686 Å (V(Å3) = 422.9), β = 90.4° (WO_3 database code ICSD 80056), and consistent with the results reported by Grimes [35] in his paper. This confirms that the TiO_x adhesion layer does not interfere with the subsequent solvothermal growth. Crystallite size is much smaller than the size of a whole lamella, being only 25.09 nm. Thus, every leaf is formed by several nanometer-sized WO_3 crystallites.

The diffraction peak at 2θ of 23.2°, 23.6° and 24.4° corresponds to the (002), (020) and (200) facets (see Figure 5). The two samples show different preferentially exposed facets which are (200) for WO_3/coll and (002) for WO_3/st. A possible mechanism for solvothermal growth that explains the preferential (002) facet exposure in WO_3/st is proposed by Gong et al. [33], who state that, during the solvothermal reaction, the Cl^- functions as a kind of capping ion analogous to what was reported for F^- [38–40]. Since the (002) facets have a larger surface energy than (020) and (200) ((002) (1.56 Jm^{-2}) > (020) (1.54 Jm^{-2}) > (002) (1.43 Jm^{-2})), Cl^- is preferably adsorbed on the (002) facets. This adsorption weakens the surface affinity for the precursor to this orientation, thus inhibiting the crystal growth rate along the perpendicular to these planes, resulting in the increased exposure of the (002) facets. It is well known that the reactivity of a photocatalyst is influenced by its surface atomic and electronic structure [40], so the preferential exposure of (002) instead of (200) family of planes could strongly influence the photocatalytic performance of these WO_3 structures. Indeed, as reported by Gong et al. [33], the preferential exposure of (002) facets was found to improve the photoelectrochemical performance of WO_3 photoanodes. According to Wang et al. [34], the preferential exposure of the highly active (001) facets that belong to the same {001} family of planes as (002) facets endows WO_3 nanocrystals with significantly enhanced photocatalytic activity, which can be attributed to the stabilization of reactive oxygen species and the reduced recombination of photogenerated electrons and holes at these interfaces.

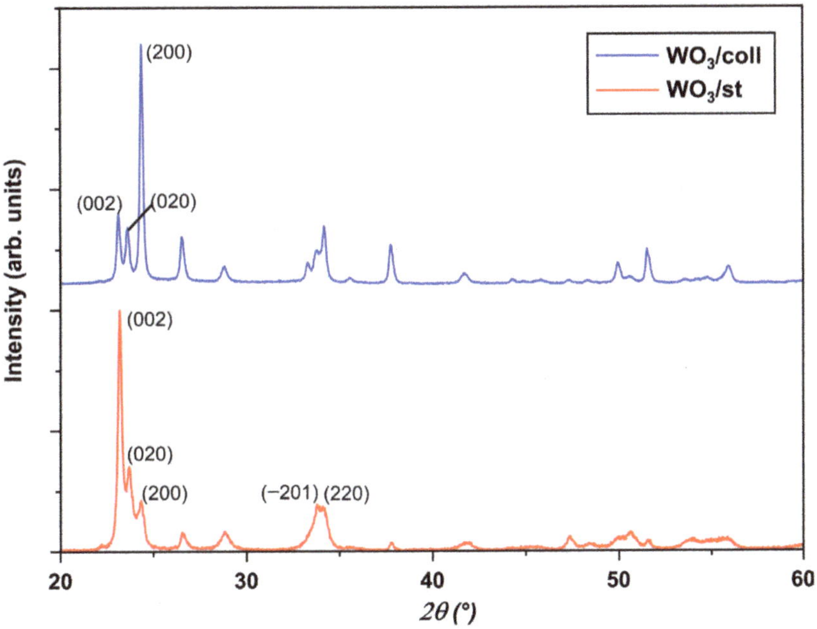

Figure 5. XRD patterns recorded on FTO substrate compared to the WO_3/coll standard of known art produced in our laboratory.

As concerns the WO_3/st-Ti grown on simple glass slides, a comparison of the collected X-ray diffraction patterns at RT is reported in Figure 5.

A qualitative phase analysis of the collected patterns was performed by means of the AXS EVA software (v.6.0.0.7) (Bruker Corporation, Billerica, MA, USA). Both samples were biphasic and composed of tungsten oxide WO_3 in two polymorph modifications, namely monoclinic (s.g. $P2_1/n$), which is largely prevalent (>95%) and hexagonal (s.g. $P6_3/mcm$) (<4%). The hexagonal phase was observed previously in materials obtained through other types of hydrothermal processes, and was found to be active in oxygen evolution processes as well [41]. The presence of hexagonal/monoclinic junctions was considered beneficial for improved charge separation by some authors [42]. Prolonged exposure of WO_3/st to pH 7 water (WO_3-washed) does not change the compositional profile substantially, indicating that both phases persist on the surface of glass supports during the employment of the photocatalyst in neutral aqueous media. The Rietveld refinement plots for both samples are shown in Figures S7 and S8. Quantitative phase analysis and unit-cell parameters, as well as the crystallite size of the main phases, are summarized in Table 1.

Table 1. Quantitative phase analysis and unit-cell parameters, as well as the crystallite size of the main phases, for the investigated samples. The R-weighted pattern refinement agreement factor was Rwp = 0.030 and 0.032 for the sample WO_3/st and WO_3/st-washed, respectively.

				Unit-Cell Parameters and Crystallite Size					
	Phase	**QPA [1] Wt%**	**Space Group**	**a (Å)**	**b (Å)**	**c (Å)**	**β (°)**	**V (Å3)**	**Crystallite Size (nm)**
WO_3	WO_3 (monoclinic)	96.0 (3)	$P2_1/n$	7.3055 (8)	7.5256 (9)	7.6761 (10)	90.43 (1)	422.01 (9)	23 (1)
	WO_3 (hexagonal)	4.0 (3)	$P6_3/mcm$	7.354 (2)	–	7.397 (6)	–	346.5 (4)	–
WO_3 washed	WO_3 (monoclinic)	99.0 (2)	$P2_1/n$	7.3106 (8)	7.5271 (9)	7.6813 (10)	89.97 (1)	422.69 (9)	21 (1)
	WO_3 (hexagonal)	1.0 (2)	$P6_3/mcm$	–	–	–	–	–	–

[1] Quantitative Phase Analysis.

The result of the XRD investigation allows us to conclude that the nature of the support (either simple glass or TCO) does not affect the crystalline nature of WO_3/st by a large margin. Coherent scattering domains are also similar (ca. 20 nm) with either SnO_2:F or borosilicate glass supports, with different surface chemistry. This is probably due to the fact that both surfaces are modified, to stabilize the adhesion of WO_3/st, with the same adhesion (TiO_x) and seed (WO_3) layers. A largely prevalent monoclinic phase is found with both substrates, which is expected to play a major role in the photoinduced processes activated upon the optical excitation of the semiconductor. Furthermore, the solvothermal coating method is surprisingly easy to extend to the functionalization of a variety of substrates of shapes other than planar (i.e., irregular porous supports, glass beads) (Figure S9).

3.2. *Photoelectrochemical Properties*

Evaluating the photoelectrochemical response of WO_3/st (Figures S10 and S11) will allow us to judge the ability of these WO_3 films to achieve efficient electron/hole separation, which is the primary goal and crucial for any solar energy conversion device, irrespective of its intended use in either solar fuel generation or environmental remediation processes. In addition, the information gained from photoelectrochemical experiments can be extended to the open circuit (i.e., photocatalytic) case where the efficiency figures are more elusive to obtain without ambiguity. The primary efficiency figure for this type of photoelectrochemical application is the magnitude of the current density, which is dependent on the rate of hole injection into the electrolyte, yielding OH radicals as the consequence of water oxidation. The generation of ·OH radicals in WO_3 aqueous suspensions under visible light illumination ($\lambda > 420$ nm) was confirmed by EPR spin-trapping measurements. Few seconds of photoirradiation of the WO_3 suspension in the presence of pbn causes the prompt formation of a triplet of doublets characterized by coupling constants aN = 15.4 G, aH = 2.7 G, which is shown in Figure S12. The coupling constant values are in agreement with the trapping of ·OH by pbn to form the paramagnetic adduct [pbn-OH]. These results confirm that photoexcited WO_3 is able to initiate water oxidation by producing OH radicals. First, we can observe that the WO_3/st-type electrode reproducibly yields (max. relative standard deviation ca. 10%) much better results with respect to the state of the art [35] from which we adapted our preparative procedure, with an average limiting photocurrent of ca. 2.2 mA/cm^2. This corresponds to a ca. 60% improvement over the reported case. WO_3/st is also a better performer compared to one of our best standard WO_3/coll (synthesis and characterization was previously published) [21] by a factor of ca. 40% (Figure S11). Interestingly, the WO_3/st J/V curve is generally steeper compared to the standards, suggesting low charge transport resistance across the oriented electrode and good interfacial charge separation efficiency. The global resistance R = (dV/dJ) figure of merit, comprising both the interfacial and the charge transport resistance, is indeed lower by a factor of 2 in the

WO_3/st material (Nyquist Plots were compared in Figure S10). This confirms that the TiO_x adhesion layer offers a stable anchor for prolonged photoelectrode operation while being sufficiently permeable to electrons to allow for their efficient collection.

Current potential results were complemented by photon-to-electron conversion spectra: IPCE and APCE (Figure 6). IPCE% of WO_3/st and WO_3/coll are compared under 1 V and 1.5 V applied potential. We note that the limiting value of WO_3/st is reached with a reduced overpotential of ca. 0.5 V compared to WO_3/coll, consistent with the lower charge transfer resistance. First, we observe that IPCE is in good agreement with the J/V results, showing that the area subtended by the WO_3/st photoconversion is ca. 30% higher than that of WO_3/coll. It is also interesting to note that most of the improvement of WO_3/st comes from a better conversion in the visible region, from 400 to 500 nm. APCE% spectra confirm, for the WO_3/st, intrinsically higher charge transfer and charge transport efficiencies with respect to WO_3/coll. For the former, we record APCE$\approx$ 70% in the near UV, meaning that 70% of the photons absorbed in such a region is converted into useful charge carriers for driving an energetically demanding redox reaction. In the blue visible region, such efficiency is still significant, with a maximum value of the order of 50%. Consistent with the gap of ca. 2.7 eV, photoconversion ceases completely around 500 nm. The energy gap estimated from the photoaction spectra was confirmed by Tauc analysis (Supplementary Materials, Figure S13), displaying a reasonably linear region and affording an indirect band gap of 2.63 eV and a direct gap of 2.88 eV.

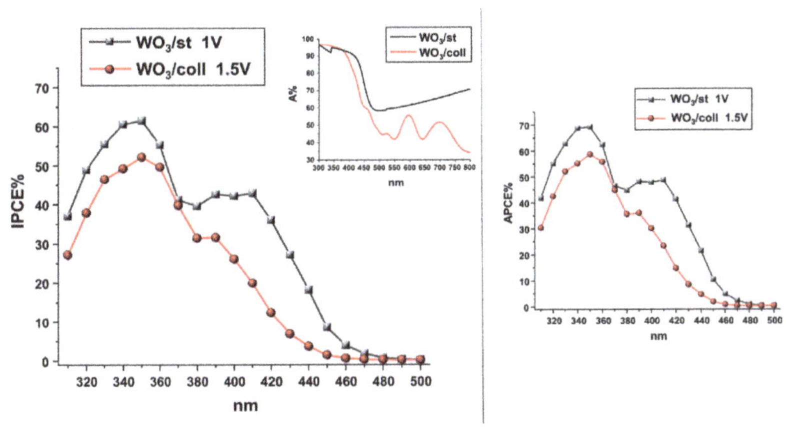

Figure 6. IPCE% and APCE% of WO_3/st and WO_3/coll under, respectively, 1 V and 1.5 V bias in 0.7 M Na_2SO_4. Their absorptance spectrum (A = 100 − R% − T%) is shown in the inset.

We probed the impact of the presence of DRO, EE2 and OZ on the J/V characteristic of WO_3 to gain direct indication of the photoelectrochemical degradation of these molecular targets. We note that the low solubility of DRO, EE2 and OZ in aqueous solution makes the direct observation of hole scavenging by these species impossible to be clearly observed by simply recording the photocurrent. For this reason, we explored the photoelectrochemical response of WO_3 in ACN/0.1 M $LiClO_4$, where concentrations of DRO, EE2 and OZ could be raised up to 0.01 M. The photocurrent recorded with blank ACN/$LiClO_4$ solutions is due to both the oxidation of the residual water present in non-anhydrous ACN and the direct oxidation of the organic electrolyte [43].

In Figure S14, it is possible to observe, that, when DRO EE2 and OZ are present in the electrolyte, a 300 mV cathodic shift of the photoanodic onset potential occurs, associated with an increased photocurrent in the 0.2–0.8 V vs. SCE interval, a region of the J/V curve which is usually affected by electron/hole recombination losses. Both these effects are consistent with the fast capture of either holes or photogenerated radicals by the selected target organic species, resulting in their irreversible oxidation. We also point out that no dark electrochemical processes resulting from the electrochemical oxidation of DRO, EE2 and OZ on the FTO/WO_3 surface are observed up to 1.8 V vs. SCE. We have thus provided direct evidence of the occurrence of the photoelectrochemical oxidation of DRO, EE2 and OZ at the illuminated WO_3 interface. Moreover, in the presence of DRO, EE2 and OZ, we generally observe a drop in the limiting photocurrent of the photoanodes, a fact that we have tentatively explained with the adsorption of the drug oxidation intermediates on the WO_3 surface, leading to a decreased active area of the photoelectrodes. In addition, EPR spin-trapping experiments were carried out on a WO_3 aqueous suspension containing DRO irradiated for few seconds in the presence of DMPO (see Section 2.3.4). It was observed the formation of the quartet (1:2:2:1, aN = aH = 14.8 G) was characteristic of the trapping of OH• radicals by DMPO (see Figure 7). After a few minutes, a triplet of doublets (aN = 15.36 G, aH = 20.28 G) appeared; these signals are possibly due to an adduct of a carbon radical originated from DRO and DMPO.

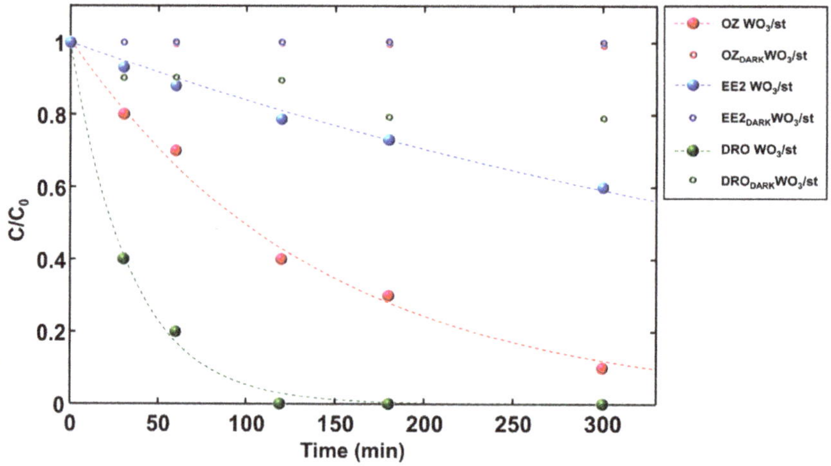

Figure 7. Photodegradation kinetics in batch system, data expressed as C/C_0 vs. time; EE2 WO_3/st, DRO WO_3/st, OZ WO_3/st represent the data of the irradiation experiments, $EE2_{DARK}WO_3$/st, $DRO_{DARK}WO_3$/st, $OZ_{DARK}WO_3$/st represent the data in dark conditions. Each solution was prepared at an initial concentration of 4 mg L^{-1}.

3.3. *Photochemical Degradation of DRO, EE2 and OZ*

Batch experiments. To study the photocatalytic removal of the selected pharmaceuticals, batch experiments with and without the WO_3/st photocatalyst, under irradiation and in dark conditions, were performed under the conditions described in Section 2.4. The concentration of the three tested drugs was determined by HPLC, as reported in Sections 2.4, 2.6 and 2.7. The determined concentration (C)was divided by the initial concentration (C/C_0). In the case of EE2 and OZ, in the absence of illumination, we demonstrated that the photocatalytic system does not contribute to drug adsorption. Indeed, the concentrations of neither EE2 nor OZ vary significantly under dark conditions, regardless of whether the catalyst is present or absent, and the ratio C/C_0 is close to 1. A reduction of approximately 20% (C/C_0 roughly 0.8) was observed only for the DRO molecule (as shown in Figure S15), and was thus likely attributable to an effective adsorption event. When comparing the results from the tests conducted in the darkness, it became evident that this specific pollutant is adsorbed onto the photocatalyst. Indeed, the data from dark conditions (in the absence of WO_3/st) revealed no alterations in the DRO concentration, whereas a significant change was noted upon the addition of the photocatalyst (Figure S15). Moreover, the outcomes of irradiation tests conducted without the WO_3/st powder indicate that the direct photochemical degradation of pollutants is in no case significant. On the contrary, a remarkable decrease in the concentration of drugs was observed only when the test solution was irradiated in the presence of WO_3/st. Figure 7 shows the degradation kinetics of the three drugs, clearly indicating the occurrence of a photocatalytic process in the presence of the WO_3/st suspension.

Flow system experiments. For each pollutant (DRO, EE2 and OZ), the contribution to adsorption was investigated by evaluating the differences in the concentration determined by the HPLC of the contaminants in solution before and after contact with the glass spheres coated with WO_3/st (WO_3/st-GS). In these experiments, 20 g of photoactive spheres were placed in a vial and kept in contact with 75 mL of aqueous solution for 420 min in the dark. The experimental data obtained showed that, for DRO, the uptake onto the WO_3/st-GS in (Figure S15) is negligible. Possibly, the difference in the adsorption of DRO onto the supported photocatalyst system WO_3/st-GS is even lower than that onto the WO_3/st due to the higher surface area of the latter one.

Experiments on the degradation kinetics by using the flow system (see Figure 3) were carried out by using 200 mL of 0.7 mM Na_2SO_4, aqueous solution containing 4 mg L^{-1} of a drug to simulate the surface water conditions for what concerns both salinity and pH (see Section 2.4) [36]. Concerning the light absorption efficiency in the flow photoreactor, we have estimated, by exploiting diffuse reflectance measurements, that an attenuation of 400 nm light of ca. 60% (Figure S16) due to absorption by WO_3 is achieved within an optical path of 1 cm. This is a similar attenuation to that observed with the WO3st suspension in static experiments (Figure S17). Considering that the diameter of the photoreactor is 2.2 cm, we obtain, along the radial coordinate, a nearly complete absorption (ca. 96%) of 400 nm light within the active volume of the reactor. In Figure 8, the results obtained using the flow system under irradiation are shown. Overall, a substantial drop in the concentrations of all pharmaceutical targets is observed upon the treatment of the test solutions in the flow reactor. However, the DRO and OZ degradation rates were smaller than those found in batch experiments. Possibly, the slower kinetics are due to the shorter contact time between the solution and the WO_3/st photocatalyst in the flow system with respect to the batch experiments. It has already been reported that, in a flow reactor, the immobilization of WO_3 on a substrate leads to a lower activity [44]. Experimental data were fitted using a pseudo-first order (PFO) (1) kinetic equation. In the literature, many cases of pseudo-first order photocatalytic decomposition of organic compounds at similar concentrations in a flow reactor are reported [45,46]. In Table 2, the results obtained from both batch and flow experiments are shown.

From the data in Table 2, it can be seen that a WO_3-based photocatalyst is able to degrade all the selected drugs. Indeed, a WO_3-based photocatalyst, much like other semiconductor catalysts, is not selective, since the degradation reaction occurs through reactions with hydroxyl radicals (OH•), which are a strong oxidant characterized by an oxidation potential equal to $E^\circ = +2.1$ V vs. NHE. In the case of EE2, the kinetic constants estimated from the flow and batch experimental data are higher than those of DRO and OZ. This could be explained by the presence, in the EE2 structure, of reactive OH groups and electron-rich carbon atoms. These functional groups likely serve as primary sites for initial reactions in the advanced oxidation process, potentially necessitating a shorter residence time within the flow reactor compared to DRO and OZ. Simultaneously, the flow configuration enhances the mass transfer of the drug within the reactor through convection, thereby increasing the likelihood of collisions between EE2 and the freely diffusing reactive oxygen species generated by the photocatalyst. Consistent with findings from batch experiments, DRO exhibits the highest degradation rate constant, leading to near-complete depletion from the test solution within 450 min. In the flow system, a WO_3-based photocatalyst operates under 400 nm irradiation, a condition under which it has already been proved that WO_3 exhibits higher degradation efficiency of organic compounds with respect to other largely employed photocatalysts such as TiO_2 [47]. The comparison of the results obtained in this study with the literature data is complicated due to the presence of numerous factors that can influence reactor performance [48]. Among these factors, the type of photocatalyst is particularly significant. Photocatalytic activity is influenced not only by chemical composition but also by structural characteristics, dimensions, surface impurities, and the surface area. Only a few studies have been carried out to investigate the efficiency of WO_3 in the photodegradation of the contaminants of emerging concern in a continuous flow reactor. Among them, Martins [44] had reported a transformation rate constant of 3.5×10^{-2} min^{-1} and 2.2×10^{-2} min^{-1} for tetracycline, obtained from two photoreactors having different configurations, which are of the same order of magnitude as those obtained in the present study.

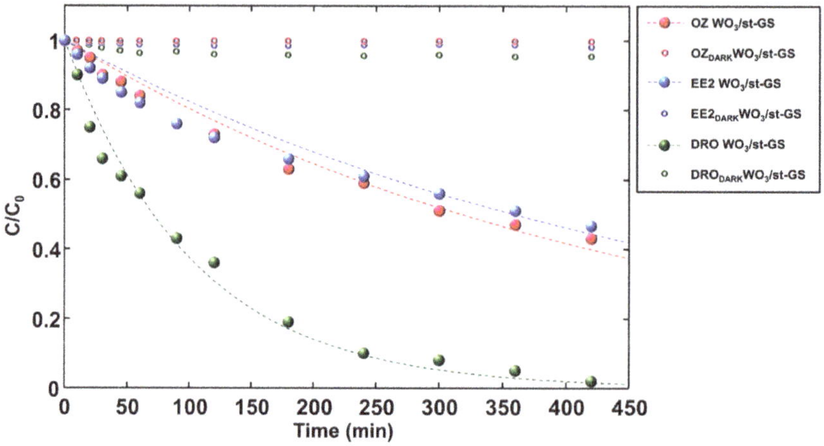

Figure 8. Drug photodegradation kinetics in the flow reactor, data expressed as C/C_0 vs. time; EE2 WO_3/st, DRO WO_3/st, OZ WO_3/st represent the data of the irradiation experiments, $EE2_{DARK}$ WO_3/st, $DRO_{DARK}WO_3$/st, $OZ_{DARK}WO_3$/st represent the data in dark conditions. Each solution was prepared at an initial concentration of 4 mg L^{-1}.

Table 2. Parameters obtained by non-linear fitting using a first-order kinetic model. The error is given as confidence interval at 95% of probability.

		k_{PFO} (min^{-1})	R^2
Batch	EE2 WO_3/st	0.0018 ± 0.0002	0.9778
	DRO WO_3/st	0.0293 ± 0.0040	0.9969
	OZ WO_3/st	0.0071 ± 0.0010	0.9936
Flow	EE2 WO_3/st-GS	0.0019 ± 0.0002	0.9313
	DRO WO_3/st-GS	0.0098 ± 0.0009	0.9880
	OZ WO_3/st-GS	0.0022 ± 0.0002	0.9694

3.4. *Characterization of the Degradation Intermediates*

After 8 h of circulation in the flow system, the EE2, DRO and OZ samples were analyzed by LC/MS in positive (DRO and OZ) and negative (EE2) ion modes. MS/MS data were acquired in order to establish the best ion transitions for use in MRM analysis. The instrument settings were adjusted to maximize the response of each precursor–product combination. Chromatographic separation of the compounds was achieved on a concentration gradient reversed-phase column with a total run time of 15 min at a flow rate of 150 μL/min. The chromatograms of the different drugs were acquired by monitoring the m/z transitions 295 > 145 and 295 > 143 for the $[EE2 - H]^-$ ion. The $[DRO + H^+]^+$ ion was monitored via the m/z transitions 367 > 97 and 367 > 91 and the $[OZ + H^+]^+$ ion via the m/z transitions 287 > 269 and 287 > 51. As mentioned earlier, the degradation of pharmaceuticals follows a scheme involving the reaction of contaminants with hydroxyl radicals (OH•), whose presence has been confirmed through EPR measurements. A general scheme is reported in Supplementary Information (Scheme S1). In the following, the degradation intermediates originating from the oxidation of the three drugs investigated are described.

Mass spectrometry data show that EE2 was transformed into intermediate products of smaller molecular weight. To effectively track the conversion, it was necessary to identify the intermediates and by-products. The mass spectra show four characteristic peaks; the precursor ions detected in the sample were at m/z of 313, 267, 249, and 148. Compared with other studies on the photo-transformation products of EE2, it is found that the conversion of EE2 by HO• attack can occur, generating the by-products reported in Scheme 1 [49,50], according to a pathway consistent with the known reactivity of WO_3 upon band gap excitation in the presence of aqueous electrolytes. Reactive oxygen species like HO• are strong enough to break the C-C, C-O, and O-H bonds in the EE2 molecule and its intermediates. The overall photocatalytic degradation process of EE2 via WO_3/st can be expressed as proposed in Scheme 1. HPLC/MS analysis confirms that, indeed, OH radicals are the main reactive oxidants in the photodegradation of EE2 in the presence of WO_3/st due to the presence of hydroxylated intermediates. Intermediate A may correspond to a hydroxylated compound following the docking of OH with the addition of a hydroxyl group, linked to the benzene ring. Product B could be formed through processes such as oxidation, carbonyl, and ring breakdown. Product B, undergoing a sequence of oxidative attacks, could produce the oxygen-rich product C. The latter, after the complete oxidation and cleavage of the partly oxidized six-membered ring-bearing carbonyl and hydroxyl groups, affords the molecular ion D seen at m/z 148 (see Table S1).

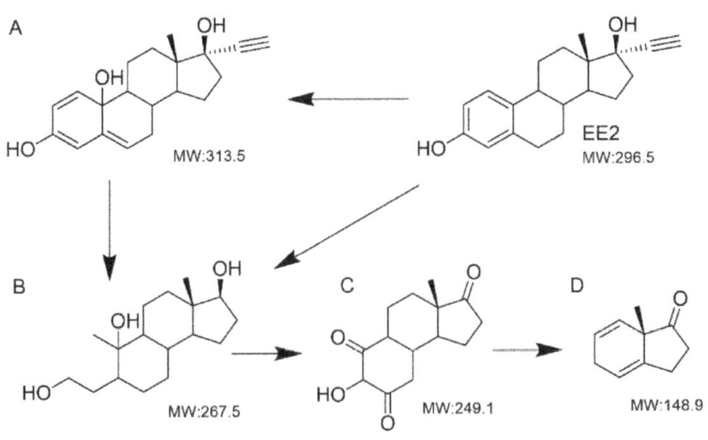

Scheme 1. Proposed EE2 degradation pathway and intermediates structures (**A**–**D**) according to LC/MS analysis.

The photodegradation of DRO leads to the appearance of 383 m/z, 343 m/z and 341 m/z peaks (see Scheme 2 and Table S2). The 383 m/z by-product could be due to the addition of OH radicals onto the five-membered ring, similar to what was reported by Huang et al. [51] where the progesterone molecule (progenitor of the class to which DRO belongs) was studied. Subsequently, the oxidation process leads to the formation of other by-products, shorter than the starting molecules II and III.

HPLC/MS analyses were also carried out to study the photodegradation intermediates of the oxazepam molecule (see Table S3). Four by-products were identified which correspond to 302 m/z (two possible structures), 225 m/z and 209 m/z peaks. The possible degradation pathway and photoproduced intermediates are shown in Scheme 3. As reported in other studies [52–54], where molecules such as alprazolam, which has similar structural characteristics as OZ and also belongs to the same benzodiazepine class, are investigated, it is hypothesized that the degradation of OZ may be mediated by photoproduced OH radicals as well.

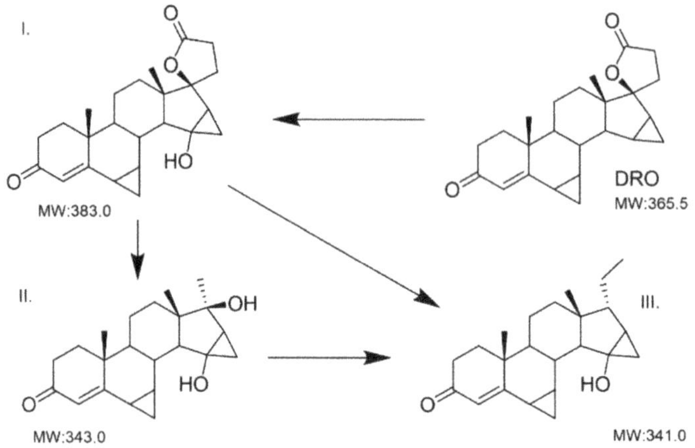

Scheme 2. DRO degradation pathway and proposed intermediate structures (**I–III**) according to LC/MS analysis.

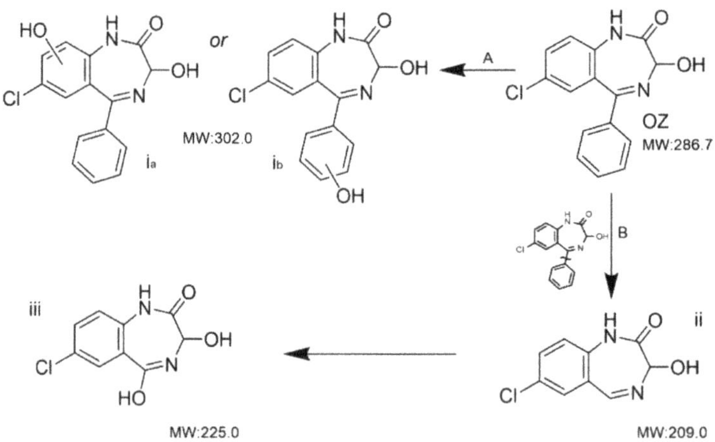

Scheme 3. Proposed OZ degradation pathways and proposed by-product structures (**i$_a$**,**i$_b$**–**iii**) according to LC/MS analysis.

From the HPLC/MS analysis, hydroxylated intermediates were, in fact, identified, indicating the attack of OH radicals (pathway A) on the aromatic ring to form the products ia and ib.

In pathway B, the oxazepam might undergo the cleavage of the bond between the nitrogen-containing heterocycle and the benzene ring, leading to detachment of the phenyl group (ii). Then, intermediate ii could be hydroxylated to product iii by further attack of OH radicals.

4. Conclusions

This work has been focused on the application of oriented WO_3 nanostructures with high photoelectrochemical activity to water remediation, exploiting advanced photocatalytic oxidation processes. Oxidative degradation of pharmaceuticals, selected among steroid hormones and psychoactive drugs classified as high-priority emerging environmental contaminants, was successfully accomplished both in batch and flow photoreactors, with the latter exploiting easy-to-handle inert glass supports coated with solvothermally grown WO_3 (WO_3/st-GS). HPLC-MS analysis allowed for the separation and the clear identification of oxidation intermediates, pointing out that photogenerated HO radicals, generated upon the band gap excitation of WO_3, are the primary agents for the advanced oxidative degradation of all the selected pharmaceutical targets. The study of the degradation kinetics has shown a different rate of drug degradation in the batch reactor compared to the flow one. In two out of three cases, the batch system displayed faster kinetics, probably due to a more efficient contact between the solution and the photocatalyst. In fact, dispersing the catalyst in the form of a fine powder leads to a greater interfacial area, resulting in a more efficient production of OH radicals. Conversely, in the flow system, WO_3/st is immobilized on a solid support composed of millimeter-sized glass spheres, resulting in a decreased active surface/volume ratio and a less intimate contact with the solution carrying the drugs. This may result in the need for a longer treatment, particularly with robust molecules characterized by high chemical stability, like the family of benzodiazepines. Nevertheless, supported WO_3 opens to easy handling and recovery of the photocatalyst and to the possibility of engineering flow systems suitable to treat larger volumes of contaminated water in a continuous fashion and with low counterpressure, while keeping the photocatalyst confined on a solid matrix.

Supplementary Materials: The following supporting information can be downloaded at: https://www.mdpi.com/article/10.3390/nano14100860/s1, Figure S1: Cyclic voltammetry of FTO, FTO-TiO_x and FTO-TiO_x-WO_3 seed layer in 0.01 M K_4FeCN_6 and 0.1 M $LiClO_4$; Figure S2: SEM image of WO_3/coll; Figure S3: (a) Solar Simulator equipped with a AM1.5G filter and 380 nm UV cut-off filter, (b) 4 µg/mL drug solution in vial under magnetic stirring; Figure S4: WO_3 films solvothermally grown on FTO before (a) and after (b) the annealing process. Electrodes equipped with the TiO_x adhesion layer are indicated by (Ti) (i.e., WO_3/st-Ti-FTO); Figure S5: SEM image of WO_3/st; Figure S6: EDS spectra of WO_3/st; Figure S7: Phase identification for WO3 specimens grown on a microscope slide surface. The powder of both samples is mainly composed of WO_3 in its monoclinic form ($P2_1/n$; green reflections). WO_3 in the hexagonal form ($P6_3/mcm$; purple reflections) is also detected as associated minor phase; Figure S8: Rietveld refinement plots of the powder diffraction pattern for the WO_3/st specimens' growth on the microscope slide surface. The experimental profile is represented by black dots and the best-fit refinement profile is the flow red line. The lower green curve is the weighted difference between observed and calculated patterns. Vertical ticks mark the position of reflections for the identified phases in the two specimens. Figure S9: Glass column filled with WO_3/st-GS; Figure S10: Nyquist plots of WO_3/coll (red) and WO_3/st (black) photoelectrode recorded at 0.6 V vs. SCE in 0.7 M Na_2SO_4 under AM1.5G illumination. 10 mV sinusoidal perturbation in the range 20000-0.1 Hz; Figure S11: Average JV performance of a batch of WO_3/st photoanodes (black line) compared to some of our best internal standards WO_3/coll (green line). 0.7 M Na_2SO_4 under AM1.5G illumination; Figure S12: EPR spectrum obtained after few seconds photoirradiation ($\lambda > 400$ nm) of aqueous suspension of WO_3 containing H_2SO_4 (0.1 M) and pbn (5×10^{-2} M); Figure S13: Tauc Plots of WO_3/st according to the direct (A) and indirect (B) transitions; Figure S14: J/V chopped curves under AM 1.5G illumination of WO_3 in 0.1 M of $LiClO_4$ in ACN solution in presence of 0.1 M of DRO (A) 0.1 M of EE2 (B) and 0.1 M of OZ; Figure S15: Comparison test (in presence of WO_3): DRO in dark (DRODARK WO_3/st)), DRO irradiated (DRO WO_3/st) and DRO in dark in batch system with spheres coated with WO_3 (DRODARK WO_3/st-GS); Figure S16: Transmission spectrum of WO_3/st loaded on glass spheres ([T% (WO_3/st-GS − T% GS)]) obtained with an integrating sphere in diffuse reflection mode along an optical path of 1 cm; Figure S17: Transmission spectra (1 cm optical path) of naked glass spheres GS, WO_3 coated glass spheres (WO_3/st-GS) and WO_3/st suspension (3.5 mg/mL) obtained in diffuse reflection mode; Scheme S1: Scheme for degradation of pollutants by WO_3; Table S1: EE2 degradation intermediates, retention times, fragment ions and proposed structures; Table S2: DRO degradation intermediates, retention times, fragment ions and proposed structures; Table S3: DRO degradation intermediates, retention times, fragment ions and proposed structures.

Author Contributions: Methodology, formal analysis, investigation, writing—original draft, M.C., V.C. and C.S.; investigation, writing—original draft, M.A.; methodology, investigation, M.O.; data curation, T.C.; funding acquisition, supervision, writing—review and editing, A.M.; conceptualization, funding acquisition, supervision, writing—review and editing, S.C. and L.P. All authors have read and agreed to the published version of the manuscript.

Funding: The contribution at this work of M.C. was supported by the program PON "Research and Innovation" 2014–2020 (PON R&I), Action IV.6 "Contratti di ricerca su tematiche Green". L.P., C.S., A.M. and S.C. acknowledge financial support from PNRR MUR project ECS_00000033_ECOSISTER. V.C. acknowledges PON Research and Innovation 2014–2020 (art. 24, comma 3, lett. (a), Law n. 240, 30 December 2010 and s.m.i. and D.M. n. 1062, 10 August 2021).

Data Availability Statement: Data are contained within the article.

Acknowledgments: The authors would like to acknowledge Negar Eftekhari and Cinzia Brenna of the Electron Microscopy Center of the University of Ferrara for additional SEM and EDS characterization.

Conflicts of Interest: The authors declare no conflicts of interest.

References

1. Richardson, S.D.; Kimura, S.Y. Water Analysis: Emerging Contaminants and Current Issues. *Anal. Chem.* **2016**, *88*, 546–582. [CrossRef]
2. Vasilachi, I.C.; Asiminicesei, D.M.; Fertu, D.I.; Gavrilescu, M. Occurrence and fate of emerging pollutants in water environment and options for their removal. *Water* **2021**, *13*, 181. [CrossRef]
3. Mezzelani, M.; Gorbi, S.; Regoli, F. Pharmaceuticals in the aquatic environments: Evidence of emerged threat and future challenges for marine organisms. *Mar. Environ. Res.* **2018**, *140*, 41–60. [CrossRef] [PubMed]
4. Lazzara, R.; Blázquez, M.; Porte, C.; Barata, C. Low environmental levels of fluoxetine induce spawning and changes in endogenous estradiol levels in the zebra mussel *Dreissena polymorpha*. *Aquat. Toxicol.* **2012**, *106*, 123–130. [CrossRef] [PubMed]
5. Kelly, S.; Davies, E.; Fearns, S.; McKinnon, C.; Carter, R.; Gerlinger, C.; Smithers, A. Effects of oral contraceptives containing ethinylestradiol with either drospirenone or levonorgestrel on various parameters associated with well-being in healthy women: A randomized, single-blind, parallel-group, multicentre study. *Clin. Drug Investig.* **2010**, *30*, 325–336. [CrossRef] [PubMed]
6. Zhao, H.N.; Tian, Z.; Kim, K.E.; Wang, R.; Lam, K.; Kolodziej, E.P. Biotransformation of Current-Use Progestin Dienogest and Drospirenone in Laboratory-Scale Activated Sludge Systems Forms High-Yield Products with Altered Endocrine Activity. *Environ. Sci. Technol.* **2021**, *55*, 13869–13880. [CrossRef] [PubMed]
7. Zhao, Y.; Castiglioni, S.; Fent, K. Environmental Progestins Progesterone and Drospirenone Alter the Circadian Rhythm Network in Zebrafish (*Danio rerio*). *Environ. Sci. Technol.* **2015**, *49*, 10155–10164. [CrossRef] [PubMed]
8. Tang, Z.; Liu, Z.H.; Wang, H.; Dang, Z.; Liu, Y. A review of 17α-ethynylestradiol (EE2) in surface water across 32 countries: Sources, concentrations, and potential estrogenic effects. *J. Environ. Manag.* **2021**, *292*, 112804. [CrossRef] [PubMed]
9. Yuan, S.F.; Liu, Z.H.; Lian, H.X.; Yang, C.T.; Lin, Q.; Yin, H.; Lin, Z.; Dang, Z. Fast trace determination of nine odorant and estrogenic chloro- and bromo-phenolic compounds in real water samples through automated solid-phase extraction coupled with liquid chromatography tandem mass spectrometry. *Environ. Sci. Pollut. Res.* **2018**, *25*, 3813–3822. [CrossRef]
10. Rocha, M.J.; Rocha, E. Synthetic Progestins in Waste and Surface Waters: Concentrations, Impacts and Ecological Risk. *Toxics* **2022**, *10*, 163. [CrossRef]
11. Avar, P.; Maasz, G.; Takács, P.; Lovas, S.; Zrinyi, Z.; Svigruha, R.; Takátsy, A.; Tóth, L.G.; Pirger, Z. HPLC-MS MS analysis of steroid hormones in environmental water samples. *Drug Test. Anal.* **2015**, *8*, 123–127. [CrossRef] [PubMed]
12. Fick, J.; Brodin, T.; Heynen, M.; Klaminder, J.; Jonsson, M.; Grabicova, K.; Randak, T.; Grabic, R.; Kodes, V.; Slobodnik, J.; et al. Screening of benzodiazepines in thirty European rivers. *Chemosphere* **2017**, *176*, 324–332. [CrossRef] [PubMed]
13. Vossen, L.E.; Červený, D.; Sarma, O.S.; Thörnqvist, P.O.; Jutfelt, F.; Fick, J.; Brodin, T.; Winberg, S. Low concentrations of the benzodiazepine drug oxazepam induce anxiolytic effects in wild-caught but not in laboratory zebrafish. *Sci. Total Environ.* **2020**, *703*, 134701. [CrossRef] [PubMed]
14. Salimi, M.; Esrafili, A.; Gholami, M.; Jafari, A.J.; Kalantary, R.R.; Farzadkia, M.; Kermani, M.; Sobhi, H.R. Contaminants of emerging concern: A review of new approach in AOP technologies. *Environ. Monit. Assess.* **2017**, *189*, 1–22. [CrossRef] [PubMed]
15. Sumpter, P.; Margiotta-Casaluci, L. Environmental Occurrence and Predicted Pharmacological Risk to Freshwater Fish of over 200 Neuroactive Pharmaceuticals in Widespread Use. *Toxics* **2022**, *10*, 233. [CrossRef] [PubMed]
16. Liu, Z.; Demeestere, K.; Hulle, S.V. Pretreatment of Secondary Effluents in View of Optimal Ozone-Based AOP Removal of Trace Organic Contaminants: Bench-Scale Comparison of Efficiency and Energy Consumption. *Ind. Eng. Chem. Res.* **2020**, *59*, 8112–8120. [CrossRef]
17. Cai, Q.Q.; Wu, M.Y.; Li, R.; Deng, S.H.; Lee, B.C.Y.; Ong, S.L.; Hu, J.Y. Potential of combined advanced oxidation—Biological process for cost-effective organic matters removal in reverse osmosis concentrate produced from industrial wastewater reclamation: Screening of AOP pre-treatment technologies. *J. Chem. Eng.* **2020**, *389*, 123419. [CrossRef]

18. Molinari, A.; Argazzi, R.; Maldotti, A. Photocatalysis with $Na_4W_{10}O_{32}$ in water system: Formation and reactivity of OH radicals. *J. Mol. Catal. A Chem.* **2013**, *372*, 23–28. [CrossRef]
19. Luong, J.H.T.; Male, K.B.; Glennon, J.D. Boron-doped diamond electrode: Synthesis, characterization, functionalization and analytical applications. *Analyst* **2009**, *134*, 1965–1979. [CrossRef]
20. Yoshihara, S.; Murugananthan, M. Decomposition of various endocrine-disrupting chemicals at boron-doped diamond electrode. *Electrochim. Acta* **2009**, *54*, 2031–2038. [CrossRef]
21. Cristino, V.; Marinello, S.; Molinari, A.; Caramori, S.; Carli, S.; Boaretto, R.; Argazzi, R.; Meda, L.; Bignozzi, C.A. Some aspects of the charge transfer dynamics in nanostructured WO_3 films. *J. Mater. Chem. A* **2016**, *4*, 2995–3006. [CrossRef]
22. Behbahani, M.A.; Ranjbar, M.; Kameli, P.; Salamati, H. Hydrogen sensing by wet-gasochromic coloring of $PdCl_2(aq)/WO_3$ and the role of hydrophilicity of tungsten oxide films. *Sens. Actuators B Chem.* **2013**, *188*, 127–136. [CrossRef]
23. Cristino, V.; Caramori, S.; Argazzi, R.; Meda, L.; Marra, G.L.; Bignozzi, C.A. Efficient photoelectrochemical water splitting by anodically grown WO_3 electrodes. *Langmuir* **2011**, *27*, 7276–7284. [CrossRef] [PubMed]
24. De Tacconi, N.R.; Chenthamarakshan, C.R.; Yogeeswaran, G.; Watcharenwong, A.; De Zoysa, R.S.; Basit, N.A.; Rajeshwar, K. Nanoporous TiO_2 and WO_3 films by anodization of titanium and tungsten substrates: Influence of process variables on morphology and photoelectrochemical response. *J. Phys. Chem. B* **2006**, *110*, 25347–25355. [CrossRef]
25. Watcharenwong, A.; Chanmanee, W.; de Tacconi, N.R.; Chenthamarakshan, C.R.; Kajitvichyanukul, P.; Rajeshwar, K. Anodic growth of nanoporous WO_3 films: Morphology, photoelectrochemical response and photocatalytic activity for methylene blue and hexavalent chrome conversion. *J. Electroanal. Chem.* **2008**, *612*, 112–120. [CrossRef]
26. Hahn, R.; Macak, J.M.; Schmuki, P. Rapid anodic growth of TiO_2 and WO_3 nanotubes in fluoride free electrolytes. *Electrochem. Commun.* **2007**, *9*, 947–952. [CrossRef]
27. Daniel, M.F.; Lassegues, J.C.; Garie, R. Infrared and Raman Spectroscopies of rf Sputtered Tungsten Oxide Films. *J. Solid State Chem.* **1988**, *73*, 127–139. [CrossRef]
28. Leftheriotis, G.; Papaefthimiou, S.; Yianoulis, P. The effect of water on the electrochromic properties of WO_3 films prepared by vacuum and chemical methods. *Sol. Energy Mater Sol. Cells* **2004**, *83*, 115–124. [CrossRef]
29. Meda, L.; Tozzola, G.; Tacca, A.; Marra, G.; Caramori, S.; Cristino, V.; Bignozzi, C.A. Photo-electrochemical properties of nanostructured WO3 prepared with different organic dispersing agents. *Sol. Energy Mater Sol. Cells* **2010**, *94*, 788–796. [CrossRef]
30. Santato, C.; Ulmann, M.; Àugustynski, J. Photoelectrochemical properties of nanostructured tungsten trioxide films. *J. Phys. Chem. B* **2001**, *105*, 936–940. [CrossRef]
31. Santato, C.; Odziemkowski, M.; Ulmann, M.; Augustynski, J. Crystallographically oriented mesoporous WO_3 films: Synthesis, characterization, and applications. *J. Am. Chem. Soc.* **2001**, *123*, 10639–10649. [CrossRef] [PubMed]
32. Park, J.H.; Bard, A.J. Photoelectrochemical tandem cell with bipolar dye-sensitized electrodes for vectorial electron transfer for water splitting. *Electrochem. Solid-State Lett.* **2006**, *9*, E5. [CrossRef]
33. Zhang, J.; Zhang, P.; Wang, T.; Gong, J. Monoclinic WO_3 nanomultilayers with preferentially exposed (002) facets for photoelectrochemical water splitting. *Nano Energy* **2015**, *11*, 189–195. [CrossRef]
34. Wang, S.; Chen, H.; Gao, G.; Butburee, T.; Lyu, M.; Thaweesak, S.; Yun, J.H.; Du, A.; Liu, G.; Wang, L. Synergistic crystal facet engineering and structural control of WO_3 films exhibiting unprecedented photoelectrochemical performance. *Nano Energy* **2016**, *24*, 94–102. [CrossRef]
35. Su, J.; Feng, X.; Sloppy, J.D.; Guo, L.; Grimes, C.A. Vertically aligned WO_3 nanowire arrays grown directly on transparent conducting oxide coated glass: Synthesis and photoelectrochemical properties. *Nano Lett.* **2011**, *11*, 203–208. [CrossRef] [PubMed]
36. Cristino, V.; Pasti, L.; Marchetti, N.; Berardi, S.; Bignozzi, C.A.; Molinari, A.; Passabi, F.; Caramori, S.; Amidani, L.; Orlandi, M.; et al. Photoelectrocatalytic degradation of emerging contaminants at $WO_3/BiVO_4$ photoanodes in aqueous solution. *Photochem. Photobiol. Sci.* **2019**, *18*, 2150–2163. [CrossRef] [PubMed]
37. Longobucco, G.; Pasti, L.; Molinari, A.; Marchetti, N.; Caramori, S.; Cristino, V.; Boaretto, R.; Bignozzi, C.A. Photoelectrochemical mineralization of emerging contaminants at porous WO_3 interfaces. *Appl. Catal. B* **2017**, *204*, 273–282. [CrossRef]
38. Yang, H.G.; Sun, C.H.; Qiao, S.Z.; Zou, J.; Liu, G.; Smith, S.C.; Cheng, H.M.; Lu, G.Q. Anatase TiO_2 single crystals with a large percentage of reactive facets. *Nature* **2008**, *453*, 638–641. [CrossRef] [PubMed]
39. Zhao, X.; Jin, W.; Cai, J.; Ye, J.; Li, Z.; Ma, Y.; Xie, J.; Qi, L. Shape- and size-controlled synthesis of uniform anatase TiO_2 nanocuboids enclosed by active {100} and {001} facets. *Adv. Funct. Mater.* **2011**, *21*, 3554–3563. [CrossRef]
40. Zhang, D.; Wang, S.; Zhu, J.; Li, H.; Lu, Y. WO_3 nanocrystals with tunable percentage of (001)-facet exposure. *Appl. Catal. B* **2012**, *123*, 398–404. [CrossRef]
41. Peng, T.; Ke, D.; Xiao, J.; Wang, L.; Hu, J.; Zan, L. Hexagonal phase WO_3 nanorods: Hydrothermal preparation, formation mechanism and its photocatalytic O2 production under visible-light irradiation. *J. Solid State Chem.* **2012**, *194*, 250–256. [CrossRef]
42. Hu, Z.; Ji, Z.; Lim, W.W.; Mukherjee, B.; Zhou, C.; Tok, E.S.; Sow, C.H. K-Enriched WO_3 nanobundles: High electrical conductivity and photocurrent with controlled polarity. *ACS Appl. Mater. Interfaces* **2013**, *5*, 4731–4738. [CrossRef] [PubMed]
43. Mi, Q.; Coridan, R.H.; Brunschwig, B.S.; Gray, H.B.; Lewis, N.S. Photoelectrochemical oxidation of anions by WO_3 in aqueous and nonaqueous electrolytes. *Energy Environ. Sci.* **2013**, *6*, 2646–2653. [CrossRef]
44. Martins, A.S.; Guaraldo, T.T.; Wenk, J.; Mattia, D.; Zanoni, M.V.B. Nanoporous WO_3 grown on a 3D tungsten mesh by electrochemical anodization for enhanced photoelectrocatalytic degradation of tetracycline in a continuous flow reactor. *J. Electroanal. Chem.* **2022**, *920*, 116617. [CrossRef]

45. Bhattacharjee, S.; Chakraborty, S.; Mandol, K.; Liu, L.; Choi, H.; Bhattacharjee, C. Optimization of process parameters during photocatalytic degradation of phenol in UV annular reactor. *Desalin. Water Treat.* **2015**, *54*, 2270–2279. [CrossRef]
46. Zheng, Q.; Aiello, A.; Choi, Y.S.; Tarr, K.; Shen, H.; Durkin, D.P.; Shuai, D. 3D printed photoreactor with immobilized graphitic carbon nitride: A sustainable platform for solar water purification. *J. Hazard. Mater.* **2020**, *399*, 123097. [CrossRef] [PubMed]
47. Mioduska, J.; Łapiński, M.S.; Karczewski, J.; Hupka, J.; Zielińska-Jurek, A. New LED photoreactor with modulated UV–vis light source for efficient degradation of toluene over WO_3/TiO_2 photocatalyst. *Chem. Eng. Res. Des.* **2022**, *193*, 145–157. [CrossRef]
48. Binjhade, R.; Mondal, R.; Mondal, S. Continuous photocatalytic reactor: Critical review on the design and performance. *J. Environ. Chem. Eng.* **2022**, *10*, 107746. [CrossRef]
49. Mazellier, P.; Méité, L.; De Laat, J. Photodegradation of the steroid hormones 17β-estradiol (E2) and 17α-ethinylestradiol (EE2) in dilute aqueous solution. *Chemosphere* **2008**, *73*, 1216–1223. [CrossRef]
50. He, H.; Lin, Y.; Yang, X.; Zhu, X.; Xie, W.; Lai, C.; Yang, S.; Zhang, Z.; Huang, B.; Pan, X. The photodegradation of 17 alpha-ethinylestradiol in water containing iron and dissolved organic matter. *Sci. Total Environ.* **2022**, *814*, 152516. [CrossRef]
51. Huang, Z.; Lin, Q.; Luo, H.; Guo, P.; Weng, Q.; Lei, Y.; Cheng, S.; Liu, S.S. Degradation of progesterone by coexisting free radical and nonradical pathways in the CuO/HNTs-PS system. *J. Chem. Eng.* **2020**, *398*, 125458. [CrossRef]
52. Mitsika, E.E.; Christophoridis, C.; Kouinoglou, N.; Lazaridis, N.; Zacharis, C.K.; Fytianos, K. Optimized Photo-Fenton degradation of psychoactive pharmaceuticals alprazolam and diazepam using a chemometric approach—Structure and toxicity of transformation products. *J. Hazard Mater.* **2021**, *403*, 123819. [CrossRef] [PubMed]
53. Calisto, V.; Domingues, M.R.M.; Esteves, V.I. Photodegradation of psychiatric pharmaceuticals in aquatic environments—Kinetics and photodegradation products. *Water Res.* **2011**, *45*, 6097–6106. [CrossRef] [PubMed]
54. You, W.D.; Ye, P.; Yang, B.; Luo, X.; Fang, J.; Mai, Z.T.; Sun, J.L. Degradation of 17 Benzodiazepines by the UV/H_2O_2 Treatment. *Front. Environ. Sci.* **2021**, *9*, 474. [CrossRef]

nanomaterials

MDPI

Article

Exploring the Synergistic Mechanisms of Nanopulsed Plasma Bubbles and Photocatalysts for Trimethoprim Degradation and Mineralization in Water

Dimitris Tsokanas [1,2] and Christos A. Aggelopoulos [1,*]

[1] Laboratory of Cold Plasma and Advanced Techniques for Improving Environmental Systems, Institute of Chemical Engineering Sciences, Foundation for Research and Technology Hellas (FORTH/ICE-HT), 26504 Patras, Greece

[2] Chemistry Department, University of Patras, 26504 Patras, Greece

* Correspondence: caggelop@iceht.forth.gr; Tel.: +30-2610965205

Abstract: In this study, the synergetic action of nanopulsed plasma bubbles (PBs) and photocatalysts for the degradation/mineralization of trimethoprim (TMP) in water was investigated. The effects of ZnO or TiO_2 loading, plasma gas, and initial TMP concentration were evaluated. The physicochemical characterization of plasma-treated water, the quantification of plasma species, and the use of appropriate plasma species scavengers shed light on the plasma-catalytic mechanism. ZnO proved to be a superior catalyst compared to TiO_2 when combined with plasma bubbles, mainly due to the increased production of ·OH and oxygen species resulting from the decomposition of O_3. The air–PBs + ZnO system resulted in higher TMP degradation (i.e., 95% after 5 min of treatment) compared to the air–PBs + TiO_2 system (i.e., 87%) and the PBs-alone process (83%). The plasma gas strongly influenced the process, with O_2 resulting in the best performance and Ar being insufficient to drive the process. The synergy between air–PBs and ZnO was more profound (SF = 1.7), while ZnO also promoted the already high O_2–plasma bubbles' performance, resulting in a high TOC removal rate (i.e., 71%). The electrical energy per order in the PBs + ZnO system was very low, ranging from 0.23 to 0.46 kWh/m^3, depending on the plasma gas and initial TMP concentration. The study provides valuable insights into the rapid and cost-effective degradation of emerging contaminants like TMP and the plasma-catalytic mechanism of antibiotics.

Keywords: cold atmospheric plasma; plasma catalysis; antibiotics; water treatment; zinc oxide; plasma bubbles

1. Introduction

The presence and persistence of antibiotics in aquatic ecosystems have become a major concern in recent years. Antibiotics are widely used as they are essential medications for combating bacterial infections in humans, animals, and plants. However, despite their effectiveness, their abuse, misuse, and improper disposal by various industries, including aquaculture, healthcare, and agriculture, pose major threats to public health and the environment, as their accumulation in water bodies has led to the emergence of antibiotic-resistant bacteria [1,2]. Among these antibiotics, trimethoprim, a widely used medication to treat bacterial infections, has been found to contribute to the contamination of aquatic ecosystems [3,4].

The removal of antibiotics from water is a major challenge due to their persistence and diverse chemical properties. Various water treatment methods have been developed to address this issue [5,6]. However, most of these methods are expensive, are not environmentally friendly, or cannot achieve high removal efficiencies. Advanced oxidation processes (AOPs) are considered more effective in oxidizing a wide range of organic compounds, including antibiotics, as well as being versatile and applicable to various types of wastewater [7]. For instance, photocatalysis has been widely applied for the degradation of antibiotics involving the use of a photocatalyst which becomes activated when exposed to ultraviolet (UV) or visible light [8]. Among the AOPs, non-thermal plasma (NTP) is an emerging and promising process that does not require the use of chemical agents [9,10]. During NTP, a plethora of long- and short-lived reactive species are generated together with UV light and highly energetic electrons that participate in electron transfer reactions with organic molecules, initiating chain reactions that lead to the degradation of pollutants [11,12]. Plasma species include reactive oxygen species (ROS) such as hydroxyl radicals, atomic oxygen, ozone and hydrogen peroxide and reactive nitrogen species (RNS) such as nitrate/nitrite anions, peroxynitrite, etc. [13,14]. Normally, after plasma treatment, the antibiotics are decomposed into less toxic and harmful intermediates, if complete mineralization cannot be achieved [15]. However, the effectiveness of most plasma reactors (e.g., gas–liquid DBD) is limited by the short lifetime of the species and their low penetration into the liquid phase [16]. Therefore, the main challenge remains the uncomplete mineralization of antibiotics resulting from the limited utilization of the UV radiation, electrons and reactive species generated during plasma discharge.

Recently, to alleviate this problem, plasma bubbles have been increasingly explored as a means of enhancing the mass transfer of plasma species from the gas to the aqueous phase [17,18] or in combination with photocatalysts (plasma catalysis) to improve the degradation rate and mineralization of the pollutant by increasing plasma energy utilization [19,20]. During plasma catalysis, the plasma-generated reactive species (electrons, ions, radicals, and UV radiation) can activate the catalyst surface and promote the generation of additional reactive species or convert certain plasma species (e.g., O_3) to species with higher oxidation potential (e.g., $\cdot OH$) [21].

In the present study, underwater plasma bubbles, energized by low-frequency high-voltage (HV) nanopulses, were combined with TiO_2 or ZnO photocatalysts for the degradation of the antibiotic trimethoprim (TMP) in water. Considering that reports on TMP degradation by plasma are scarce [22,23], a thorough investigation was conducted to provide insightful information. Thus, a point-to-point comparison was made between the two catalysts in terms of plasma-catalytic synergy during TMP degradation, species formation and physicochemical properties of the plasma-treated water. In this context, the effect of various parameters in the absence and presence of the two catalysts, such as treatment time, catalyst loading and plasma gas, was investigated. Furthermore, the role of the main plasma species in TMP degradation was evaluated through scavenging experiments, while the process energy efficiency (electrical energy per order) and TMP mineralization were assessed. The strategy of combining underwater plasma bubbles with two well-known photocatalysts could contribute to the understanding of the plasma-catalytic mechanisms that take place during the degradation of antibiotics in water and to the establishment of a highly effective, energy-efficient and rapid process for the destruction of this class of emerging pollutants.

2. Materials and Methods

2.1. Materials

All reagents were of analytical grade: trimethoprim (TMP, $C_{14}H_{18}N_4O_3$, M = 290.32 g/mol), ZnO (zinc oxide nanopowder, <100 nm particle size), TiO_2 (titanium (IV) oxide nanopowder, 21 nm particle size), D-mannitol (D-man), p-benzoquinone (BQ), potassium phosphate monobasic (MP), sodium pyruvate (SP), terephthalic acid (TA), and titanium oxysulfate were purchased from Sigma-Aldrich (St. Louis, MO, USA). Solutions were prepared using Milli-Q water with a resistivity of 18.2 MΩ·cm, obtained by the purification of deionized water, unless otherwise stated. Compressed dry air, oxygen and argon were supplied by Linde (Athens, Greece).

2.2. Experimental Setup, Electrical Diagnostics and Treatment Conditions

The experimental setup used to treat the TMP-polluted water and characterize the plasma is illustrated in Figure 1. It consisted of (i) a DBD-based plasma bubble reactor driven by (ii) a nanosecond pulsed voltage delivered by a power supply (NPG-18/3500), (iii) a digital oscilloscope (Rigol MSO2302A) connected to voltage (Tektronix P6015A, 0–75 MHz) and current probes (Pearson electronics 2877, 300 Hz–200 MHz) for the discharge power calculation, (iv) an optical characterization arrangement, and (v) a feeding gas system. Details regarding the calculation of the power dissipated in the DBD-based plasma reactor can be found elsewhere [24].

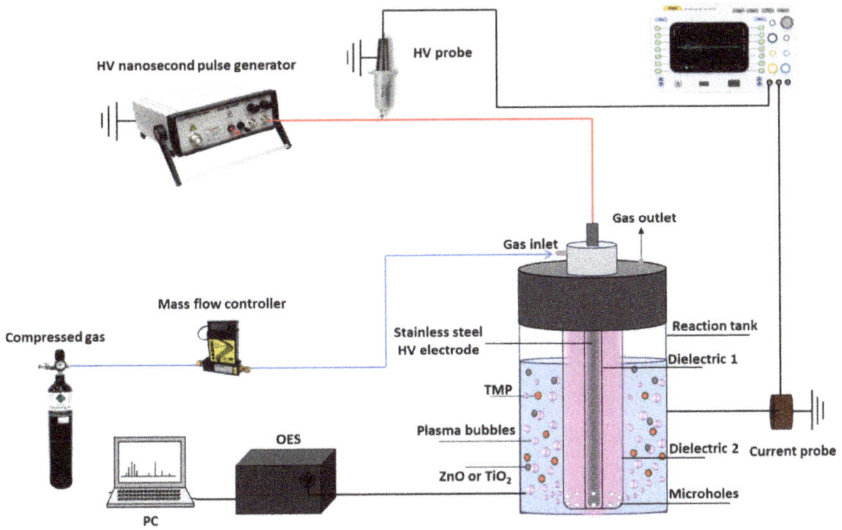

Figure 1. The experimental setup used to treat water samples contaminated by trimethoprim (TMP) and characterize the plasma.

The plasma reactor, based on a coaxial DBD column (Figure 1) capable of delivering plasma bubbles directly into the aqueous solution, was similar to that previously reported [17,18]. Briefly, it consisted of a high voltage stainless-steel rod (5 mm diameter) covered by an inner quartz tube (8.0 mm external diameter) forming a coaxial DBD column with an outer quartz tube (12.0 mm internal diameter). The latter was constructed with 10 microholes uniformly distributed around its base so that the plasma gas, flowing between these dielectric tubes at a constant flow rate (3.0 L/min), could provide plasma bubbles in the solution. The microholes were constructed with a diameter of 400–500 μm to ensure a small bubble size, and therefore a large gas–liquid interfacial area, as a means to increase the mass transfer of short-lived plasma species from the gas to the liquid phase [25]. This coaxial DBD configuration was placed in the center of a quartz reaction tank (46 mm inner diameter, 2 mm wall thickness, 70 mm height), which was filled with the water sample to be treated. A copper tape was attached to the outer surface of the reaction tank and acted as a grounded electrode.

An amount of 70 mL of TMP-polluted water was exposed to plasma treatment (20 s to 5 min) with an initial pollutant concentration ranging from 5 to 20 mg/L. Pulse voltage and frequency were equal to 25.6 kV and 200 Hz, respectively. For the plasma-catalytic experiments, the catalyst loading ranged from 0 to 0.4 g/L. The solutions containing the catalyst were previously sonicated for 20 min at room temperature. To investigate the effect of plasma chemistry on TMP degradation, different feed gases were introduced into the reactor (air, oxygen and argon). All experiments were carried out in duplicate with excellent reproducibility.

2.3. Chemical Analysis of Water Samples

The concentration of TMP was evaluated by high-performance liquid chromatography (HPLC, Shimadzu LC-2050C) equipped with a C18 column (4.6 mm × 50 mm, 3 μm particle size) and a UV detector. According to Liang et al. [23], Αναφορές the temperature was set to 30 °C, detection wavelength to 271 nm and pressure to 100 bar. The flow rate was set to 1.0 mL/min and injection volume to 20 μL. The mobile phase was isocratic and consisted of 0.2 mol/L NaH_2PO_4 solution and methanol (70%/30% v/v). The TMP degradation efficiency, $D(\%)$, was calculated as follows:

$$D(\%) = \left[\frac{C_0 - C_t}{C_0}\right] \times 100 \tag{1}$$

where C_0 is the initial TMP concentration and C_t is its concentration after plasma treatment for a given time.

The degradation rate was described by a pseudo-first-order reaction with respect to the pollutant concentration, and thus, the experimental data were fitted to the following equation:

$$In\left(\frac{C_0}{C_t}\right) = k_{app}t \tag{2}$$

where k_{app} (min^{-1}) represents the apparent rate constant.

The synergetic factor, namely the synergetic effect between plasma and catalysis, was calculated by the following equation:

$$SF = \frac{k_{plasma+catalyst}}{k_{plasma} + k_{catalyst}} \tag{3}$$

where $k_{plasma+catalyst}$ is the rate constant of TMP degradation in the plasma-catalytic system, k_{plasma} is the rate constant in the plasma-alone system, and $k_{catalyst}$ is the rate constant of TMP adsorption onto the catalyst.

The electrical energy per order is the energy required for the decrease in pollutant concentration by one order of magnitude in a unit volume of water and defined as follows:

$$E_{EO} = \frac{Pt}{Vlog\left(\frac{C_0}{C_t}\right)} \tag{4}$$

where P is the power dissipated in the plasma reactor, V stands for the volume of the treated solution, and t refers to the treatment time.

Finally, a total organic carbon analyzer (Shimadzu, TOC-VCSH) was used to quantify the mineralization of plasma-treated samples. The TOC removal efficiency was calculated by the following equation:

$$TOC(\%) = \left[\frac{TOC_0 - TOC_f}{TOC_0}\right] \times 100 \tag{5}$$

where TOC_0 and TOC_f indicate the TOC concentration (mg/L) before treatment and after plasma treatment, respectively.

2.4. Identification of Reactive Species in Gas and Aqueous Phase and Use of Scavengers

The critical short- and long-lived reactive oxygen species in the absence and presence of both catalysts were determined. The concentration of hydroxyl radicals (·OH) was detected using a photoluminescence spectrometer (Hitachi F2500) Αναφορές [26]. More specifically, the fluorescent properties of 2-hydroxyterephthalic acid (HTA) were monitored, which are produced from the reaction between ·OH and terephthalic acid (TA). When the sample was irradiated by UV-A light of 310 nm wavelength, an intensity peak at 425 nm was observed. Furthermore, ozone (O_3) was quantified by a Hanna multiparameter photometer (HI83399–2), using calibrated reagent kits. Concerning hydrogen peroxide (H_2O_2), the titanium oxysulfate method was used [27]. Briefly, 1 mL of titanium oxysulfate solution in 2 M sulfuric acid (25 g/L) was added to 2 mL of each sample, and the absorption peak at 407 nm in the UV-Vis spectrum was quantified (Shimadzu UV-1900). The main atomic and molecular excited plasma species in the gas phase were identified through optical emission spectroscopy (OES) measurements by using a fiber optic spectrometer (AvaSpec-ULS2048CL-EVO, Avantes) in the range of 200–900 nm.

The transient evolution of the physicochemical characteristics of the aqueous solution during plasma treatment was evaluated. The conductivity and pH of the water samples were measured using a Consort multiparameter (Benchtop meter C1020) and a QUANTOFIX® Relax unit (Macherey-Nagel, GmbH), respectively. The initial conductivity of the ultrapure water was equal to 4.6 μScm^{-1} and the pH was 6.5.

To investigate the role of plasma species on the degradation of TMP, scavenging experiments were also conducted. Sodium pyruvate (SP), D-mannitol (D-man), monopotassium phosphate (MP), and p-benzoquinone (BQ) were selected as scavenging agents of H_2O_2, $\cdot OH$, e_{aq}^- and $\cdot O_2^-$, respectively. The concentration of each scavenger varied from 0.2 to 6 mM to ensure the maximum inhibition of TMP degradation. The action efficiency of plasma species was calculated as follows:

$$\varphi = \left[\frac{D_0 - D_1}{D_0}\right] \times 100\% \tag{6}$$

where D_1 and D_0 is the TMP degradation in the presence and absence of a scavenger, respectively.

2.5. Catalyst Characterization

The crystallographic and pore structure of the catalysts were examined before and after plasma treatment. Plasma-treated catalysts were collected after their exposure to plasma bubbles in the absence of pollutant. The plasma-treated water including the catalyst remained at 80 °C until complete evaporation. The resulting plasma-treated solid (ZnO, TiO_2) was collected and used for characterization purposes.

The crystal structure of pristine and plasma-treated ZnO and TiO_2 nanopowder was measured by X-ray powder diffraction (XRD) analysis in the range of 5–90° (2θ). A Bruker D8 Advance diffractometer was used, applying Cu K-alpha radiation (X-ray wavelength λ = 0.15418 nm). The specific surface area and pore size distribution of the pristine and plasma-treated catalysts were measured using a Quantochrome Autosorb IQ-C-MP apparatus. The samples were first degassed at 150 °C for 2 h, and afterwards, nitrogen adsorption and desorption isotherms were obtained at relative pressure P/P0 = 0.05–0.99, where P represents the absolute pressure and P0 the saturation vapor pressure. Using the Brunauer–Emmett–Teller (BET) method, the specific surface area was determined from the nitrogen adsorption (P/P0 in the range 0.05–0.25). Finally, the density functional theory (DFT) method was employed to obtain the volume-based pore size distributions from the whole nitrogen adsorption and desorption isotherms.

3. Results

3.1. TiO_2 and ZnO Characterization (XRD, BET) before and after Plasma Treatment and TEM Images

The catalysts were characterized through XRD before and after the plasma bubble treatment. TiO_2 showed two crystalline phases with a characteristic tetragonal crystal structure, rutile and anatase, as expected. Anatase was the dominant phase (~90%) with apparent peaks at 2theta values of 25.32°, 37.86°, 48.06°, 53.97°, and 55.09°, which are attributed to the (101), (004), (200), (105), and (211) crystal planes, respectively. The rutile phase was observed at minor peaks like 27.45°, 35.98°, and 41.25°, assigned to the (110), (101), and (111) crystal planes, respectively. After plasma treatment, there were no alterations in the structure of TiO_2 (Figure S1). The XRD pattern of ZnO confirmed its hexagonal wurtzite structure, with the main diffraction peaks at 31.72°, 34.44°, and 36.26°

attributed to the (100), (002), and (101) crystal planes, respectively. Similar to TiO_2, the underwater plasma bubble treatment did not result in changes in the crystal structure of ZnO (Figure S1).

The specific surface area of each catalyst after treatment with nanopulsed plasma bubbles was almost identical to that of the pristine catalysts, with the surface area of TiO_2 being ~4 times higher compared to that of ZnO (Table 1). Nevertheless, as will be presented below (Section 3.3), the combination of plasma bubbles with ZnO was more effective in degrading TMP than the combination of plasma bubbles with TiO_2. Furthermore, the plasma-catalytic synergy between the plasma bubbles and catalysts could not be attributed to the modification of the ZnO or TiO_2 pore structure during treatment, as the plasma bubbles did not induce noticeable changes in the pore volume and pore size distribution of the catalysts (Table 1 and Figure S2).

Table 1. Specific surface area and pore volume of TiO_2 and ZnO before and after plasma bubble treatment (pulse voltage: 25.6 kV, pulse frequency: 200 Hz, plasma gas: air, gas flow rate: 3 L/min, treatment time: 5 min, catalyst loading: 0.2 g/L).

	Specific Surface Area (m^2/g)/Pore Volume (cm^3/g)	
Catalyst	**Pristine**	**Plasma-Treated**
TiO_2	54.96/0.121	54.92/0.160
ZnO	13.18/0.024	13.01/0.026

Figure S3 displays representative TEM images of the TiO_2 and ZnO samples, providing insight into their respective sizes. The TiO_2 nanoparticles exhibit a nearly spherical shape, with a mean diameter of approximately 20 nm. Conversely, the ZnO particles display a hexagonal wurtzite structure and possess a mean size of less than 100 nm.

3.2. Electrical and Optical Plasma Characterization

The electrical power dissipated in the reactor is a measure to evaluate the energy efficiency of the present remediation process. Typical voltage and current nanopulses monitored during TMP treatment by plasma bubbles and the resulting instantaneous power are depicted in Figure 2a. The waveforms initially consist of a high-amplitude pulse followed by smaller-amplitude pulses. The pulse rising time was approximately 4 ns, while the current peak was ~50 A. Despite the applied voltage being a unipolar pulse, the current exhibited bipolar characteristics, with positive pulses (primary discharge) and negative pulses (secondary discharge), attributed to both conduction and displacement current. Although the resulting instantaneous power peak (corresponding to the primary discharge) was very high and equal to ~1.2 MW, the average power dissipated in the plasma bubble reactor was quite low and equal to ~0.8 W due to the very low duty cycle. This is the feature responsible for the better energy efficiency obtained when HV nanosecond pulses are used to energize plasma reactors compared to HV microsecond pulses or sinusoidal HV [18].

Optical emission spectroscopy (OES) was conducted to detect the main excited molecular and atomic plasma species under air, oxygen, and argon atmospheres (Figure 2b). Under O_2–plasma, the main species identified include the following ROS: an ·OH emission peak at 309 nm [28], ·O emissions at 777 and 844 nm [15], and weak O_2^+ emissions [29]. Under argon–plasma, the spectrum is dominated by multiple argon transition lines (696 to 852 nm), ·OH, and weak ·O emissions [30]. The OES spectrum obtained during air–plasma treatment is dominated by nitrogen molecules (N_2—second positive system), nitrogen cation (N_2^+—first negative system), ·OH, and weak ·O and O_2^+ emissions [31,32]. The different OES spectra between the various plasma gases result in the generation of different species within the aqueous solution, affecting the degradation of TMP in water (see the results below in Section 3.5).

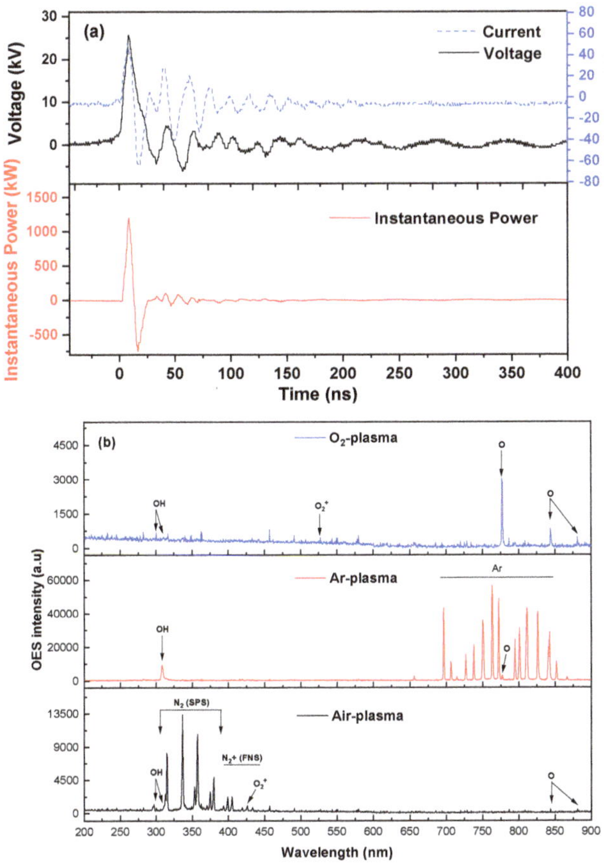

Figure 2. (**a**) Instantaneous voltage, current, and power waveforms during TMP treatment by nanopulsed air–plasma bubbles and (**b**) optical emission spectra of oxygen, argon, and air–plasma in the plasma bubble reactor (pulse voltage: 25.6 kV, pulse frequency: 200 Hz, gas flow rate: 3 L/min).

3.3. *Effect of Catalyst Addition on Plasma-Treated Water Characteristics*

3.3.1. Plasma Species Formation

In dielectric barrier discharge (DBD) plasma, electrons are primarily generated through photoionization and collisional ionization processes in the gas phase. Therefore, an electron colliding with a neutral gas molecule transfers enough energy to ionize the molecule, resulting in the creation of additional electrons and positively charged ions. Subsequently, free radicals capture electrons from neighboring atoms or molecules, leading to the generation of several new chemically active species. The degradation of organic pollutants by plasma lies mainly in the parallel action of reactive oxygen and nitrogen species (RONS). Both short- and long-lived species can be crucial for TMP degradation, and therefore, the concentrations of H_2O_2, O_3, and ·OH in the plasma-treated water were measured. The concentration of nitrogen-related species (nitrate, nitrite) was negligible, as previously reported for plasma-treated water inside a DBD-based plasma bubble reactor driven by HV nanopulses [18].

H_2O_2 is considered representative among the long-lived plasma species, being mostly formed by recombination reactions between radicals. It is noteworthy that H_2O_2 could also lead to the formation of additional oxygen species (R1–R4). A detailed discussion and reactions on the formation of plasma species during water treatment can be found elsewhere [14]:

$$H_2O + e^- \rightarrow \cdot H + \cdot OH + e_{aq}^- \quad (R1)$$

$$\cdot OH + \cdot OH \rightarrow H_2O_2 \quad (R2)$$

$$\cdot HO_2 + \cdot HO_2 \rightarrow H_2O_2 + O_2 \quad (R3)$$

$$e^- + H_2O_2 \rightarrow \cdot OH + OH^- \quad (R4)$$

Compared to the plasma bubbles, the concentration of H_2O_2 was decreased in the presence of ZnO and increased substantially in the presence of TiO_2 (Figure 3a). In particular, after 5 min of treatment with air–plasma bubbles, the H_2O_2 concentration was 0.25 mg/L (plasma bubbles), 0.06 mg/L (plasma bubbles + ZnO), and 1.2 mg/L (plasma bubbles + TiO_2). Therefore, TiO_2 promoted the reactions that led to the formation of H_2O_2, in contrast to ZnO.

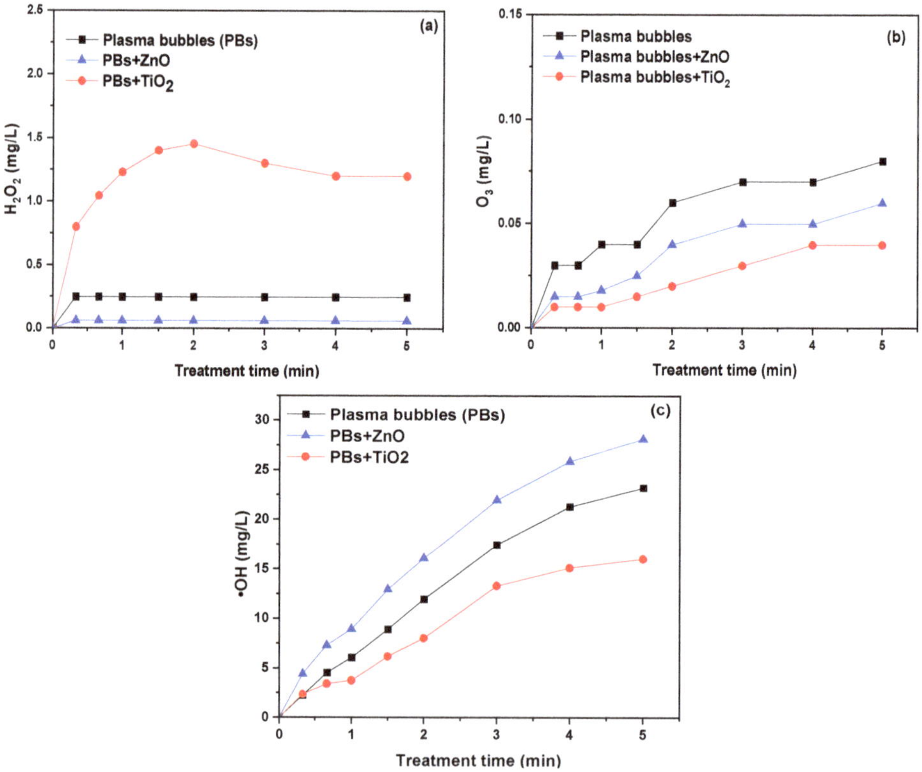

Figure 3. Concentration of plasma species in the absence and presence of TiO_2 or ZnO. (**a**) H_2O_2, (**b**) O_3, and (**c**) $\cdot OH$ (pulse voltage: 25.6 kV, pulse frequency: 200 Hz, plasma gas: air, flow rate: 3 L/min, catalyst loading: 0.2 g/L).

Ozone plays a major role in the degradation of pollutants and, therefore, its formation should be quantified. In all cases, O_3 increased with the treatment time almost linearly, while the presence of both catalysts in the plasma seemed to promote the decomposition of O_3 (Figure 3b). After 5 min of treatment, the O_3 concentration decreased from 0.08 mg/L (plasma bubbles) to 0.06 mg/L and 0.04 mg/L in the presence of ZnO and TiO_2, respectively. The presence of a catalyst may promote the decomposition of O_3 through various mechanisms: (i) O_3 is adsorbed on the active sites of ZnO or TiO_2 and decomposed by the oxygen defects into highly reactive species, like atomic (R5) and singlet oxygen [33,34]; (ii) plasma radiation (hv) results in the decomposition of O_3 into hydrogen peroxide (R6); and (iii) the electrons generated by the irradiation of the photocatalysts by the UV emitted by the plasma can react with O_3 and reduce it to short-lived and highly reactive $\cdot O_3^-$, $\cdot O_4^-$, $^1O_2$, $\cdot O_2^-$, and $\cdot OH$ [21].

$$O_3 + {}^* \rightarrow \cdot O + O_2 \quad \text{(R5)}$$

where * symbolizes the active sites of the catalyst

$$O_3 + \text{plasma } (hv) + H_2O \rightarrow H_2O_2 + O_2 \quad \text{(R6)}$$

Among the various plasma species, short-lived $\cdot OH$ is one of the most critical species that interact rapidly with organic pollutants [35,36]. It is directly formed through water dissociation by plasma electrons as well as from the consumption of the long-lived O_3 and H_2O_2:

$$H_2O + e^- \rightarrow \cdot H + \cdot OH + e_{aq}^- \quad \text{(R7)}$$

$$O_3 + H_2O_2 \rightarrow \cdot OH + \cdot HO_2 + O_2 \quad \text{(R8)}$$

$$H_2O_2 + \text{plasma } (hv) \rightarrow 2\cdot OH \quad \text{(R9)}$$

Compared to plasma alone, the presence of ZnO enhanced the production of $\cdot OH$, in contrast to TiO_2, which had the opposite effect (Figure 3c). For instance, after 5 min of treatment, the concentration of $\cdot OH$ in the aqueous phase treated by plasma bubbles was 23 mg/L and increased to 28 mg/L in the presence of ZnO (plasma bubbles + ZnO), but it decreased to 16 mg/L in the presence of TiO_2 (plasma bubbles + TiO_2). The increase in $\cdot OH$ production under ZnO-catalyzed conditions may be attributed to the reaction between the UV-plasma generated holes with water molecules (R10 and R11):

$$\text{photocatalyst} + \text{plasma } (hv) \rightarrow e_{cb}^- + h_{vb}^+ \quad \text{(R10)}$$

$$h_{vb}^+ + H_2O \rightarrow \cdot OH + H^+ \quad \text{(R11)}$$

On the other hand, the decrease in $\cdot OH$ production under TiO_2-catalyzed conditions may be attributed to its recombination to form H_2O_2 (R2). It is worth reminding here that the concentration of H_2O_2 increased in the presence of TiO_2 (Figure 3a). A possible explanation could be as follows: Given that TiO_2 has almost quadruple the specific surface area of ZnO and its size is four times smaller than ZnO (Table 1, Figure S3), it is plausible to assume that the active sites in TiO_2 are more densely packed, making the recombination of $\cdot OH$ to form H_2O_2 (R2) more likely than in ZnO. In other words, in plasma-catalytic wastewater treatment, the use of a photocatalyst with an increased surface area and very small size may potentially lead to the increased production of H_2O_2 through the recombination of $\cdot OH$. This could explain why H_2O_2 increases and $\cdot OH$ decreases under TiO_2-catalyzed conditions, while the opposite occurs with ZnO. However, in plasma-catalytic wastewater treatment, the goal is to optimize the conditions that minimize radical recombination to H_2O_2, thus maximizing pollutant degradation.

In conclusion, both catalysts promoted the decomposition of the plasma-generated O_3, probably to other oxygen radicals. The presence of TiO_2 resulted in a decreased $\cdot OH$ concentration and more H_2O_2 compared to plasma alone. In contrast, regarding ZnO, the $\cdot OH$ concentration was enhanced and less H_2O_2 was produced compared to plasma alone. It is worth noting that the oxidation potential of H_2O_2 is approximately +1.78 volts (V) under standard conditions, whereas the oxidation potential of $\cdot OH$ is considerably higher, often exceeding +2.8 volts (V) or even higher. These results could contribute to understanding the degradation mechanism of TMP in the presence of either ZnO or TiO_2. Additionally, identifying the key species involved in degrading specific pollutants can help in selecting the most suitable catalyst for each pollutant.

3.3.2. Physicochemical Water Properties

The aqueous solution's pH is considered a critical parameter for the effectiveness of plasma in water treatment. Neutral or alkaline pH levels increase the production of ·OH due to O_3 and H_2O_2 decomposition, thereby enhancing or reducing pollutant degradation depending on the critical species for its degradation [37,38]. Additionally, alkaline conditions may decrease the oxidation potential of O_3. Significantly, no substantial fluctuations were observed in the pH during the plasma bubble treatment in the presence or absence of ZnO or TiO_2 (Figure S4a). This pH stability (~6.5) was further supported by the electrical conductivity of the aqueous solutions (Figure S4b), which showed an insignificant gradual increase with treatment time (from 5 to 20 μS/cm) in all cases.

3.4. *The Effect of Catalyst Loading on the Degradation of TMP*

Numerous studies have investigated the potential synergy between plasma and catalysts for antibiotic degradation [39–42]. In this study, to harness the UV radiation emitted by plasma discharge, two photocatalysts, ZnO and TiO_2, were selected. Comparisons of these two catalysts are well documented in other applications and remediation methods [43–46]. The effect of catalyst loading on TMP degradation was thoroughly investigated by repeating the plasma experiments in the presence of different loadings of ZnO or TiO_2 (Figure 4). Initially, different loadings of TiO_2 nanopowder were added (0.1, 0.2, 0.3, and 0.4 g/L). The TiO_2 catalyst did not appear to significantly enhance the degradation efficiency of TMP (Figure 4a). However, slight changes were observed with varying loadings and when compared to the plasma bubbles-alone system. In particular, there was no notable effect on the degradation of TMP with the addition of 0.1 g/L TiO_2, which removed approximately ~83.2% of the pollutant in 5 min, similar to the plasma bubbles-alone system. The other three loadings exhibited better degradation efficiency, as depicted in Figure 4a, with the pseudo-first-order degradation kinetics displayed in Figure 4c. Almost 87.4% degradation was achieved, with an apparent rate constant of 0.42 min^{-1}. It is worth noting that in terms of TMP removal, 0.2 g/L TiO_2 was the optimum loading, as higher loadings led to similar TMP degradation. This may be due to agglomerations between TiO_2 nanoparticles and the potential coverage of its active sites.

Similar experiments were conducted in the presence of ZnO, revealing a different behavior, as ZnO exhibited significant synergy with the plasma bubbles. All ZnO loadings resulted in an enhancement in TMP degradation (Figure 4b). In particular, with the addition of 0.1 and 0.2 g/L ZnO, TMP degradation increased from ~83.2% (plasma bubbles) to ~93.7% and ~94.8%, respectively, after 5 min of treatment. Higher ZnO loadings resulted in a less profound impact, reaching ~93.3% (0.3 g/L) and ~92.8% (0.4 g/L) degradation of TMP. This could be attributed to the agglomeration of the ZnO particles, which are not able to interact properly with the plasma bubbles. The corresponding pseudo-first-order degradation kinetic constants increased from 0.35 min^{-1} (plasma bubbles) to 0.56, 0.59, 0.54, and 0.53 min^{-1}, respectively (Figure 4d), indicating that 0.2 g/L is the ideal ZnO loading.

A direct comparison between the two catalysts under their optimal loading is presented in Figure 5a. It is evident that TMP degradation under plasma bubbles + ZnO was higher compared to the plasma bubbles + TiO_2 system and plasma bubbles-alone process. For instance, TMP degradation at the end of treatment (i.e., 5 min) was ~83.2% (plasma bubbles), ~94.8% (plasma bubbles + ZnO), and ~87.4% (plasma bubbles + TiO_2). Additionally, negligible TMP adsorption onto both catalysts was observed under no plasma conditions (only bubbles with a catalyst).

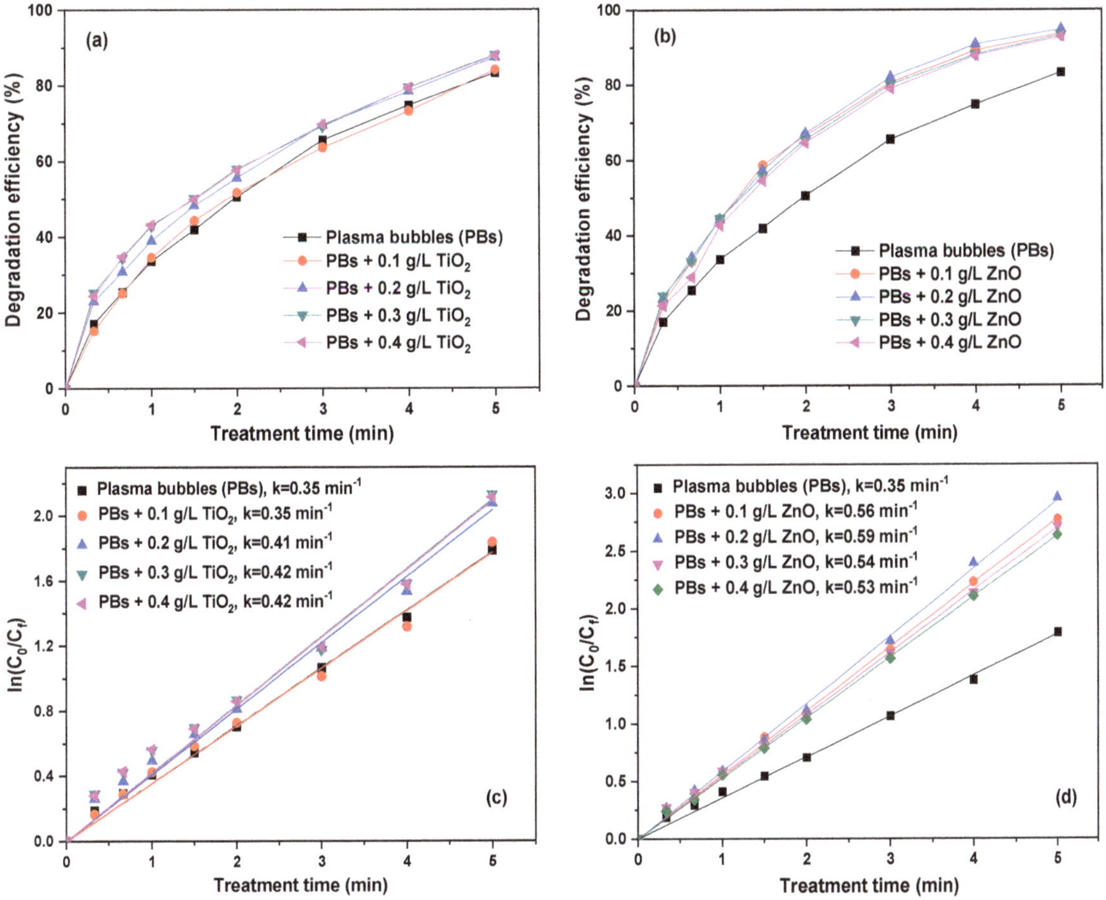

Figure 4. TMP degradation efficiency at various (**a**) TiO_2 and (**b**) ZnO loadings. Pseudo-first-order degradation kinetics of (**c**) TiO_2 and (**d**) ZnO (pulse voltage: 25.6 kV, pulse frequency: 200 Hz, plasma gas: air, flow rate: 3 L/min, initial TMP concentration: 20 mg/L).

Utilizing the previously calculated apparent rate constant, the synergistic factor (SF) was deduced for each catalyst at every loading, as depicted in Figure 5b. The corresponding SF in the presence of ZnO was 1.56, 1.70, 1.54, and 1.53 for 0.1, 0.2, 0.3, and 0.4 g/L, respectively, while the corresponding values for TiO_2 were 1, 1.14, 1.17, and 1.17. Therefore, the highest SF achieved was 1.7 with 0.2 g/L ZnO, which is approximately 1.5 times higher than TiO_2 at the same loading. Consequently, ZnO appears to be much more beneficial when combined with plasma bubbles for the degradation of TMP. Interestingly, ZnO exhibited increased plasma-catalytic performance compared to TiO_2 despite its lower surface area (Table 1). The superiority of ZnO can be rationalized by the vital role of plasma species in the TMP degradation process (see discussion below in Section 3.7) and their relative concentration, which in turn depends on the type of the catalyst (as discussed in Section 3.3.1).

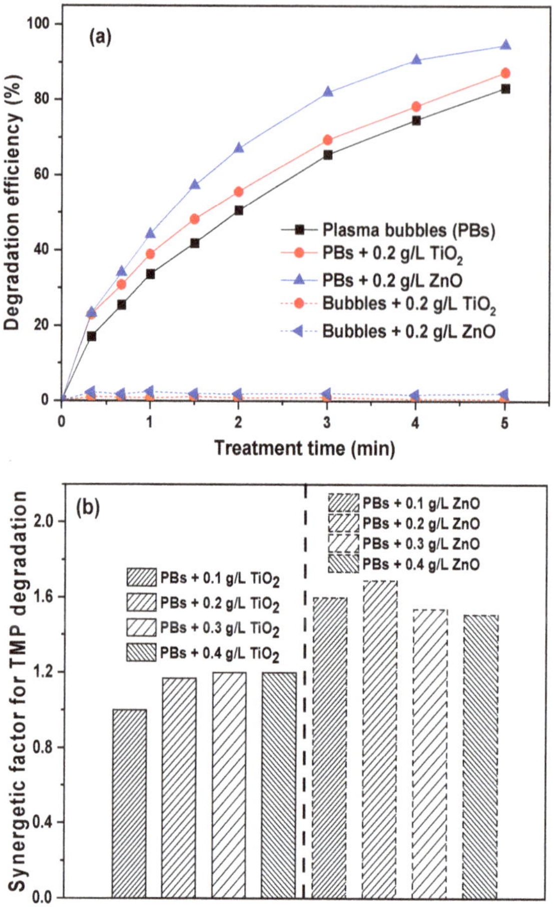

Figure 5. (**a**) Comparison between the TMP degradation efficiencies of plasma bubbles in the presence of TiO_2 or ZnO (catalyst loading: 0.2 g/L). (**b**) Synergetic factor at various ZnO and TiO_2 loadings (pulse voltage: 25.6 kV, pulse frequency: 200 Hz, plasma gas: air, flow rate: 3 L/min, initial TMP concentration: 20 mg/L).

3.5. The Effect of Plasma Gas and Initial TMP Concentration on Its Degradation by Plasma Bubbles in the Absence and Presence of ZnO

The effectiveness of plasma for water treatment depends on numerous parameters related to plasma conditions and chemistry [10]. Plasma chemistry is related to the nature of the produced reactive oxygen and nitrogen species (RONS), which depend on the gas used for the discharge formation [47]. Altering the plasma gas can result in the production of different types and concentrations of RONS, which play a major role in pollutant degradation. For this reason, the effect of plasma gas (air, oxygen, and argon) on TMP degradation was investigated under the conditions of plasma bubbles in the absence and presence of ZnO, which proved to be a more effective catalyst for TMP degradation (Figure 6). When oxygen and argon were used instead of air as the feeding gas, a different TMP degradation profile was observed. Under O_2–plasma bubbles, very high TMP degradation (i.e., ~94%) was recorded after only 2 min of treatment, whereas the addition of ZnO further promoted this performance to ~96% (Figure 6a). This higher performance of oxygen is related to the enhanced reactive oxygen species (ROS) production under an O_2-containing atmosphere, indicating the significance of ROS over reactive nitrogen species (RNS) in the TMP degradation process.

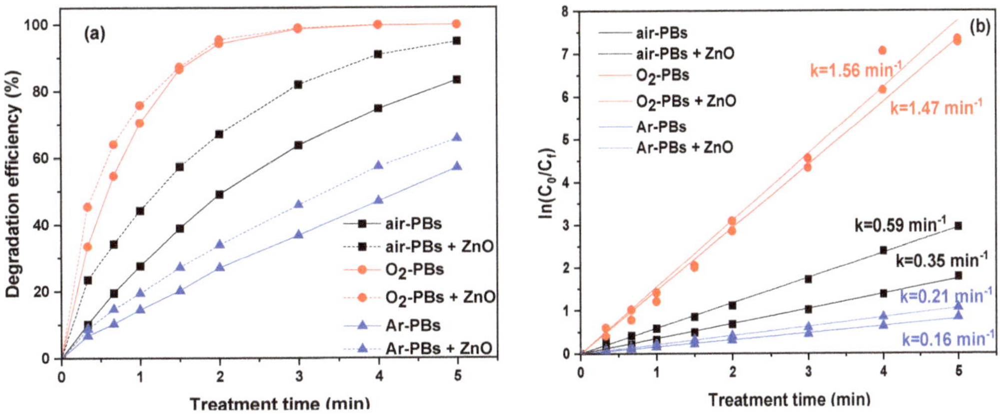

Figure 6. (**a**) TMP degradation as a function of treatment time under air–plasma bubbles, O_2–plasma bubbles, and Ar–plasma bubbles in the presence and absence of ZnO and (**b**) corresponding pseudo-first-order degradation kinetics (pulse voltage: 25.6 kV, pulse frequency: 200 Hz, flow rate: 3 L/min, initial TMP concentration: 20 mg/L, ZnO loading: 0.2 g/L).

On the other hand, under argon–plasma bubbles, a lower TMP degradation was achieved (~57%) even after 5 min of treatment, while the addition of ZnO increased the TMP degradation to 65% (Figure 6a). It could, therefore, be assumed that the reactions and species promoted by argon, such as electrons, argon ions, and ·OH, are also involved in TMP degradation but not as the major species. It is also interesting to note that the impact of ZnO with the different gases was not the same. The apparent rate constant of air–plasma bubbles + ZnO was ~1.6 times higher compared to that of air–plasma bubbles, being 0.59 and 0.35 min^{-1}, respectively (Figure 6b). Concerning argon, the apparent rate constant increased from 0.16 to 0.21 min^{-1} (~1.3 times higher). In addition, considering the already very high performance of O_2–plasma bubbles, the presence of ZnO did not result in a significant enhancement in the degradation rate of the parent TMP molecule (1.47 to 1.56 min^{-1}). Nevertheless, the TOC measurements presented below (Section 3.8) are indicative of the plasma-catalytic synergy between ZnO and O_2–plasma bubbles. Based on the apparent rate constants, the synergistic factor (SF) values for air–plasma bubbles + ZnO and argon–plasma bubbles + ZnO were 1.70 and 1.31, respectively. A possible explanation for why the effect of ZnO was more intense in the case of air could lie in the fact that the absorption spectrum of ZnO (UVA/UVB region) overlaps with the corresponding radiation emitted by air–plasma. On the other hand, the radiation emitted from argon is mainly in the VUV/visible light region, meaning the photoactivation of the catalyst is less effective.

The effect of the initial pollutant concentration on its degradation was extensively investigated. Nonetheless, from a practical standpoint, with the inclusion of this analysis in the present study, we aimed to assess the efficacy of the current approach across a range of pollutant concentrations in water, encompassing both higher and lower (and more realistic) levels of TMP contamination in water. Therefore, the air–plasma bubbles process was tested under various initial TMP concentrations, both with and without ZnO. For this purpose, experiments were conducted using initial TMP concentrations of 20 mg/L, 10 mg/L, and 5 mg/L, with the results depicted in Figure 7. Evidently, the lower the initial TMP concentration, the higher its degradation efficiency. For 20 mg/L TMP, 83.2% degradation was achieved after 5 min of treatment, while complete degradation (99.9%) was achieved at an initial concentration of 10 mg/L. For the lowest concentration (5 mg/L), complete degradation was also achieved but with a lower treatment time (4 min). The addition of ZnO apparently enhanced the TMP degradation, as reflected in the apparent degradation rate constants (Figure 7b). In particular, for 20 mg/L TMP, the rate increased by ~1.7 times from 0.35 min^{-1} to 0.59 min^{-1}. At a lower concentration, 10 mg/L, it increased from 0.82 min^{-1} to 0.95 min^{-1}, whereas for an even lower concentration of TMP, 5 mg/L, the constant was ~1.3 times higher with the addition of ZnO.

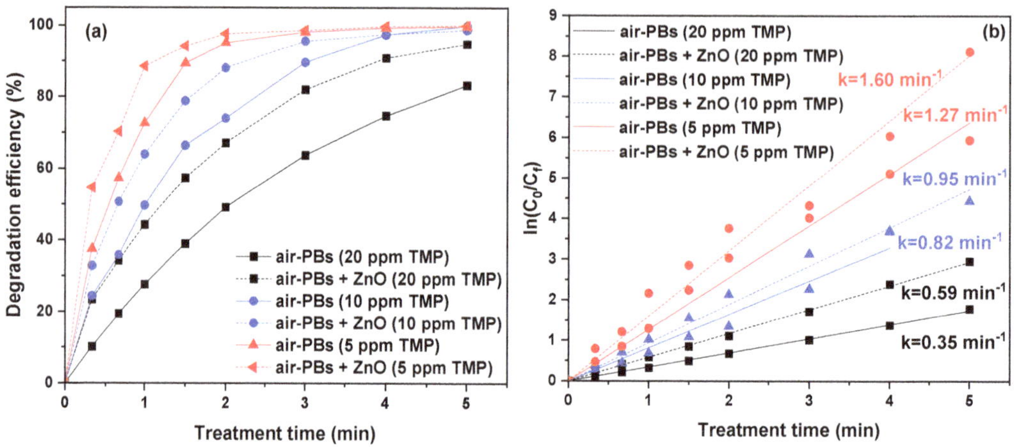

Figure 7. (**a**) TMP degradation as a function of treatment time under air–plasma bubbles at various initial TMP concentrations in water in the presence and absence of ZnO and (**b**) corresponding pseudo-first-order degradation kinetics (pulse voltage: 25.6 kV, pulse frequency: 200 Hz, flow rate: 3 L/min, ZnO loading: 0.2 g/L).

3.6. Process Energy Efficiency

The energy efficiency of a process is a critical factor in determining its broader applicability as a treatment method. One measure used to evaluate process effectiveness is electrical energy per order (E_{EO}), which considers the discharge power, treated solution volume, and treatment time needed to reduce the pollutant concentration by at least one order of magnitude. In this study, the E_{EO} for TMP degradation at two different initial concentrations in water (5 and 20 mg/L) under air– and oxygen–plasma bubbles, in the absence and presence of ZnO, is depicted in Figure 8.

For 20 mg/L TMP, the E_{EO} under air– and O_2–plasma bubbles combined with ZnO was 0.46 and 0.28 kWh/m^3, respectively. The air–plasma bubbles-alone system failed to achieve degradation higher than 90%, while the E_{EO} under O_2–plasma bubbles was also very low (i.e., 0.31 kWh/m^3). Even lower energy requirements were observed for an initial TMP concentration of 5 mg/L under air–plasma bubbles combined with ZnO (0.23 kWh/m^3), which is significant considering its closer resemblance to real-life conditions. The E_{EO} of this study is compared to other AOPs (for TMP degradation) and to other plasma-catalytic studies (for the treatment of water polluted by antibiotics) in Table 2, as this study is the first attempt to combine plasma with catalysis for TMP degradation. Specifically, Chen et al. investigated the degradation of amoxicillin by DBD plasma combined with a $Ce_{0.5}Bi_{0.5}VO_4$ catalyst on a honeycomb ceramic plate, reporting an increase in amoxicillin degradation from 30% (DBD alone) to 65%, with the corresponding E_{EO} being 79.4 kWh/m^3 [48]. Xiao et al. combined DBD with a ZnO/cellulose acetate film catalyst for the degradation of sulfadiazine, reporting an E_{EO} of 15.4 kWh/m^3 [49]. Regarding TMP degradation, the E_{EO} achieved by the combination of plasma bubbles with ZnO is significantly lower compared to other AOPs (Table 2), highlighting the potential of this combination for the rapid and effective degradation of antibiotics in water.

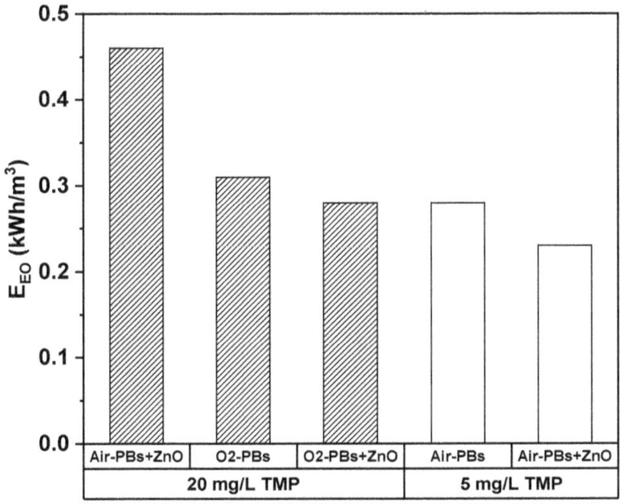

Figure 8. Electrical energy per order (E_{EO}) for the TMP degradation in water under air– and oxygen–plasma bubbles in the absence/presence of ZnO (pulse voltage: 25.6 kV, pulse frequency: 200 Hz, flow rate: 3 L/min, initial TMP concentration: 5 or 20 mg/L, ZnO loading: 0.2 g/L).

Table 2. Comparison of this study with other AOPs for the degradation of antibiotics in water.

Plasma Process	Pollutant	Degradation Efficiency (%)	Treatment Time (min)	E_{EO} (kWh/m^3)	Ref.
DBD plasma + $Ce_{0.5}Bi_{0.5}VO_4$	Amoxicillin 50 mg/L	94.5	30	79.4	[47]
DBD plasma + ZnO on cellulose acetate films	Sulfadiazine 20 mg/L	97.2	60	15.4	[48]
Plasma + graphene–TiO_2–Fe_3O_4	Oxytetracycline 40 mg/L	98.1	60	-	[49]
DBD + Cu-CeO_2@CA	Ciprofloxacin 200 mg/L	89.5	40	-	[50]
UV–chloride (1 mM)	Trimethoprim 2.9 mg/L	91	20	3.6	[51]
Photocatalysis (TiO_2 film irradiated with simulated solar radiation)	Trimethoprim 10 mg/L	90	101.5	8441.4	[52]
Nanopulsed air–plasma bubbles + ZnO	Trimethoprim 20 mg/L	90.9	4	0.46	This study
Nanopulsed O_2–plasma bubbles	Trimethoprim 20 mg/L	94.3	2	0.31	This study
Nanopulsed O_2–plasma bubbles + ZnO	Trimethoprim 20 mg/L	95.5	2	0.28	This study
Nanopulsed air–plasma bubbles + ZnO	Trimethoprim 5 mg/L	94.2	1.5	0.23	This study

3.7. The Role of Plasma Species in the Degradation of TMP by Plasma Bubbles

To understand the mechanism of TMP degradation by plasma bubbles, it was essential to determine the contribution of the main plasma species to its degradation. Therefore, the role of $\cdot OH$, $\cdot O_2^-$, e_{aq}^-, and H_2O_2 was explored by using D-mannitol (D-man), p-benzoquinone (BQ), monopotassium phosphate (MP), and sodium pyruvate (SP), respectively, as suitable scavengers. The selected optimum concentration of each scavenger was the minimum concentration at which no further inhibition of TMP degradation was observed.

It is evident that the presence of all scavengers inhibited the degradation of TMP, indicating that all these species contributed to its degradation (Figure 9a). For instance, after 4 min of plasma bubble treatment, the degradation of TMP decreased from 74.7% (no scavenger) to 69.5%, 59.4%, 39.7%, and 26.7% in the presence of MP, SP, D-man, and BQ, respectively. Therefore, hydrated electrons and H_2O_2 contributed the least to TMP degradation, while $\cdot OH$ proved to play a considerable role in the process, with the most significant contribution coming from $\cdot O_2^-$. Depending on the treatment time, the calculated action efficiency of $\cdot O_2^-$ ranged between ~53 and 74%, while the corresponding action efficiencies of $\cdot OH$, H_2O_2, and e_{aq}^- ranged between ~43 and 70%, ~9 and 26%, and ~3 and 15%, respectively (Figure 9b). The significantly more pronounced role of $\cdot OH$ in the degradation of TMP compared to H_2O_2 justifies the superior plasma-catalytic performance of ZnO compared TiO_2. It is worth reminding here that in the presence of TiO_2, the concentration of H_2O_2 was increased, while the concentration of $\cdot OH$ was decreased compared to plasma alone. Conversely, with ZnO, the presence of this substance led to a decrease in H_2O_2 concentration and an increase in $\cdot OH$ concentration (Figure 3a,c).

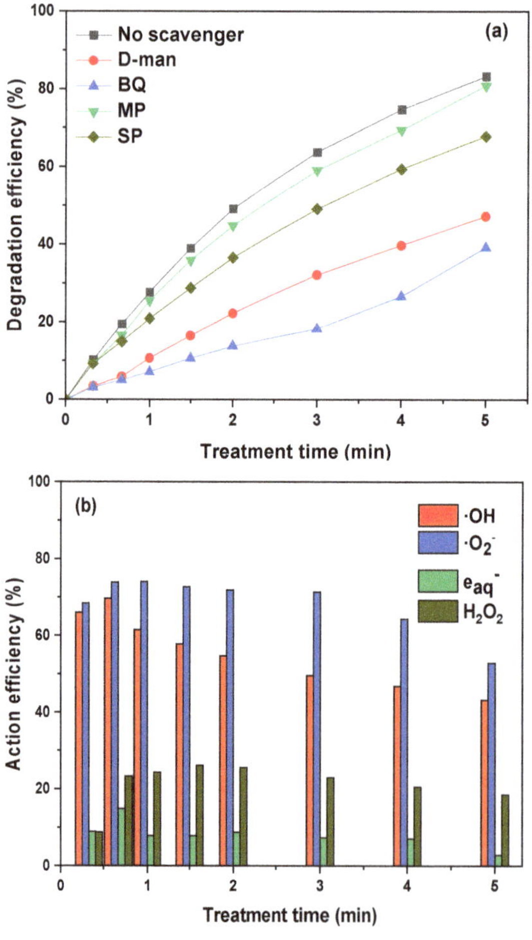

Figure 9. (**a**) Effect of scavengers on TMP degradation by plasma bubbles and (**b**) action efficiency of plasma species (pulse voltage: 25.6 kV, pulse frequency: 200 Hz, plasma gas: air, flow rate: 3 L/min, initial TMP concentration: 20 mg/L, D-man concentration: 6 mM, BQ concentration: 1 mM, MP concentration: 1 mM, SP concentration: 1 mM).

3.8. The Mineralization of TMP under Plasma Bubbles in the Absence and Presence of ZnO

The effectiveness of this remediation approach for TMP degradation was explored in terms of total organic carbon (TOC) removal. The samples tested were from the following systems: air–plasma bubbles, air–plasma bubbles + ZnO, O_2–plasma bubbles, and O_2–plasma bubbles + ZnO for 10, 20, and 30 min of plasma treatment. ZnO was beneficial to TOC removal for both the air–plasma bubbles and O_2–plasma bubbles treatments (Figure 10). In particular, the TOC removal rate for air–plasma bubbles was 27% after 30 min, while in the presence of ZnO, it increased significantly to 35% with a shorter treatment time (i.e., 20 min). In the case of the O_2–plasma bubbles treatment, the TOC removal rate remained stable at ~30% regardless of the treatment time, while the addition of ZnO resulted in a gradual increase with treatment time, reaching an impressive TOC removal rate of 71% after 30 min of treatment. Therefore, ZnO-catalyzed plasma bubbles resulted in the complete removal of the TMP parent molecule but also in noticeable TOC removal.

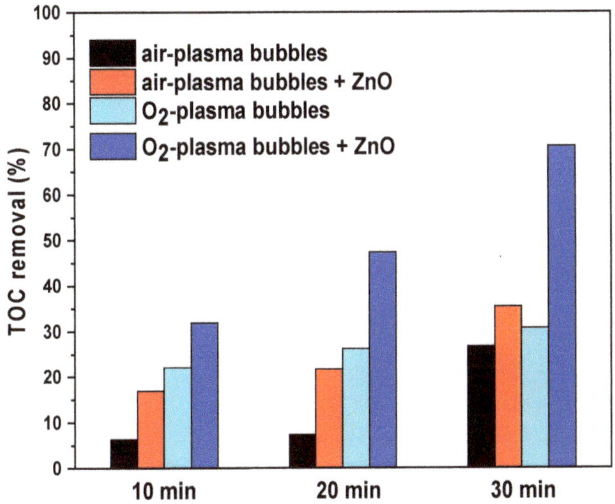

Figure 10. TOC removal during TMP degradation by air– and O_2–plasma bubbles in the presence/absence of ZnO (pulse voltage: 25.6 kV, pulse frequency: 200 Hz, flow rate: 3 L/min, initial TMP concentration: 20 mg/L, ZnO loading: 0.2 g/L).

4. Conclusions

In this study, a novel plasma-catalytic process was proposed for degrading trimethoprim (TMP) in water by combining underwater plasma bubbles with ZnO or TiO_2 photocatalysts. The plasma bubbles were energized by high-voltage nanosecond pulses to enhance the production yield of plasma species. Both catalysts maintained their pore structure after plasma bubble treatment, and the physicochemical properties of water (pH and electrical conductivity) remained unaffected.

ZnO and TiO_2 both facilitated the decomposition of plasma-generated O_3 in water to other oxygen radicals. However, TiO_2 led to a decrease in ·OH concentration and an in-crease in H_2O_2, while ZnO resulted in an increased ·OH concentration and a decrease in H_2O_2. ZnO-catalyzed plasma bubbles outperformed TiO_2-catalyzed ones in TMP degradation, achieving 94.8% degradation compared to 87.4% with TiO_2 after 5 min of treatment. The synergetic factor was 1.70 for ZnO and 1.17 for TiO_2, indicating that the synergy was ~1.5 times higher for ZnO.

TMP degradation was higher under oxygen– and air–plasma bubbles compared to argon–plasma bubbles, with ZnO-catalyzed air–plasma bubbles showing effectiveness across various initial TMP concentrations. The plasma-catalytic synergy was higher in air–plasma bubbles compared to argon–plasma bubbles (with synergistic factors of 1.70 and 1.31, respectively), which aligns with the wavelength of UV light produced by each plasma gas and the UV absorption range of ZnO. $\cdot O_2^-$ and $\cdot OH$ were the primary contributors to TMP degradation, with $\cdot OH$ playing a more significant role than H_2O_2. This justified the superior plasma-catalytic performance of ZnO compared to TiO_2.

The combination of HV nanopulses with plasma bubbles and ZnO proved to be optimal for minimizing energy costs, with the calculated electrical energy per order (E_{EO}) ranging from 0.23 to 0.46 kWh/m^3. The ZnO-catalyzed O_2–plasma bubbles achieved a remarkable 71% removal rate of TOC after 30 min of treatment. These findings underscore the potential of this study in contributing to the development of highly effective processes for antibiotic degradation in water.

Supplementary Materials: The following supporting information can be downloaded at: https://www.mdpi.com/article/10.3390/nano14100815/s1, Figure S1. The XRD patterns of (a) TiO_2 and (b) ZnO before and after plasma bubble treatment (pulse voltage: 25.6 kV, pulse frequency: 200 Hz, plasma gas: air, flow rate: 3 L/min, treatment time: 5 min, catalyst loading: 0.2 g/L). Figure S2. Pore size distribution of (a) ZnO and (b) TiO_2 before and after plasma bubble treatment (pulse voltage: 25.6 kV, pulse frequency: 200 Hz, plasma gas: air, flow rate: 3 L/min, treatment time: 5 min, catalyst loading: 0.2 g/L). Figure S3. TEM images of (a) TiO_2 nanoparticles and (b) ZnO nanopowder. Figure S4. (a) pH and (b) conductivity of plasma-treated water as a function of plasma bubble treatment time in the absence or the presence of catalysts (pulse voltage: 25.6 kV, pulse frequency: 200 Hz, plasma gas: air, flow rate: 3 L/min, catalyst loading: 0.2 g/L).

Author Contributions: C.A.A.: conceptualization, methodology, investigation, visualization, supervision, resources, writing, funding acquisition. D.T.: investigation, visualization, writing—original draft. All authors have read and agreed to the published version of the manuscript.

Funding: This research was funded by the Hellenic Foundation for Research and Innovation (HFRI) under grant agreement No [3560].

Data Availability Statement: No new data were created or analyzed in this study. Data sharing is not applicable to this article.

Conflicts of Interest: The authors declare no conflicts of interest.

References

1. Musie, W.; Gonfa, G. Fresh water resource, scarcity, water salinity challenges and possible remedies: A review. *Heliyon* **2023**, *9*, e18685. [CrossRef] [PubMed]
2. Syeed, M.M.; Hossain, M.S.; Karim, M.R.; Uddin, M.F.; Hasan, M.; Khan, R.H. Surface water quality profiling using the water quality index, pollution index and statistical methods: A critical review. *Environ. Sustain. Indic.* **2023**, *18*, 100247. [CrossRef]
3. Martins, T.S.; Bott-Neto, J.L.; Oliveira, O.N., Jr.; Machado, S.A.S. Paper-based electrochemical sensors with reduced graphene nanoribbons for simultaneous detection of sulfamethoxazole and trimethoprim in water samples. *J. Electroanal. Chem.* **2021**, *882*, 114985. [CrossRef]
4. Yue, X.; Li, Z.; Zhao, S. A new electrochemical sensor for simultaneous detection of sulfamethoxazole and trimethoprim antibiotics based on graphene and ZnO nanorods modified glassy carbon electrode. *Microchem. J.* **2020**, *159*, 105440. [CrossRef]
5. Mpatani, F.M.; Aryee, A.A.; Kani, A.N.; Han, R.; Li, Z.; Dovi, E.; Qu, L. A review of treatment techniques applied for selective removal of emerging pollutant-trimethoprim from aqueous systems. *J. Clean. Prod.* **2021**, *308*, 127359. [CrossRef]
6. Dolar, D.; Vuković, A.; Ašperger, D.; Košutić, K. Efficiency of RO/NF membranes at the removal of veterinary antibiotics. *Water Sci. Technol.* **2012**, *65*, 317–323. [CrossRef] [PubMed]
7. Ma, D.; Yi, H.; Lai, C.; Liu, X.; Huo, X.; An, Z.; Li, L.; Fu, Y.; Li, B.; Zhang, M.; et al. Critical review of advanced oxidation processes in organic wastewater treatment. *Chemosphere* **2021**, *275*, 130104. [CrossRef] [PubMed]
8. Wei, Z.; Liu, J.; Shangguan, W. A review on photocatalysis in antibiotic wastewater: Pollutant degradation and hydrogen production. *Chin. J. Catal.* **2020**, *41*, 1440–1450. [CrossRef]
9. Kyere-Yeboah, K.; Bique, I.K.; Qiao, X.-c. Advances of non-thermal plasma discharge technology in degrading recalcitrant wastewater pollutants. A comprehensive review. *Chemosphere* **2023**, *320*, 138061. [CrossRef] [PubMed]
10. Aggelopoulos, C.A. Recent advances of cold plasma technology for water and soil remediation: A critical review. *Chem. Eng. J.* **2022**, *428*, 131657. [CrossRef]

11. Aggelopoulos, C.; Meropoulis, S.; Hatzisymeon, M.; Lada, Z.G.; Rassias, G. Degradation of antibiotic enrofloxacin in water by gas-liquid nsp-DBD plasma: Parametric analysis, effect of H_2O_2 and CaO_2 additives and exploration of degradation mechanisms. *Chem. Eng. J.* **2020**, *398*, 125622. [CrossRef]
12. Gao, X.; Huang, K.; Zhang, A.; Wang, C.; Sun, Z.; Liu, Y. Simultaneous degradation of glucocorticoids and sterilization using bubbling corona discharge plasma based systems: A promising terminal water treatment facility for hospital wastewater. *Chem. Eng. J.* **2021**, *430*, 132845. [CrossRef] [PubMed]
13. Magureanu, M.; Bilea, F.; Bradu, C.; Hong, D. A review on non-thermal plasma treatment of water contaminated with antibiotics. *J. Hazard. Mater.* **2021**, *417*, 125481. [CrossRef] [PubMed]
14. Mouele, E.S.M.; Tijani, J.O.; Badmus, K.O.; Pereao, O.; Babajide, O.; Fatoba, O.O.; Zhang, C.; Shao, T.; Sosnin, E.; Tarasenko, V.; et al. A critical review on ozone and co-species, generation and reaction mechanisms in plasma induced by dielectric barrier discharge technologies for wastewater remediation. *J. Environ. Chem. Eng.* **2021**, *9*, 105758. [CrossRef]
15. Meropoulis, S.; Giannoulia, S.; Skandalis, S.; Rassias, G.; Aggelopoulos, C.A. Key-study on plasma-induced degradation of cephalosporins in water: Process optimization, assessment of degradation mechanisms and residual toxicity. *Sep. Purif. Technol.* **2022**, *298*, 121639. [CrossRef]
16. Schneider, M.; Rataj, R.; Bláha, L.; Kolb, J.F. Experimental review of different plasma technologies for the degradation of cylindrospermopsin as model water pollutant. *Chem. Eng. J.* **2023**, *451*, 138984. [CrossRef]
17. Zhang, T.; Zhou, R.; Wang, P.; Mai-Prochnow, A.; McConchie, R.; Li, W.; Zhou, R.; Thompson, E.W.; Ostrikov, K.; Cullen, P.J. Degradation of cefixime antibiotic in water by atmospheric plasma bubbles: Performance, degradation pathways and toxicity evaluation. *Chem. Eng. J.* **2021**, *421*, 127730. [CrossRef]
18. Meropoulis, S.; Aggelopoulos, C.A. Plasma microbubbles vs. gas-liquid DBD energized by low-frequency high voltage nanopulses for pollutants degradation in water: Destruction mechanisms, composition of plasma-activated water and energy assessment. *J. Environ. Chem. Eng.* **2023**, *11*, 109855. [CrossRef]
19. Zhou, R.; Zhou, R.; Alam, D.; Zhang, T.; Li, W.; Xia, Y.; Mai-Prochnow, A.; An, H.; Lovell, E.C.; Masood, H.; et al. Plasmacatalytic bubbles using CeO_2 for organic pollutant degradation. *Chem. Eng. J.* **2021**, *403*, 126413. [CrossRef]
20. Ansari, M.; Sharifian, M.; Ehrampoush, M.H.; Mahvi, A.H.; Salmani, M.H.; Fallahzadeh, H. Dielectric barrier discharge plasma with photocatalysts as a hybrid emerging technology for degradation of synthetic organic compounds in aqueous environments: A critical review. *Chemosphere* **2021**, *263*, 128065. [CrossRef] [PubMed]
21. Aggelopoulos, C.A.; Dolinski, O. A comprehensive insight on plasma-catalytic degradation of organic pollutants in water: Comparison between ZnO and TiO_2. *Chemosphere* **2024**, *347*, 140667. [CrossRef] [PubMed]
22. Kim, K.-S.; Yang, C.-S.; Mok, Y.S. Degradation of veterinary antibiotics by dielectric barrier discharge plasma. *Chem. Eng. J.* **2013**, *219*, 19–27. [CrossRef]
23. Liang, J.-P.; Zhou, X.-F.; Zhao, Z.-L.; Yang, D.-Z.; Wang, W.-C. Degradation of trimethoprim in aqueous by persulfate activated with nanosecond pulsed gas-liquid discharge plasma. *J. Environ. Manag.* **2021**, *278*, 111539. [CrossRef] [PubMed]
24. Hatzisymeon, M.; Daletou, M.K.; Rassias, G.; Aggelopoulos, C.A. Degradation of organic pollutants combining plasma discharges generated within soil with TiO_2 and ZnO catalysts: Comparative analysis, optimization and mechanisms. *Sep. Purif. Technol.* **2023**, *320*, 124119. [CrossRef]
25. Hong, J.; Zhang, T.; Zhou, R.; Zhou, R.; Ostikov, K.; Rezaeimotlagh, A.; Cullen, P.J. Plasma bubbles: A route to sustainable chemistry. *AAPPS Bull.* **2021**, *31*, 26. [CrossRef]
26. Li, S.; Timoshkin, I.V.; Maclean, M.; Macgregor, S.J.; Wilson, M.P.; Given, M.J.; Wang, T.; Anderson, J.G. Fluorescence detection of hydroxyl radicals in water produced by atmospheric pulsed discharges. *IEEE Trans. Dielectr. Electr. Insul.* **2015**, *22*, 1856–1865. [CrossRef]
27. O'Sullivan, D.W.; Tyree, M. The kinetics of complex formation between Ti(IV) and hydrogen peroxide. *Int. J. Chem. Kinet.* **2007**, *39*, 457–461. [CrossRef]
28. Jamróz, P.; Pohl, P.; Żyrnicki, W. Spectroscopic evaluation of a low power atmospheric pressure mixed argon–helium microwave induced plasma combined with the chemical generation of volatile species for the optical emission spectrometric determination of arsenic, antimony and mercury. *J. Anal. At. Spectrom.* **2012**, *27*, 1772–1779. [CrossRef]
29. Rezaei, F.; Abbasi-Firouzjah, M.; Shokri, B. Investigation of antibacterial and wettability behaviours of plasma-modified PMMA films for application in ophthalmology. *J. Phys. D Appl. Phys.* **2014**, *47*, 085401. [CrossRef]
30. Rezaei, F.; Nikiforov, A.; Morent, R.; De Geyter, N. Plasma Modification of Poly Lactic Acid Solutions to Generate High Quality Electrospun PLA Nanofibers. *Sci. Rep.* **2018**, *8*, 2241. [CrossRef] [PubMed]
31. Zhou, R.; Zhou, R.; Wang, P.; Luan, B.; Zhang, X.; Fang, Z.; Xian, Y.; Lu, X.; Ostrikov, K.K.; Bazaka, K. Microplasma Bubbles: Reactive Vehicles for Biofilm Dispersal. *ACS Appl. Mater. Interfaces* **2019**, *11*, 20660–20669. [CrossRef] [PubMed]
32. Chang, T.; Shen, Z.; Huang, Y.; Lu, J.; Ren, D.; Sun, J.; Cao, J.; Liu, H. Post-plasma-catalytic removal of toluene using MnO_2–Co_3O_4 catalysts and their synergistic mechanism. *Chem. Eng. J.* **2018**, *348*, 15–25. [CrossRef]
33. Wu, K.; Sun, Y.; Liu, J.; Xiong, J.; Wu, J.; Zhang, J.; Fu, M.; Chen, L.; Huang, H.; Ye, D. Nonthermal plasma catalysis for toluene decomposition over $BaTiO_3$-based catalysts by Ce doping at A-sites: The role of surface-reactive oxygen species. *J. Hazard. Mater.* **2021**, *405*, 124156. [CrossRef] [PubMed]
34. Yu, X.; Li, S.; Dang, X.; Wang, H.; Qu, J.; Zheng, H. Facile fabrication of three-dimensional MnO_2 for trichloroethylene degradation by plasma catalysis. *Sep. Purif. Technol.* **2023**, *325*, 124680. [CrossRef]

35. Deng, Y.; Zhao, R. Advanced Oxidation Processes (AOPs) in Wastewater Treatment. *Curr. Pollut. Rep.* **2015**, *1*, 167–176. [CrossRef]
36. Huang, C.P.; Dong, C.; Tang, Z. Advanced chemical oxidation: Its present role and potential future in hazardous waste treatment. *Waste Manag.* **1993**, *13*, 361–377. [CrossRef]
37. Marotta, E.; Ceriani, E.; Schiorlin, M.; Ceretta, C.; Paradisi, C. Comparison of the rates of phenol advanced oxidation in deionized and tap water within a dielectric barrier discharge reactor. *Water Res.* **2012**, *46*, 6239–6246. [CrossRef] [PubMed]
38. Kovačević, V.; Dojcinovic, B.; Jović, M.; Roglic, G.; Obradović, B.; Kuraica, M. Measurement of reactive species generated by dielectric barrier discharge in direct contact with water in different atmospheres. *J. Phys. D Appl. Phys.* **2017**, *50*, 19. [CrossRef]
39. Gong, S.; Sun, Y.; Zheng, K.; Jiang, G.; Li, L.; Feng, J. Degradation of levofloxacin in aqueous solution by non-thermal plasma combined with Ag_3PO_4/activated carbon fibers: Mechanism and degradation pathways. *Sep. Purif. Technol.* **2020**, *250*, 117264. [CrossRef]
40. Guo, H.; Jiang, N.; Wang, H.; Lu, N.; Shang, K.; Li, J.; Wu, Y. Pulsed discharge plasma assisted with graphene-WO3 nanocomposites for synergistic degradation of antibiotic enrofloxacin in water. *Chem. Eng. J.* **2019**, *372*, 226–240. [CrossRef]
41. Guo, H.; Jiang, N.; Wang, H.; Shang, K.; Lu, N.; Li, J.; Wu, Y. Enhanced catalytic performance of graphene-TiO_2 nanocomposites for synergetic degradation of fluoroquinolone antibiotic in pulsed discharge plasma system. *Appl. Catal. B Environ.* **2019**, *248*, 552–566. [CrossRef]
42. Li, Y.; Xie, W.; Hu, X.; Shen, G.; Zhou, X.; Xiang, Y.; Zhao, X.; Fang, P. Comparison of Dye Photodegradation and its Coupling with Light-to-Electricity Conversion over TiO_2 and ZnO. *Langmuir* **2010**, *26*, 591–597. [CrossRef]
43. Liao, Y.; Xie, C.; Liu, Y.; Chen, H.; Li, H.; Wu, J. Comparison on photocatalytic degradation of gaseous formaldehyde by TiO_2, ZnO and their composite. *Ceram. Int.* **2012**, *38*, 4437–4444. [CrossRef]
44. Sakthivel, S.; Neppolian, B.; Shankar, M.V.; Arabindoo, B.; Palanichamy, M.; Murugesan, V. Solar photocatalytic degradation of azo dye: Comparison of photocatalytic efficiency of ZnO and TiO_2. *Solar Energy Mater. Sol. Cells* **2003**, *77*, 65–82. [CrossRef]
45. Chantes, P.; Jarusutthirak, C.; Danwittayakul, S. A comparison study of photocatalytic activity of TiO_2 and ZnO on the degradation of real batik wastewater. In Proceedings of the International Conference on Biological, Environment and Food Engineering (BEFE-2015), Singapore, 15–16 May 2015.
46. Murugesan, P.; Evanjalin Monica, V.; Moses, J.A.; Anandharamakrishnan, C. Water decontamination using non-thermal plasma: Concepts, applications, and prospects. *J. Environ. Chem. Eng.* **2020**, *8*, 104377. [CrossRef]
47. Chen, Y.; Sun, X.; Zheng, L.; Liu, Y.; Zhao, Y.; Huang, S.; Li, S. Synergistic catalysis degradation of amoxicillin by DBD plasma-catalyst system constructed by DBD plasma and $Ce_{0.5}Bi_{0.5}VO_4$/HCP coating. *Process Saf. Environ. Prot.* **2024**, *181*, 416–428. [CrossRef]
48. Xiao, S.; Shen, Z.; Song, S.; Han, S.; Du, Y.; Wang, H. Enhanced sulfadiazine degradation in a multi-electrode paralleling DBD plasma system coupled with ZnO/cellulose acetate films. *J. Environ. Chem. Eng.* **2023**, *11*, 109063. [CrossRef]
49. Guo, H.; Wang, Y.; Yao, X.; Zhang, Y.; Li, Z.; Pan, S.; Han, J.; Xu, L.; Qiao, W.; Li, J.; et al. A comprehensive insight into plasma-catalytic removal of antibiotic oxytetracycline based on graphene-TiO_2-Fe_3O_4 nanocomposites. *Chem. Eng. J.* **2021**, *425*, 130614. [CrossRef]
50. Hu, K.; Xie, Q.; Wang, H.; Zhang, B.; Huang, Y.; Song, S.; Zhang, H.; Ding, Y.; Huang, H.; Wu, C. Synergistic catalysis of Cu-CeO_2@CA composite film in a circulating DBD plasma system and its effect on ciprofloxacin degradation. *Chem. Eng. J.* **2023**, *455*, 140895. [CrossRef]
51. Ye, W.-K.; Tian, F.-X.; Chen, C.; Ye, J.; Liu, F.-w.; Wang, B.; Hu, X.-J.; Xu, B. Performance evaluation of the UV activated chlorite process on trimethoprim: Degradation efficiency, energy consumption and disinfection by-products formation. *Chemosphere* **2023**, *327*, 138540. [CrossRef] [PubMed]
52. Ljubas, D.; Juretić, H.; Badrov, A.; Biošić, M.; Babić, S. Photocatalytic Degradation of Pharmaceutical Trimethoprim in Aqueous Solution over Nanostructured TiO_2 Film Irradiated with Simulated Solar Radiation. *Appl. Sci.* **2023**, *13*, 5681. [CrossRef]

nanomaterials

MDPI

Article

Preparation of Heterogeneous Fenton Catalysts Cu-Doped MnO_2 for Enhanced Degradation of Dyes in Wastewater

Xiao Liu [1,2,3,*], Lu Wang [1], Jiran Li [1], Rong Li [1], Runze He [1], Wanglong Gao [1] and Neng Yu [4]

1 School of Materials Science & Engineering, North Minzu University, Yinchuan 750021, China; luwang111222333@163.com (L.W.); 18314293192@163.com (J.L.); lr157821@163.com (R.L.); 15591779243@163.com (R.H.); chenyielegant@163.com (W.G.)
2 Key Laboratory of Polymer Materials and Manufacturing Technology, North Minzu University, Yinchuan 750021, China
3 International Scientific and Technological Cooperation Base of Industrial Waste Recycling and Advanced Materials, Yinchuan 750021, China
4 Huadian Electric Power Research Institute Company, Hangzhou 310012, China; neng-yu@chder.com
* Correspondence: plzxlx@163.com

Abstract: Herein, a series of heterogeneous Fenton catalysts, Cu doped MnO_2 (CDM), with different Cu/Mn molar ratios were prepared via a hydrothermal reaction. Meanwhile, detailed characterizations were used to study the structures of CDM, and it is amazing that the morphology of CDM changed from nanowires to nanoflowers with an increasing amount of Cu doped. Apart from this, both the specific surface area and oxygen vacancy increased obviously with the increasing Cu/Mn molar ratio. Then, the degradation of different dyes was utilized to evaluate the catalytic activity of different CDM with H_2O_2 used as the oxidizing agent, and the 50%-CDM with the highest content of Cu doped displayed the best catalytic activity. Herein, the degradation efficiency (D%) of Congo red (CR) solution with low concentration (60 mg/L) reached 100% in 3 min, while the D% of CR solution with a high concentration (300 mg/L) reached 99.4% after 5 min with a higher dosage of H_2O_2. Additionally, the 50%-CDM also displayed excellent reusability, for which the D% values were still higher than 90% after the 14th cycles. Based on the structure characteristics and mechanism analysis, the excellent catalytic capacity of 50%-CDM was due to the combined influence of large specific surface area and abundant oxygen vacancy. Thus, a promising heterogeneous Fenton catalyst was developed in this study, which proved the treatment efficiency of actual dye wastewater.

Keywords: Cu-doped MnO_2; nanocatalyst; Congo red degradation; heterogeneous Fenton reaction

1. Introduction

With the continuous development of industrialization, a large amount of dye wastewater containing refractory organics is produced. It has been shown that more than 15% of dyes are discharged into the natural environment directly, and these have brought serious environmental problems and health risks [1,2]. In particular, some high-use dyes, such as CR, methyl blue (MBE), rhodamine B (Rh B), and methylene blue (MB), have more stable, nonbiodegradable, aromatic molecular structures that pose a serious threat to the environment [3]. Thus, the development of efficient and sustainable technology for dye wastewater treatment is very urgent [4,5]. Among the various technologies utilized until now, the heterogeneous Fenton reaction is regarded as an efficient and promising approach. The heterogeneous Fenton reaction is a kind of advanced oxidation process (AOP) [6], which can generate •OH and $•O_2^-$ from H_2O_2 under the catalysis of nanocatalysts and can cause the degradation of organic pollutants further [7–9]. Compared with the traditional homogeneous Fenton reaction, the heterogeneous Fenton reaction displays smaller quantities of sludge, wider pH applicability, higher catalytic performance, and so on, for which it has attracted much attention [10,11].

In the development of a heterogeneous Fenton reaction, the key problem is to develop high-activity nanocatalysts. MnO_2, as a metal oxide, has been used widely as a nanocatalyst [12] due to its unique properties such as multiple valences of manganese, low toxicity, large surface area, primary adaptability, strong adsorption, and catalytic performance [13,14]. For example, Li et al. prepared highly porous α-MnO_2 nanorods by selective acid etching from Mn-containing spinel, and 90.9% of 4-chlorophenol can be degraded within 12 min by catalytic ozonation in a wide range of pH of 4.5–10.5 [15]. Wang et al. synthesized three-dimensional hierarchical (3D) MnO_2 via the calcination of hydrothermal products and utilized MnO_2 to degrade acid orange 7, and the 3D MnO_2 delivered great catalytic performance due to its high specific surface area [16]. Lu et al. prepared MnO_2 with different morphologies, including nanorod, nanoflower, nanowire, and nanourchin, and the results proved that the nanoflower MnO_2 displayed the highest catalytic activity. Therefore, this is more due to better oxidative and easier oxygen migration rather than just a high specific surface area [17]. He et al. evaluated the catalytic activity of MnO_2 with different crystal phases and the following results were obtained: α-MnO_2 > β-MnO_2 > γ-MnO_2 [18]. However, some different results about the structure–activity relationship of MnO_2 have been reported in other studies [1,19,20], so He et al. indicated that it is still hard to confirm the remarkable factors influencing the catalytic activity of MnO_2 due to the different preparation methods of catalysts [18].

However, the results obtained in previous studies indicated that the crystalline phase and morphologies played important roles in the catalytic activity of MnO_2. As a result, MnO_2 nanocatalysts with higher activity could be produced by novel preparation methods, and intercalation chemistry is a highly effective method to improve the catalytic performance of MnO_2 [21]. Wang et al. compared the activity of a Co(II)-intercalated δ-MnO_2 (Co-δ-MnO_2) catalyst and original δ-MnO_2 catalyst; the degradation efficiency of δ-MnO_2/PMS was 60.1%, while that of Co-δ-MnO_2/PMS reached 100% in the same reaction conditions [22]. Yu et al. synthesized Cu-doped δ-MnO_2@diatomite to degrade the MB, and the catalyst had excellent oxidation ability to dissociate H_2O_2 to a hydroxyl radical [23]. Yang et al. prepared metal-doped amorphous MnO_2 (M-AMO, M = Fe, Co, Ni, and Cu) for organic oxidation with PMS, and the Cu-AMO displayed the highest activity compared to the other metal-doped MnO_2, for which the rate constant reached 3.5 times as high as that of pure MnO_2 [24]. According to the above reports, the high catalytic activity could be attributed to two respects: one is that the metal doping had a significant impact on the surface electron and charge transfer by the creation of more oxygen vacancies, the other is that the change in catalyst surface morphology promotes the interaction between the catalyst, the oxidizing species, and the organic pollutant. Although many studies about metal-doped MnO_2 have been investigated, deep research about the effect of Cu doping on the structure and catalytic activity of MnO_2 has been reported rarely, and the mechanism of its structure–activity relationship also needs to be analyzed further.

In this study, various CDM with different contents of Cu were prepared by a simple hydrothermal reaction, and a series of characterizations were utilized to investigate the structures of CDM. To evaluate the catalytic activity of CDM, a CR degradation reaction was taken out by a heterogeneous Fenton reaction with activating H_2O_2. Finally, the catalytic mechanisms of CDM were discussed according to the experimental results of a free radical scavenging test and EPR measurement. On account of the large specific surface area and abundant oxygen vacancy, the Cu-doped MnO_2 displayed an outstanding degradation capacity of CR. Thus, a heterogeneous catalyst with high catalytic activity and good operation performance was obtained herein, which proved the treatment efficiency of a catalytic oxidation system for the actual dye wastewater.

2. Materials and Methods

2.1. Materials

Potassium permanganate ($KMnO_4$) and acetic acid were purchased from Shanghai Wokai Biotechnology Co. LTD (Shanghai, China), copper sulfate ($CuSO_4$), ammonia, CR, methyl blue (MBE), rhodamine B (Rh B), and methylene blue (MB) were obtained from Guangzhou Chemical Reagent Factory (Guangzhou, China). All other chemicals used in the experiments were purchased from China National Pharmaceutical Group Corporation (Beijing, China), and all of the reagents were chemical grade.

2.2. Synthesis Procedures

A simple hydrothermal reaction was utilized to prepare a CDM nanocatalyst according to our previous study [12]. Firstly, 0.6 g of $KMnO_4$ was dissolved into 30 mL 0.4 M acetic acid with stirring at room temperature and a certain amount of $CuSO_4 \cdot 5H_2O$ was dissolved in 10 mL of distilled water. Then, the two solutions were mixed thoroughly and transferred to a 50 mL Teflon-lined autoclave with 40 μL ammonia added. Finally, the CDM nanocatalyst with different Cu/Mn molar ratios was obtained after a hydrothermal reaction at 140 °C for 12 h. Additionally, a pure MnO_2 nanocatalyst was also prepared according to the above method without $CuSO_4 \cdot 5H_2O$. Thus, a series of nanocatalysts with different compositions were prepared, as shown in Table 1.

Table 1. Series of nanocatalysts.

Samples	0%-CDM	10%-CDM	20%-CDM	30%-CDM	40%-CDM	50%-CDM
Cu:Mn	1:0	1:10	1:5	1:3.3	1:2.5	1:2

Note: Cu:Mn means the Cu/Mn molar ratio.

2.3. Characterization

The morphologies and elemental mapping of series catalysts were observed by scanning electron microscopy (SEM, Zeiss, sigma500, Jena, Germany) with an Oxford Ultim Max Large Area SDD EDS detector. The size of catalysts was measured by a laser particle size analyzer (Malvern, MS-2000, Malvern, UK). The crystallographic structures of catalysts were characterized by a high-resolution transmission electron microscope (HRTEM, JEM-2100, Tokyo, Japan) and X-ray diffraction (XRD, Rigaku Corporation, XRD-6000, Tokyo, Japan). The specific surface area and pore size distribution of samples were analyzed by Brunauer–Emmett–Teller (BET) and Barrett–Joyner–Halenda (BJH) methods, respectively. A Fourier-transform infrared spectrophotometer (FT-IR, American Nicolet Corp. Model 170-SX, Green Bay, WI, USA) was used to investigate the chemical structures of the catalyst. X-ray photoelectron spectroscopy (XPS, Thermo Fisher Scientific, ESCALAB250Xi, Waltham, MA, USA) was utilized to study the surface chemical composition of catalysts. To study the catalytic mechanism of CDM, an electron paramagnetic resonance spectrometer (Bruker ELEXSYS E500, Karlsruhe, Germany) was used to obtain the EPR signals. The leaching of Cu and Mn in wastewater was detected by an inductively coupled plasma spectrometer (Agilent 5800 ICP-OES, Santa Clara, CA, USA).

2.4. Heterogeneous Fenton Degradation of Dyes

To study the catalytic ability of CDM that were prepared, dye degradation experiments were taken out, and different dyes were selected, including MBE, CR, Rh B, and MB. In a typical experiment, 0.05 g of CDM and 2 mL of 30% H_2O_2 were added into 50 mL of dye solution (60 mg/L), and the reaction was taken out in a shaker (150 rpm/min) for a certain time. When the degradation reaction finished, the remaining dye concentration was calculated by measuring the absorbance of dye solution using a UV-vis spectrophotometer at a certain wavelength (668 nm for MBE, 498 nm for CR, 552 nm for Rh B, and 664 nm for MB). And D% was calculated using the following equations [12]:

$$D\% = \frac{C_o - C_t}{C_t} \times 100 \quad (1)$$

Herein, C_0 and C_t (mg/L) are the initial and instantaneous concentrations of dye wastewater, respectively. All of the degradation experiments were tested in triplicate to reduce experimental error.

To investigate the optimal conditions for degradation reactions, the influences of the initial CR concentration, H_2O_2 dosage, pH of the solution, and temperature on the degradation of CR were investigated. Additionally, the reusability of catalysts was also studied as follows: after one CR degradation reaction, the catalyst was separated and used for the subsequent CR degradation experiment without any treatment. Finally, the D% of each degradation experiment was calculated.

2.5. Kinetics of Dye Degradation

In order to better analyze the kinetics for the removal of CR, the pseudo-first-order and pseudo-second-order kinetic models were utilized to analyze the data, using the following Equations (2) and (3):

$$\log(Q_{e1} - Q_t) = logQ_{e1} - \frac{k_1 t}{2.303} \quad (2)$$

$$\frac{t}{Q_t} = \frac{1}{k_2 Q_{e2}^2} + \frac{t}{Q_{e2}} \quad (3)$$

Herein, Q_e (mg/g) and Q_t (mg/g) mean the CR removal amount at equilibrium and time t (min); k_1 and k_2 mean the kinetic constant of pseudo-first-order kinetic models and pseudo-second-order kinetic models, respectively.

3. Results and Discussion

3.1. Characterization of CDM

The morphologies of CDM catalysts were investigated and are shown in Figure 1. It can be seen that 0%-CDM contained nanowires with a diameter of about 20 nm, but the structure of the nanowires began to become irregular when Cu was introduced, just as shown in Figure 1b. When the Cu/Mn molar ratio increased to 20%, there were some irregular clumps that appeared in addition to the nanowires, and these clumps were bound to the nanowires. More surprisingly, a new flower-like morphology consisting of shorter nanowires, generated as the Cu/Mn molar ratio, increased to 30%. As the content of Cu increased continually, the flower-like morphology became more regular and the "petal" parts appeared thinner, and there were no obvious nanowires that could be seen in the flower-like nanocatalyst (Figure 1e,f). In addition, the EDS elemental maps proved that O, Mn, and Cu elements were evenly distributed in the CDM and the content of Cu in catalysts increased significantly from 10%-CDM to 50%-CDM, and the concrete content of Cu is given in Table S1. Therefore, the morphologies of nanocatalysts were affected greatly by the copper doping.

In order to investigate the morphologies and crystal structures of CDM, TEM and SAED were measured and are given in Figure 2. In terms of morphologies, the results are consistent with the previous SEM characterization results. With increasing Cu content, the morphology of CDM changes from nanowires to flower-like materials, and the greater the content of Cu, the more regular the shape of the flower-like material. In Figure 2(a3,b3), it is seen that the resolved lattice spacings of 0.69 nm, 0.48 nm, and 0.21 nm belong to the crystal plane (110), (200), (301) of α-MnO_2. In addition to the lattice structure of α-MnO_2, the crystal plane structure of CuO was also observed in Figure 2(c3–f3) with an increasing content of Cu; for example, a lattice spacing of 0.14 nm corresponded to CuO (310) and a lattice spacing of 0.25 nm corresponded to CuO (002) [25]. Additionally, some diffraction points were detected by SAED characterizations and are shown in Figure 2(a4–c4), confirming that the 0%-CDM, 10%-CDM, and 20%-CDM are single crystal structures. As shown in Figure 2(d4), both diffraction points and diffraction rings were observed, which means a mixed state of single crystals and polycrystals was formed in 30%-CDM. Furthermore, only diffraction rings appeared in Figure 2(e4,f4), and this proved that only polycrystals were generated in 40%-CDM and 50%-CDM. Moreover, the diffraction rings of 40%-CDM and 50%-CDM displayed a little fuzz, which proves that there were also amorphous structures in these two nanocatalysts, aside from polycrystalline structures. Likewise, the similar amorphous structures also could be found in the thin "petal" parts according to the HRTEM images in Figure 2(e3,f3). Based on the above results, the copper was successfully doped into MnO_2 and coexisted with MnO_2 as CuO, and the crystal structure of the catalyst changed from a single crystal to the coexistence of polycrystalline and amorphous structures with an increasing content of copper.

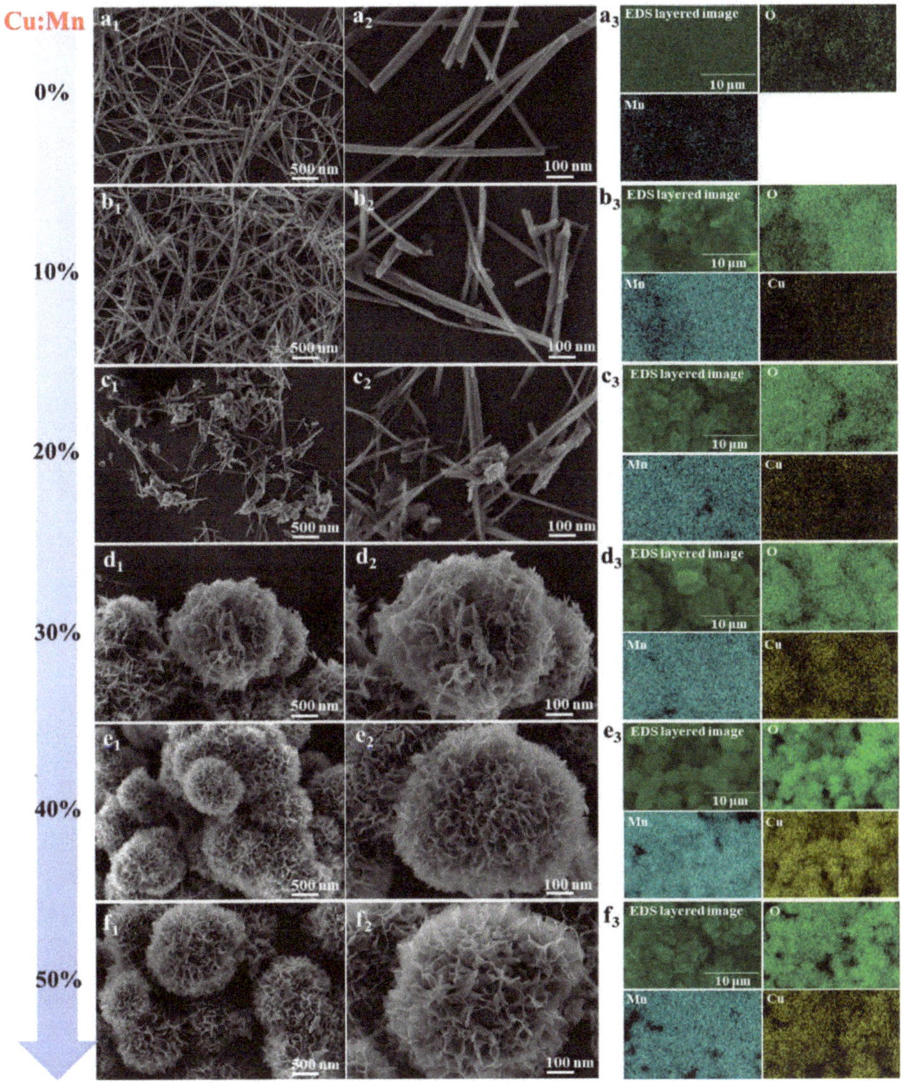

Figure 1. SEM images and EDS elemental maps of CDM catalysts: (**a**) 0%-CDM, (**b**) 10%-CDM, (**c**) 20%-CDM, (**d**) 30%-CDM, (**e**) 40%-CDM, and (**f**) 50%-CDM.

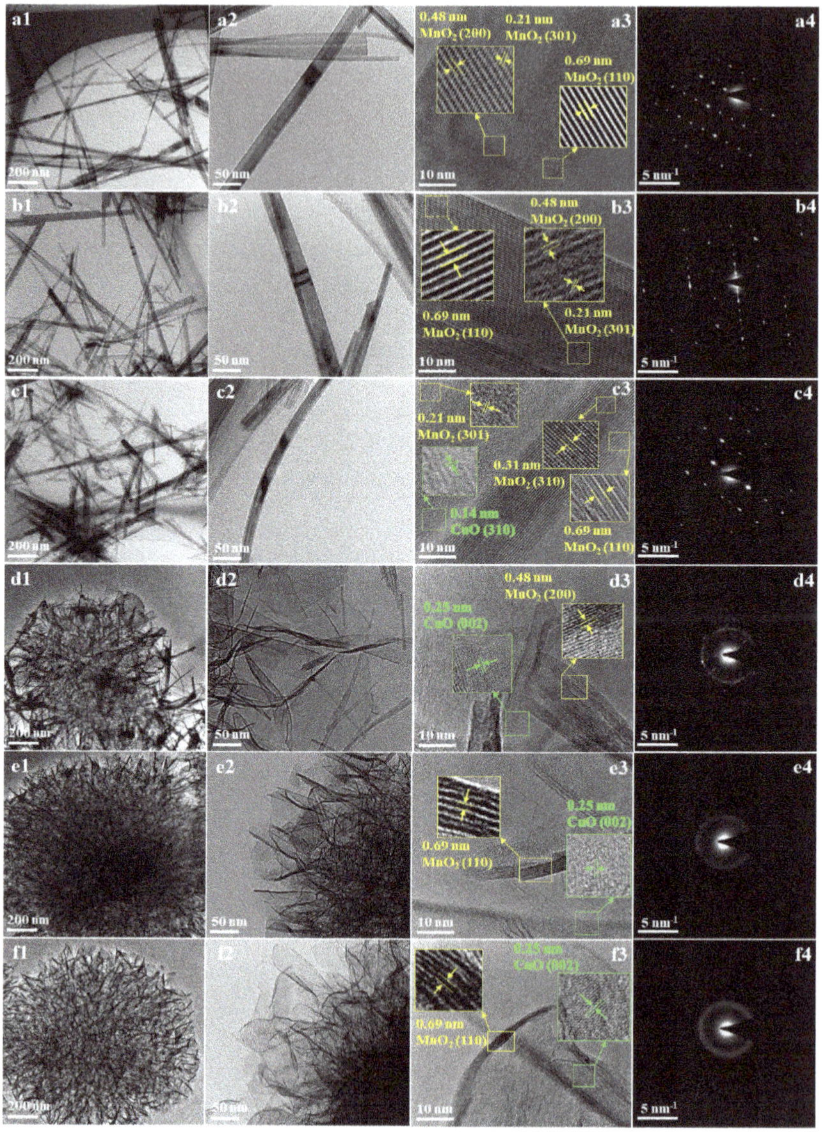

Figure 2. TEM images (**a1–f1**, **a2–f2**), HRTEM images (**a3–f3**) and SAED patterns (**a4–f4**) of CDM catalyst: (**a**) 0%-CDM, (**b**) 10%-CDM, (**c**) 20%-CDM, (**d**) 30%-CDM, (**e**) 40%-CDM, (**f**) 50%-CDM.

A laser particle size analyzer was used to characterize the size of CDM and the results are given in Figure 3a. Among them, three samples, containing 0%-CDM, 40%-CDM, and 50%-CDM, showed unimodal particle size distribution, while the other three samples (10%-CDM, 20%-CDM, and 30%-CDM) showed bimodal particle size distribution. These results were consistent with the morphological characterization from SEM and TEM, with the content of doped copper being increased. Some irregular mixed nanomaterials were formed during the transformation of CDM from nanowires to nanoflowers. For the regular nanowires (0%-CDM) and regular nanoflowers (40%-CDM and 50%-CDM), the particle size distributions were relatively uniform.

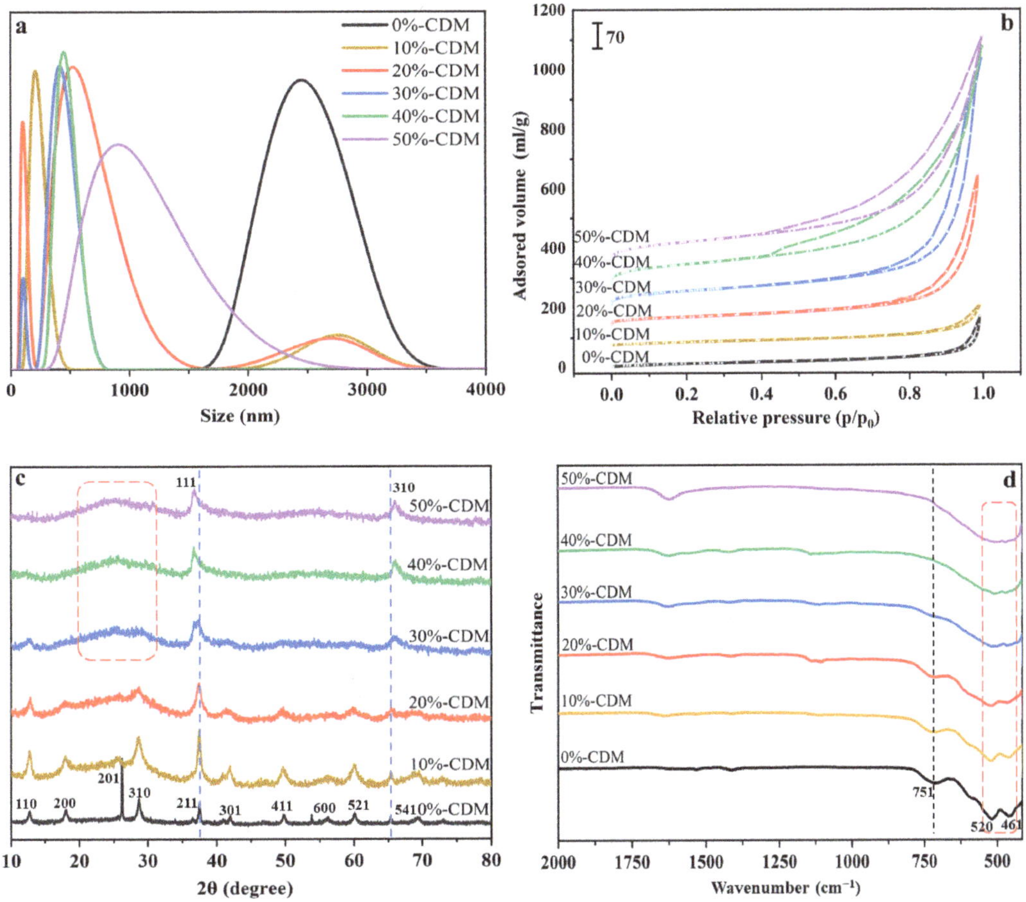

Figure 3. (**a**) Size distribution curve, (**b**) N_2 adsorption–desorption isotherms, (**c**) XRD patterns, and (**d**) FT-IR spectra of as-prepared CDM.

The specific surface area and pore size distributions of series CDM were investigated and are shown in Figure 3b and Figure S1, respectively. According to Figure 3b, all six kinds of catalysts displayed typical type IV isotherms with an evident H3-type hysteresis loop, which is a typical feature of mesoporous materials [20,26]. Moreover, type I adsorption isotherms could be found in the isotherms of 20%-CDM, 30%-CDM, 40%-CDM, and 50%-CDM, for which a steep increase in the isotherm at a low pressure range ($P/P_0 = 10^{-7}$–0.01) was detected, so some micropores were proved to exist in these catalysts [27]. Remarkably, hysteresis loops of the catalysts became larger with increasing contents of doped Cu, demonstrating increased mesopore porosity. Combined with the S_{BET} shown in Table 2, 50%-CDM really displayed the highest specific surface area, which was as high as 259.89 m^2/g, while that of 0%-CDM was only 45.85 m^2/g. In addition, the total pore volume of the catalyst became larger after the Cu doping; therefore, more available reactive region space and more effective substrate transportation could be provided, contributing to enhanced catalytic activity [22,28]. Based on the data shown in Table 3 and Figure S1, the most probable particle size Dp* also changed obviously with the doping of Cu; 20%-CDM and 30%-CDM displayed large Dp* (18.00 and 22.05 nm), while the other catalysts displayed smaller Dp* values. The change in pore size is due to the change in crystalline grain stacking structure during the formation of nanocatalysts.

Table 2. Texture parameters from BET and BJH method.

Samples	S_{BET} (m^2/g)	Vp (cm^3/g)	Dp (nm)	Dp* (nm)
0%-CDM	45.85	0.26	16.50	2.04
10%-CDM	57.12	0.22	11.97	2.04
20%-CDM	107.83	0.78	20.90	18.00
30%-CDM	176.89	1.28	20.05	22.05
40%-CDM	243.79	0.93	8.90	3.62
50%-CDM	259.89	1.18	11.37	3.55

Note: S_{BET} means the specific surface area calculated by BET multi-point method; Vp, Dp, and Dp* were calculated by the BJH method, Vp means the total pore volume, Dp means the mean diameter, and Dp* means the most probable particle size, which corresponds to the peak position of the pore size distribution curve.

Table 3. Kinetic parameters of CR degradation.

Model	Constant	0 mL H_2O_2	6 mL H_2O_2
Pseudo-first-order model	R^2	0.9448	0.9608
	$Q_{e\ cal}$ (mg g^{-1})	86.90	146.28
	K_1 (min^{-1})	0.0290	0.0210
Pseudo-second-order model	R^2	0.9958	0.9999
	$Q_{e\ cal}$ (mg g^{-1})	175.44	294.12
	K_2 (min^{-1})	0.0054	0.0033

Note: $Q_{e\ cal}$ means the maximum removal amount calculated by kinetic models.

Figure 3c showed the XRD patterns of series CDM. For the pure α-MnO_2 (0%-CDM), the typical diffraction peaks of crystalline α-MnO_2 could be found at 2θ = 12.8, 18.1, 26.2, 28.7, 37.6, 41.8, 49.8, 56.4, 60.4, and 69.5, which corresponded to the (110), (200), (201), (310), (211), (301), (411), (600), (521), and (541) planes (PDF#00-044-0141) [29,30]. After being Cu doped, the diffraction peaks of crystalline α-MnO_2 became weaker, and part of the diffraction peaks of α-MnO_2 even disappeared when the content of Cu was high enough, such as in 40%-CDM and 50%-CDM. Meanwhile, the characteristic diffraction peaks of CuO appeared obviously at 36.7 and 66.4, corresponding to the (111) and (310) planes (PDF#00-048-1548) [25,31]. In addition to the above sharp crystal diffraction peaks, an amorphous structure also appeared with an increasing content of copper, especially for 50%-CDM, 40%-CDM, and 30%-CDM. Coincidentally, the same results were also obtained in the SAED patterns before.

The chemical structures of catalysts were tested by FT-IR and are given in Figure 3d. Herein, three characteristic absorption peaks could be found at 715 cm^{-1}, 520 cm^{-1}, and 461 cm^{-1}, which corresponded to the stretching vibrations of Mn-O [18]. With an increasing content of Cu, the peak at 715 cm^{-1} disappeared and a new broad peak appeared in the range of 430–550 cm^{-1}, especially in the spectra of 40%-CDM and 50%-CDM, which could be due to the Cu-O bonds formed in the catalysts [32,33]. In addition, the peak at 1625 cm^{-1} was attributed to the -OH group combined on the surface of catalysts.

XPS measurement was used to study the surface constituents of CDM. According to the XPS broad survey spectra shown in Figure 4a, Mn and O could be found in all of the six CDM, while the characteristic peak of Cu, at the range of 933.9 eV to 962.8 eV, appeared gradually with the increasing Cu/Mn molar ratio; this is to say, it could be found in the survey scan of 20%-CDM, 30%-CDM, 40%-CDM, and 50%-CDM, but not in 0%-CDM and 10%-CDM. Figure 4b shows the spectrum of Mn 2p in 50%-CDM. The peaks appearing at 653.0 eV and 641.3 eV correspond to Mn $2p_{1/2}$ and Mn $2p_{3/2}$ [34,35], and the spin separation energy with 11.7 eV indicates the existence of the MnO_2 phase in 50%-CDM. The Cu 2p spectrum of 50%-CDM is given in Figure 4c, and two obvious peaks appeared at 953.8 eV and 933.9 eV, corresponding to Cu $2p_{1/2}$ and Cu $2p_{3/2}$ [36,37], respectively. Additionally, two associated satellite peaks were found at 962.0 eV and 943.2 eV, which proved the presence of Cu^{2+} [25,38]. The O 1s spectrum of 50%-CDM is shown in Figure 4d. Here, two characteristic peaks could be found at 529.6 eV and 531.2 eV, corresponding to lattice O and adsorbed O [39], respectively. Furthermore, the HR-XPS spectra of O 1s for series CDM are illustrated in Figure S2, and the contents of lattice O and adsorbed O in different CDM are given in Table S2. It could be found that the contents of adsorbed O increased obviously with an increasing content of Cu, which signified that greater oxygen vacancy was formed [40], and that the content of oxygen vacancy would affect the catalytic activity of CDM significantly [20]. In particular, the content of adsorbed O in 50%-CDM was a little lower than that in 40%-CDM, meaning that the oxygen vacancy would not be increased if the content of Cu increased continually.

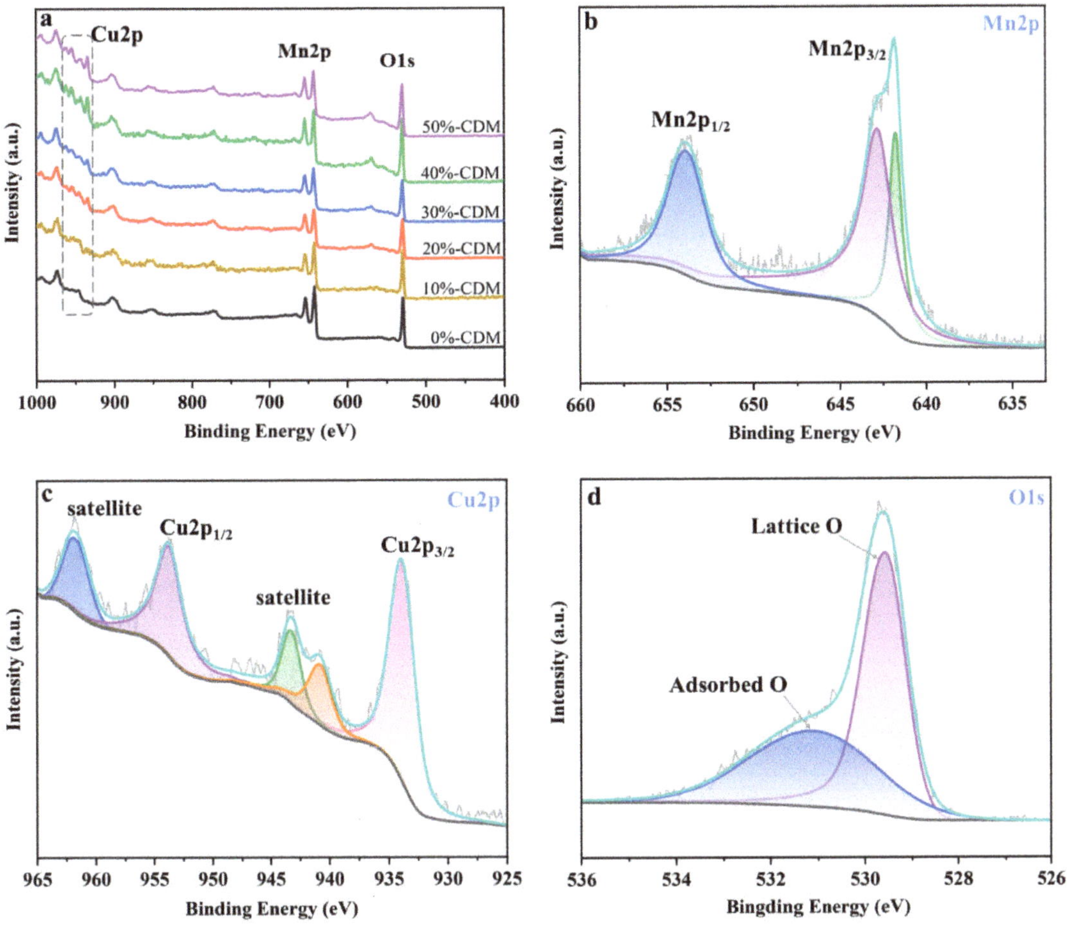

Figure 4. XPS spectra of CDM: (**a**) six CDM; (**b**) Mn 2p on 50%-CDM; (**c**) Cu 2p on 50%-CDM; (**d**) O 1s on 50%-CDM.

3.2. Catalytic Performance of CDM

The series of CDM were used to catalytically degrade CR, and the results are shown in Figure 5a. Firstly, it could be found that the degradation efficiency was as low as 15.7% when there was only H_2O_2 added without any catalyst. When CDM and H_2O_2 were added simultaneously, it was obvious that 0%-CDM displayed the lowest degradation efficiency, which only reached 63.4% after 120 min. Meanwhile, the degradation efficiency reached 100% after 5 min when 30%-CDM, 40%-CDM, and 50%-CDM were used, and the 50%-CDM displayed the fastest catalytic rate according to the partially enlarged illustration in Figure 5a. Therefore, the catalytic activity of CDM was significantly improved with an increasing amount of Cu doped. This could be explained by the following reasons: firstly, the 50%-CDM displayed the largest specific surface area, which could supply the maximum adsorption and reaction sites for CR; secondly, a greater oxygen vacancy was formed in 50%-CDM according to the XPS results of O1s, although it was lower than that in 40%-CDM. As a result, the catalytic activity of CDM was affected by the specific surface area and oxygen vacancy, simultaneously. Thus, 50%-CDM was selected as the best catalyst and used for the subsequent experiments.

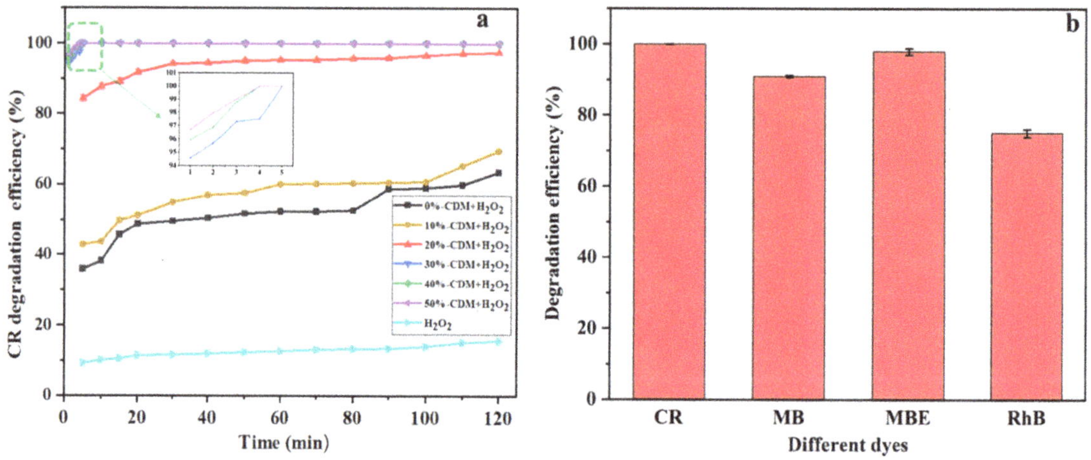

Figure 5. (**a**) Degradation of CR by different CDM catalysts (C_o = 60 mg/L, pH = 7, 2 mL 30% H_2O_2); (**b**) catalytic performance of M-5 for different dyes (C_o = 60 mg/L, pH = 7, 5 min, 2 mL 30% H_2O_2).

To evaluate the catalytic performance of 50%-CDM further, degradation experiments of different dyes (CR, MB, MBE, and RhB) were carried out, as displayed in Figure 5b. After 5 min of the reaction, the D% values of CR and MBE reached 100% and 98%, respectively, while the D% values of MB and RhB appeared to be relatively lower and only reached 90% and 75%. The reason may be that CR and MBE are anionic dyes while MB and RhB are cationic dyes, and the 50%-CDM has a positive electrical surface as a metal chemical compound. Therefore, there is a good electrostatic adsorption between anionic dyes (CR, MBE) and 50%-CDM, but there is no such electrostatic adsorption between cationic dyes (MB, RhB) and 50%-CDM. Thus, the 50%-CDM displayed better catalytic activity for CR and MBE. In the subsequent experiment, CR was selected as a target degradation substance.

3.3. Optimum Catalytic Conditions of 50%-CDM

To study the optimum catalytic conditions of CDM, the effects of CR concentration, H_2O_2 dosage, pH value, and temperature were investigated, as shown in Figure 6a–d. For a CR solution with a low concentration (60 mg/L), the D% could reach 100% only in 3 min. While the CR concentration increased above 210 mg/L, the D% could not reach 100% in 120 min. Thus, both the CR degradation rate and degradation efficiency decreased when the initial concentration of CR increased. Figure 6b shows the effect of H_2O_2 dosage, where it can be seen that the D% of 300 mg/L CR solution only reached 61.7% after 120 min without H_2O_2, while it increased to 94.5% when 6 mL H_2O_2 was added. In addition, both the degradation rate and degradation efficiency increased obviously with an increasing H_2O_2 dosage. When there is no H_2O_2, the removal process of CR is only based on physical adsorption. As H_2O_2 is added, in addition to physical adsorption, the degradation of CR by a Fenton-like reaction also plays an important role. The improvement in H_2O_2 dosage could bring out more active free radicals, which causes a quicker degradation rate and a higher degradation efficiency [23,36]. The pH value always has an important effect on the activity of catalysts, and the influences of the pH value from 3 to 11 on the degradation of CR were studied and are shown in Figure 6c. Herein, the highest catalytic capacity of 50%-CDM was obtained when the pH was 3.0, for which the D% of CR was tested as 99.4% after 5 min, and this is a fairly fast catalytic rate for highly concentrated CR solution. However, the degradation efficiency decreased significantly with an increasing pH value, and the D% decreased as low as 64.7%. The reason for this has been explained in many previous studies, as more free radicals from H_2O_2 in a Fenton-like system are generated in acidic conditions [1,36]. The effect of temperature on the degradation efficiency of CR was also investigated. As shown in Figure 6d, the degradation efficiency increased with elevated temperature. However, the differences in D% at different temperatures were not too large, and the difference between the maximum D% and the minimum D% was only 0.4% after 120 min. Thus, the temperature had no significant effect on the catalytic activity of the catalysts.

To evaluate the reusability of 50%-CDM, fifteen consecutive catalytic degradations of CR were carried out. After each run, the catalyst was separated from the solution by centrifugation, and then it was used directly for the next round of catalytic degradation without any more treatments. As shown in Figure 6e, all of the D% values were higher than 99% in the first nine catalytic degradation experiments, while the D% began to decrease from the 10th reuse and the D% dropped to 87% in the 15th reuse. However, all of the D% values were higher than 90% in the first 14 cycles. As a result, 50%-CDM displayed excellent reusability.

In order to estimate the stability of 50%-CDM, the leaching amount of Cu and Mn in wastewater after the reusability reaction with the 14th reuse was tested by ICP-OES, and the contents of Cu and Mn were detected as 0.0023 mg/L and 0.094 mg/L, which were lower than the concentrations specified in the national standard GB5749-2022 [41] in China ($C_{Cu} < 1$ mg/L, $C_{Mn} < 0.1$ mg/L). In addition, SEM and XPS were also used to evaluate the stability of 50%-CDM and the results are given in Figures S3 and S4. It was found that the morphology of the 50%-CDM still maintained a flower-like morphology, but some of the microspheres' structures collapsed and were no longer regular flower-shaped structures. According to the XPS broad survey spectra, Mn, O, and Cu are all found in Figure S4a. The characteristic peaks of Cu and Mn almost remained consistent with those of the original 50%-CDM in Figure 4b,c. Notably, the O 1s spectrum changed obviously and two new characteristic peaks appeared at 531.9 eV and 533.6 eV, which correspond to the organic C-O and organic C=O, and these may be from the degradation products of CR. Based on the above results, the 50%-CDM displayed good stability during the application.

To better analyze the kinetics for the removal of CR, two kinetic models including a pseudo-first-order model and pseudo-second-order model were utilized to fit the experimental data with H_2O_2 and without H_2O_2, as shown in Figure 7a,b, and the kinetic model rate constants are listed in Table 3. Obviously, the R^2 of the pseudo-first-order was relatively low (0.9448, 0.9608), whether with or without H_2O_2, while that of the pseudo-second-order model (0.9958, 0.9999) was higher. Thus, the removal process of CR was more consistent with the pseudo-second-order kinetics model. Additionally, the Qe calculated in the pseudo-second-order model was closer to the Qe obtained in experiments (171.15 $mg \cdot g^{-1}$ without H_2O_2 and 298.46 $mg \cdot g^{-1}$ with 6 mL H_2O_2).

In Table 4, the catalytic performances of other catalysts reported in previous studies are compared with that of 50%-CDM. Obviously, the catalyst prepared here displayed an excellent degradation efficiency, for either low-concentration CR solution or high-concentration CR solution. It is noteworthy that such an impressive catalytic performance has been rarely reported in previous studies.

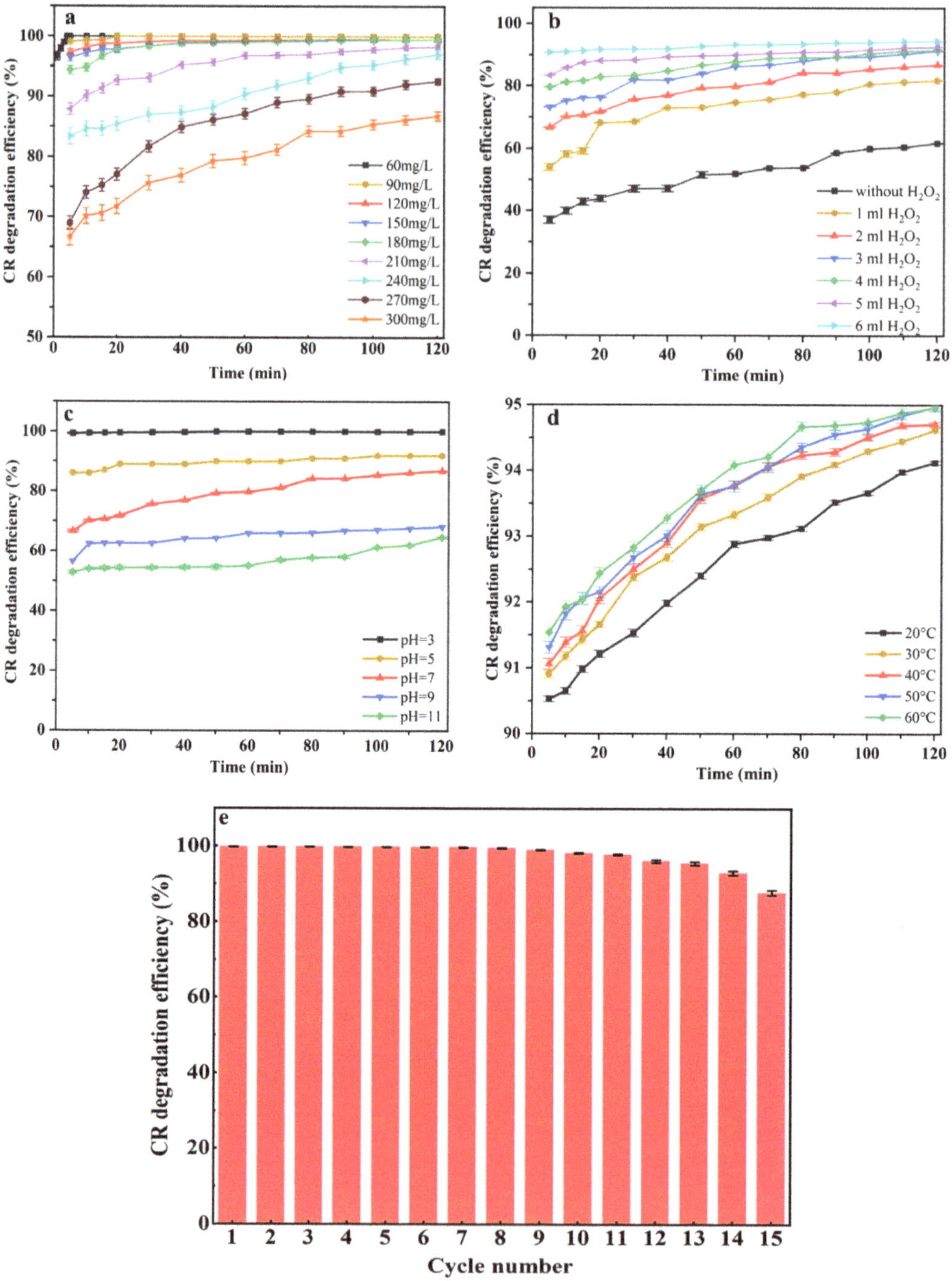

Figure 6. (**a**) Effect of CR concentrations (pH = 7, 2 mL 30% H_2O_2); (**b**) effect of dosage of H_2O_2 (C_o = 300 mg/L, pH = 7); (**c**) effect of pH (C_o = 300 mg/L, 2 mL 30% H_2O_2); (**d**) effect of temperature (C_o = 300 mg/L, pH = 7, 6 mL 30% H_2O_2); and (**e**) the reusability of 50%-CDM (C_o = 300 mg/L, pH = 3, 6 mL 30% H_2O_2, 5 min).

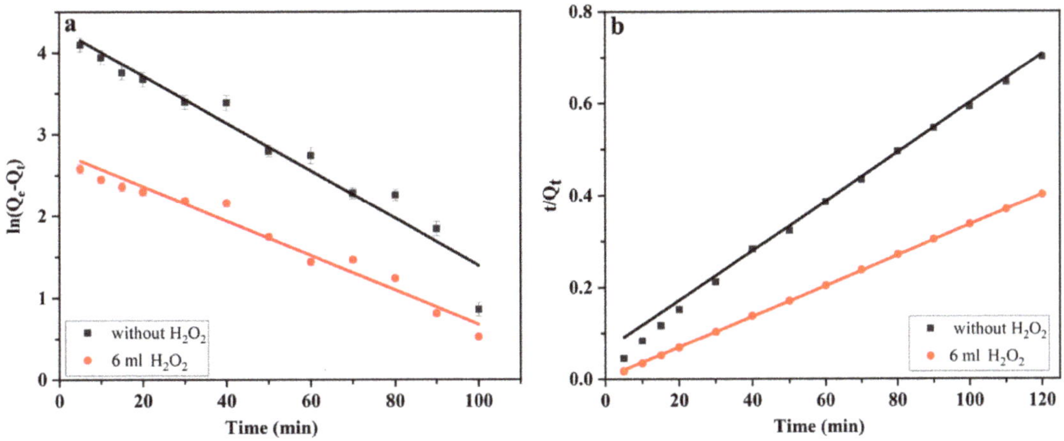

Figure 7. (**a**) Pseudo-first-order kinetics model and (**b**) pseudo-second-order kinetics model with different H_2O_2 dosages.

Table 4. Comparison of the performance of catalysts in previous studies.

Catalyst.	Dye	C_o (mg/L)	Degradation Method	Time (min)	D_{dye} (%)	Ref.
Biochar-CuO	MB	10	Periodate-AOP	30	100	[42] 2023
CuO	RhB	50	PMS-AOP	60	100	[43] 2023
Fe_3O_4/Biochar	CR	100	PMS-AOP	60	94.3	[44] 2023
Fe_3O_4/biochar	AO 7	200	photo-Fenton	120	100	[45] 2023
RuSA-N-C	AO 7	90	Fenton-like	30	100	[46] 2023
3D-MnO_2	AO 7	50	PMS-AOP	5	98.3	[16] 2022
ZnS	CR	50	photocatalytic	120	94.8	[47] 2022
ZnO/ZnS@AWA	CR	40	photocatalytic	150	98.8	[48] 2022
SnO_2-CdS	CR	10	photocatalytic	40	83.0	[49] 2022
MnO_2-ceramsite	AO 7	100	Fenton-like	60	85.2	[50] 2021
Cu doped MnO_2	CR	60 300	Fenton-like	3 5	100 99.4	This work

Note: AO 7 means acid orange 7.

3.4. Mechanism

In order to investigate the catalytic mechanism of CDM, free radical scavenging experiments were carried out by using p-benzoquinone (BQ) and isopropyl alcohol (IPA) as the scavengers for HOO• and HO•, respectively. As shown in Figure 8a, the D% decreased from 99.6% to 55.7% when BQ was added, while the D% only decreased to 94.9% when IPA was added. As a result, both HOO• and HO• played a role during the degradation process, but HOO• played a greater role apparently.

Furthermore, the EPR measurement was utilized to detect the existence of HOO• and HO• in the Fenton-like reaction. As shown in Figure 8b,c, both DMPO-HOO• (1:1:1:1) signal peaks [51] and DMPO-HO• (1:2:2:1) signal peaks [52,53] were obtained with the existence of 50%-CDM and H_2O_2, but no peaks appeared as only H_2O_2 exist but there was an absence of H_2O_2. However, the signal intensity of HO• was significantly lower than that of HOO• under the same test conditions. Thus, HOO• played a greater role in the Fenton-like reaction, which was consistent with the results obtained in the above free radical scavenging experiments.

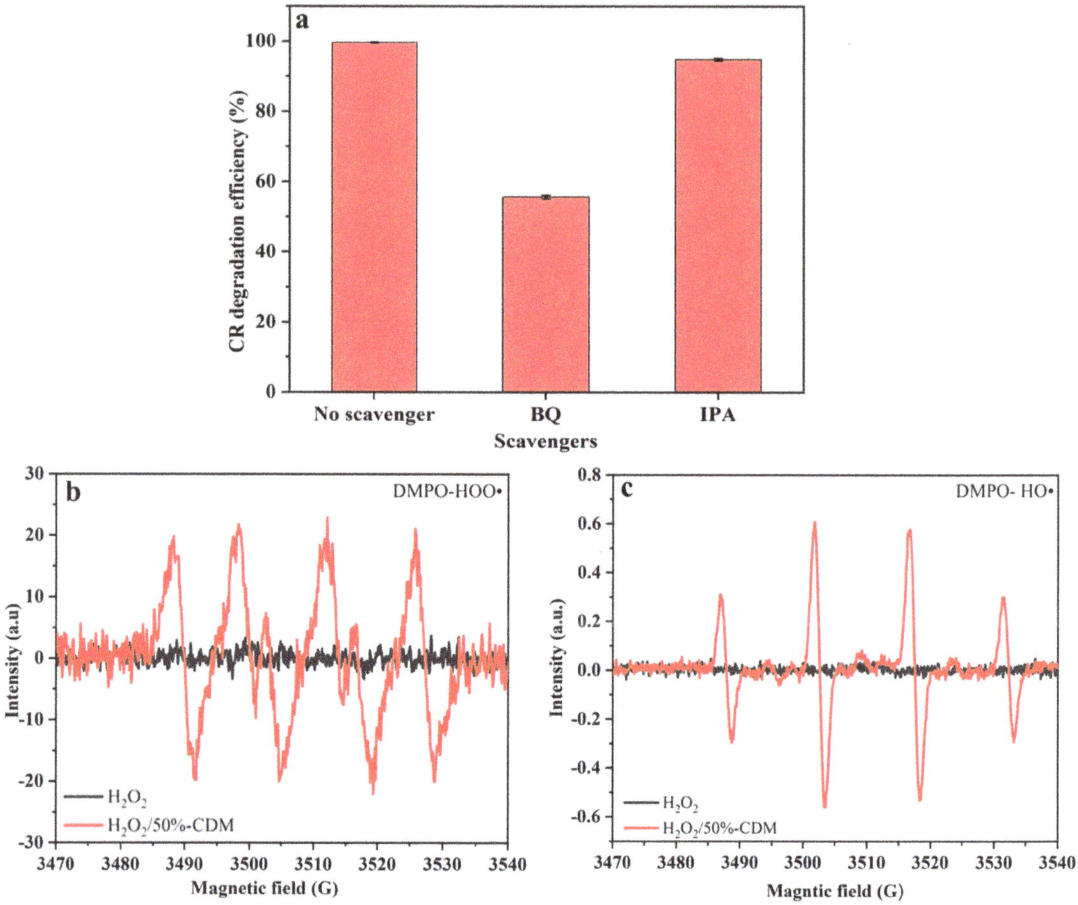

Figure 8. (**a**) The degradation efficiency of CR in the presence of scavengers (C_o = 300 mg/L, pH = 3, 6 mL 30% H_2O_2, 5 min), (**b**) EPR signals of DMPO-HO• adducts, and (**c**) EPR signals of DMPO-HOO• adducts.

According to the above results, the probable degradation mechanism of CR in the 50%-CDM/H_2O_2 catalytic system is illustrated in Figure 9. Based on the large specific surface area of the 50%-CDM (259.89 m^2/g), CR could be adsorbed onto a flower-like catalyst at first. As shown in Equations (4) and (5), H_2O_2 was decomposed to HO• and HOO• along with the valence change in manganese ions and copper ions; thus, CR was oxidized and degraded by the free radical and some non-toxic or low-toxic products such as CO_2, and H_2O and other small molecules were produced. Herein, the two metal ions could catalyze H_2O_2 to generate free radicals in the 50%-CDM/H_2O_2 catalytic system, which is also one of the reasons as to why the 50%-CDM displayed the highest catalytic activity. Additionally, copper ions and manganese ions can also promote a mutual valence transition after low-covalent ions (Mn^{3+} and Cu^+) are generated, as shown in Equations (6) and (7), for which the content of HO• detected is lower that of HOO•, and this theoretical inference is consistent with previous results obtained from EPR and radical scavenger experiments. Finally, the valence changes in metal ions through different methods ensured the regeneration of 50%-CDM, and led to the continual degradation of CR.

$$M_n^{4+}/Cu^{2+} + H_2O_2 \rightarrow M_n^{3+}/Cu^{+} + HOO\cdot + H^{+} \quad (4)$$

$$M_n^{3+}/Cu^{+} + H_2O_2 \rightarrow M_n^{4+}/Cu^{2+} + HO\cdot + OH^{-} \quad (5)$$

$$M_n^{3+} + Cu^{2+} \rightarrow M_n^{4+} + Cu^{+} \quad (6)$$

$$M_n^{4+} + Cu^{+} \rightarrow M_n^{3+} + Cu^{2+} \quad (7)$$

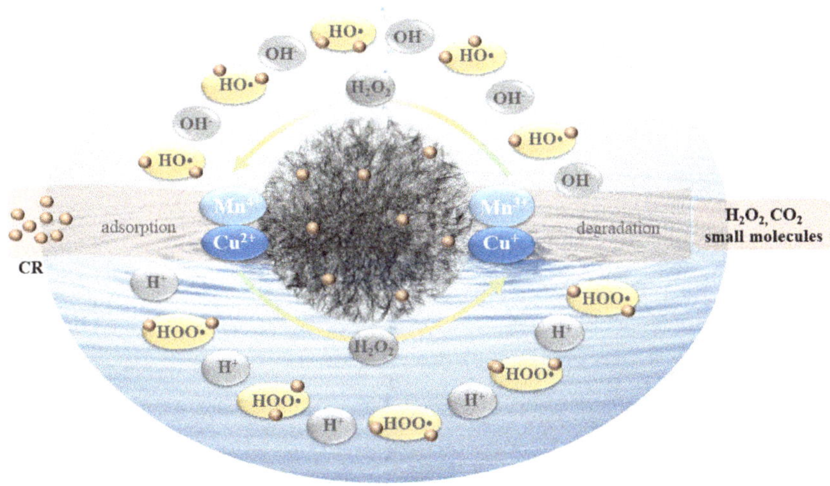

Figure 9. The probable degradation mechanism of CR in the 50%-CDM/H_2O_2 catalytic system.

4. Conclusions

In this study, Cu-doped manganese dioxide nanocatalysts with different Cu/Mn molar ratios were prepared with a simple hydrothermal reaction. Evidently, the Cu/Mn molar ratio had a significant effect on the structure and catalytic performance of CDM, whereas 50%-CDM, a flower-like nanocatalyst, with the highest content of Cu doped displayed the best catalytic activity due to the largest specific surface area and abundant oxygen vacancy. When 50%-CDM was used to degrade CR wastewater, an impressive degradation efficiency was obtained; for example, the D% reached 100% for 60 mg/L CR in 3 min at neutral conditions and the D% also reached 99.4% for 300 mg/L CR in 5 min at acidic conditions. Additionally, 50%-CDM also displayed excellent reusability, and the D% remained higher than 99% in nine cycles and higher than 90% in fourteen cycles. Finally, combing the high specific surface area and oxygen vacancy of the flower-like catalyst, a highly reactive and reusable heterogeneous Fenton catalyst was obtained in this work, which could be promising for practical dye wastewater treatment.

Supplementary Materials: The following supporting information can be downloaded at: https://www.mdpi.com/article/10.3390/nano14100833/s1, Table S1. Elemental content of catalyst obtained from EDS; Figure S1. Pore size distributions of series CDM: (**a**) 0%-CDM; (**b**) 10%-CDM; (**c**) 20%-CDM; (**d**) 30%-CDM; (**e**) 40%-CDM; (**f**) 50%-CDM; Figure S2. HR-XPS spectra of O 1s for series CDM; Table S2. The atomic% of lattice O and adsorbed O obtained from XPS; Figure S3. SEM images and EDS elemental maps of 50%-CDM after reused 14th; Figure S4. XPS spectra of 50%-CDM after reused 14th.

Author Contributions: Investigation, X.L., L.W., J.L., R.L., R.H. and W.G.; Writing—original draft, X.L. and L.W.; Writing—review & editing, X.L. and N.Y.; Funding acquisition, X.L. All authors have read and agreed to the published version of the manuscript.

Funding: This research was funded by the Key Scientific Research Project Foundation of Ningxia (No. 2022BDE03001) and the Ningxia College Student Innovation Project (S202311407039).

Data Availability Statement: Data are contained within the article and Supplementary Materials.

Conflicts of Interest: Author Neng Yu was employed by the company Huadian Electric Power Research Institute Company. The remaining authors declare that the research was conducted in the absence of any commercial or financial relationships that could be construed as a potential conflict of interest.

References

1. Yang, R.; Fan, Y.; Ye, R.; Tang, X.; Cao, X.; Yin, Z.; Zeng, Z. MnO_2-Based Materials for Environmental Applications. *Adv. Mater.* **2021**, *33*, 2004862. [CrossRef] [PubMed]
2. Wu, T.; Liang, Q.; Tang, L.; Tang, J.; Wang, J.; Shao, B.; Gong, S.; He, Q.; Pan, Y.; Liu, Z. Construction of a novel S-scheme heterojunction piezoelectric photocatalyst V-$BiOIO_3$/FTCN and immobilization with floatability for tetracycline degradation. *J. Hazard. Mater.* **2023**, *443*, 130251. [CrossRef] [PubMed]
3. Wang, F.; Gao, Y.; Liu, S.; Yi, X.; Wang, C.; Fu, H. Fabrication strategies of metal-organic frameworks derivatives for catalytic aqueous pollutants elimination. *Chem. Eng. J.* **2023**, *463*, 142466. [CrossRef]
4. Wang, F.; Zhang, Z.; Wang, C. Selective oxidation of aqueous organic pollutants over MOFs-based catalysts: A mini review. *Chem. Eng. J.* **2023**, *459*, 141538. [CrossRef]
5. Li, R.; Tang, X.; Wu, J.; Zhang, K.; Zhang, Q.; Wang, J.; Zheng, J.; Zheng, S.; Fan, J.; Zhang, W.; et al. A sulfonate-functionalized covalent organic framework for record-high adsorption and effective separation of organic dyes. *Chem. Eng. J.* **2023**, *464*, 142706. [CrossRef]
6. Costa-Serge, N.D.; Gonçalves, R.G.L.; Ramirez-Ubillus, M.A.; Li, C.; Hammer, P.; Chiron, S.; Nogueira, R.F.P. Pupo Nogueira, Effect of the interlamellar anion on CuMgFe-LDH in solar photo-Fenton and Fenton-like degradation of the anticancer drug 5-fluorouracil. *Appl. Catal. B-Environ. Energy* **2022**, *315*, 121537. [CrossRef]
7. Ren, Y.; Yu, J.; Zhang, J.; Lv, L.; Zhang, W. An in-situ strategy to analyze multi-effect catalysis in iron-copper bimetals catalyzed Fenton-like processes. *Appl. Catal. B-Environ.* **2021**, *299*, 120697. [CrossRef]
8. Zhao, Z.; Zhou, W.; Lin, D.; Zhu, L.; Xing, B.; Liu, Z. Construction of dual active sites on diatomic metal (FeCo-N/C-x) catalysts for enhanced Fenton-like catalysis. *Appl. Catal. B-Environ.* **2022**, *309*, 121256. [CrossRef]
9. Lai, C.; Yang, L.; Li, L.; Ma, D.; Cheng, M.; Liu, S.; Zhang, M.; Yan, H.; Tang, C.; Chen, Z.; et al. Design of a highly efficient Cu-based catalyst with two functional areas: The role of Cu0 and oxygen vacancies in Fenton-like system. *Chem. Eng. J.* **2023**, *464*, 142420. [CrossRef]
10. Nie, X.; Li, G.; Li, S.; Luo, Y.; Luo, W.; Wan, Q.; An, T. Highly efficient adsorption and catalytic degradation of ciprofloxacin by a novel heterogeneous Fenton catalyst of hexapod-like pyrite nanosheets mineral clusters. *Appl. Catal. B-Environ.* **2022**, *300*, 120734. [CrossRef]
11. Wang, F.; Xu, J.; Wang, Z.; Lou, Y.; Pan, C.; Zhu, Y. Unprecedentedly efficient mineralization performance of photocatalysis-self-Fenton system towards organic pollutants over oxygen-doped porous g-C_3N_4 nanosheets. *Appl. Catal. B-Environ.* **2022**, *312*, 121438. [CrossRef]
12. Liu, X.; Zhou, Z.; Wang, L.; Wang, P.; Zhang, X.; Luo, K.; Li, J.R. A general and programmable preparation of α-MnO_2/GO/CS aerogels used for efficient degradation of MB in wastewater. *Sep. Purif. Technol.* **2022**, *301*, 122034. [CrossRef]
13. Niu, M.; Yang, H.; Zhou, H.; Yi, X.; Zhou, X.; Zhan, J.; Liu, Y. Synergy of the successive modification of cryptomelane MnO_2 by potassium insertion and nitrogen doping for catalytic formaldehyde oxidation. *Chem. Eng. J.* **2022**, *431*, 133928. [CrossRef]
14. Sun, S.; Wang, Y.; Zhou, L.; Wang, X.; Kang, C. Enhanced degradation mechanism of tetracycline by MnO_2 with the presence of organic acids. *Chemosphere* **2022**, *286*, 131606. [CrossRef] [PubMed]
15. Li, P.; Zhan, S.; Yao, L.; Xiong, Y.; Tian, S. Highly porous alpha-MnO_2 nanorods with enhanced defect accessibility for efficient catalytic ozonation of refractory pollutants. *J. Hazard. Mater.* **2022**, *437*, 129235. [CrossRef] [PubMed]
16. Wang, Y.; Chen, D.; Zhang, Z.; Zhou, T.; Zou, J. Singlet oxygen-dominated activation of peroxymonosulfate by 3D hierarchical MnO_2 nanostructures for degradation of organic pollutants in water: Surface defect and catalytic mechanism. *Sep. Purif. Technol.* **2022**, *303*, 122177. [CrossRef]
17. Lu, J.; Guo, Z.; Li, M.; Dai, P.; He, M.; Kang, Y.; Sun, B.; Zhang, J. The increased oxygen vacancy by morphology regulation of MnO_2 for efficient removal of PAHs in aqueous solution. *Chemosphere* **2023**, *318*, 137966. [CrossRef] [PubMed]
18. He, Y.; Wang, L.; Chen, Z.; Shen, B.; Wei, J.; Zeng, P.; Wen, X. Catalytic ozonation for metoprolol and ibuprofen removal over different MnO_2 nanocrystals: Efficiency, transformation and mechanism. *Sci. Total Environ.* **2021**, *785*, 147328. [CrossRef] [PubMed]
19. Gong, P.; He, F.; Xie, J.; Fang, D. Catalytic removal of toluene using MnO_2-based catalysts: A review. *Chemosphere* **2023**, *318*, 137938. [CrossRef]
20. Wu, S.; Liu, H.; Huang, Z.; Xu, H.; Shen, W. O-vacancy-rich porous MnO_2 nanosheets as highly efficient catalysts for propane catalytic oxidation. *Appl. Catal. B-Environ.* **2022**, *312*, 121387. [CrossRef]

21. Chen, L.; Jia, J.; Ran, R.; Song, X. Nickel doping MnO_2 with abundant surface pits as highly efficient catalysts for propane deep oxidation. *Chem. Eng. J.* **2019**, *369*, 1129–1137. [CrossRef]
22. Wang, Y.; Sun, Y.; Wang, R.; Gao, M.; Xin, Y.; Zhang, G.; Xu, P.; Ma, D. Activation of peroxymonosulfate with cobalt embedded in layered delta-MnO_2 for degradation of dimethyl phthalate: Mechanisms, degradation pathway, and DFT calculation. *J. Hazard. Mater.* **2023**, *451*, 130901. [CrossRef] [PubMed]
23. Xiao, Y.; Huo, W.; Yin, S.; Jiang, D.; Zhang, Y.; Zhang, Z.; Liu, X.; Dong, F.; Wang, J.; Li, G.; et al. One-step hydrothermal synthesis of Cu-doped MnO_2 coated diatomite for degradation of methylene blue in Fenton-like system. *J. Colloid Interface Sci.* **2019**, *556*, 466–475. [CrossRef] [PubMed]
24. Yang, Y.; Zhang, P.; Hu, K.; Duan, X.; Ren, Y.; Sun, H.; Wang, S. Sustainable redox processes induced by peroxymonosulfate and metal doping on amorphous manganese dioxide for nonradical degradation of water contaminants. *Appl. Catal. B-Environ.* **2021**, *286*, 119903. [CrossRef]
25. Xu, Y.; Wen, Y.; Ren, T.; Yu, H.; Deng, K.; Wang, Z.; Li, X.; Wang, L.; Wang, H. Engineering the surface chemical microenvironment over CuO nanowire arrays by polyaniline modification for efficient ammonia electrosynthesis from nitrate. *Appl. Catal. B-Environ.* **2023**, *320*, 121981. [CrossRef]
26. Ma, M.; Li, W.; Tong, Z.; Yang, Y.; Ma, Y.; Cui, Z.; Wang, R.; Lyu, P.; Huang, W. 1D flower-like Fe_3O_4@SiO_2@MnO_2 nanochains inducing RGO self-assembly into aerogels for high-efficient microwave absorption. *Mater. Des.* **2020**, *188*, 108462. [CrossRef]
27. Bo, Z.; Yi, K.; Yang, H.; Guo, X.; Huang, Z.; Zheng, Z.; Yan, J.; Cen, K.; Ostrikov, K. More from Less but Precise: Industry-relevant Pseudocapacitance by Atomically-precise Mass-loading MnO_2 within Multifunctional MXene Aerogel. *J. Power Sources* **2021**, *492*, 229639. [CrossRef]
28. Wang, Y.; Tong, Y.; Chen, D.; Zhou, T.; Zhang, Q.; Zou, J. Activation of peroxymonosulfate by g-C_3N_4/ε-MnO_2 microspheres for nonradical pathway degradation of organic pollutants in water: Catalytic mechanism and degradation path. *Chem. Eng. J.* **2023**, *459*, 141643. [CrossRef]
29. Wei, W.; Wang, A.; Guo, K.; He, S.; Li, A.; Kang, X.; Fang, J. Enhanced degradation of micropollutants by UV/freshly formed colloidal MnO_2: Reactive species, kinetics and pathways. *Appl. Catal. B-Environ.* **2022**, *313*, 121441. [CrossRef]
30. Gu, W.; Li, C.; Qiu, J.; Yao, J. Facile fabrication of flower-like MnO_2 hollow microspheres as high-performance catalysts for toluene oxidation. *J. Hazard. Mater.* **2021**, *408*, 124458. [CrossRef]
31. Rajendran, K.; Pandurangan, N.; Vinod, C.P.; Khan, T.S.; Gupta, S.; Haider, M.A.; Jagadeesan, D. CuO as a reactive and reusable reagent for the hydrogenation of nitroarenes. *Appl. Catal. B-Environ.* **2021**, *297*, 120417.
32. Francis, M.M.; Thakur, A.; Balapure, A.; Dutta, J.R.; Ganesan, R. Fabricating effective heterojunction in metal-organic framework-derived self-cleanable and dark/visible-light dual mode antimicrobial CuO/AgX (X = Cl, Br, or I) nanocomposites. *Chem. Eng. J.* **2022**, *446*, 137363. [CrossRef]
33. Kumar, V.G.D.; Kumari, S.; Balaji, K.R.; Khan, A.A.; Ravikumar, C.R.; Basavaraja, B.M.; Santosh, M.S.; Rtimi, S. Singlet oxygen driven enhanced photocatalytic degradation of 1,3,7-trimethylpurine-2,6-dione using surfactant mediated PVA-CuO nanocomposites: Combining physical adsorption and photocatalysis. *Chem. Eng. J.* **2023**, *462*, 142187. [CrossRef]
34. Hsieh, M.C.; Su, Y.H.; Hsu, M.H.; Lin, A.Y.C. Enhanced MnO_2 oxidation of methotrexate through self-sensitized photolysis. *J. Hazard. Mater.* **2022**, *438*, 129494. [CrossRef] [PubMed]
35. Shokry, R.; Abd El Salam, H.M.; Aman, D.; Mikhail, S.; Zaki, T.; El Rouby, W.M.A.; Farghali, A.A.; Al, W.; Ko, Y. MOF-derived core–shell MnO@Cu/C as high-efficiency catalyst for reduction of nitroarenes. *Chem. Eng. J.* **2023**, *459*, 141554. [CrossRef]
36. Ma, G.; Syzgantseva, O.A.; Huang, Y.; Stoian, D.; Zhang, J.; Yang, S.; Luo, W.; Jiang, M.; Li, S.; Chen, C.; et al. A hydrophobic Cu/Cu_2O sheet catalyst for selective electroreduction of CO to ethanol. *Nat. Commun.* **2023**, *14*, 501. [CrossRef] [PubMed]
37. He, B.; Song, L.; Zhao, Z.; Liu, W.; Zhou, Y.; Shang, J.; Cheng, X. $CuFe_2O_4$/CuO magnetic nano-composite activates PMS to remove ciprofloxacin: Ecotoxicity and DFT calculation. *Chem. Eng. J.* **2022**, *446*, 137183. [CrossRef]
38. Suresh, R.; Karthikeyan, N.S.; Gnanasekaran, L.; Rajendran, S.; Soto-Moscoso, M. Facile synthesis of CuO/g-C_3N_4 nanolayer composites with superior catalytic reductive degradation behavior. *Chemosphere* **2023**, *315*, 137711. [CrossRef] [PubMed]
39. Luo, D.; Zhu, P.; Duan, M.; Liu, M.; Lu, H.; Huang, Z. Mechanism and degradation pathways insight of photocatalytic oxidation antibiotics by geometrical Ag/$AgNbO_3$/$BiVO_4$ plasmon Z-type heterojunction. *Sep. Purif. Technol.* **2023**, *311*, 123287. [CrossRef]
40. Wei, Z.; Wang, W.; Li, W.; Bai, X.; Zhao, J.; Tse, E.C.M.; Phillips, D.L.; Zhu, Y. Steering Electron-Hole Migration Pathways Using Oxygen Vacancies in Tungsten Oxides to Enhance Their Photocatalytic Oxygen Evolution Performance. *Angew. Chem. Int. Ed.* **2021**, *60*, 8236–8242. [CrossRef]
41. *GB5749-2022*; Hygienic Standard for Drinking Water. National Health Commission of the People's Republic of China: Beijing, China, 2022.
42. Zhang, X.; Verbist, M.; Kamali, M.; Xue, Y.; Liu, Y.; Jin, P.; Costa, M.E.V.; Appels, L.; Cabooter, D.; Dewil, R. Activation of periodate with pinewood biochar-CuO composite for the removal of recalcitrant organic pollutants—Mechanisms and degradation products. *Chem. Eng. J.* **2023**, *465*, 142916. [CrossRef]
43. Davarazar, M.; Kamali, M.; Venâncio, C.; Gabriel, A.; Aminabhavi, T.M.; Lopes, I. Activation of persulfate using copper oxide nanoparticles for the degradation of Rhodamine B containing effluents: Degradation efficiency and ecotoxicological studies. *Chem. Eng. J.* **2023**, *453*, 139799. [CrossRef]

44. Sun, Y.; Wang, T.; Han, C.; Bai, L.; Sun, X. One-step preparation of lignin-based magnetic biochar as bifunctional material for the efficient removal of Cr(VI) and Congo red: Performance and practical application. *Bioresour. Technol.* **2023**, *369*, 128373. [CrossRef] [PubMed]
45. Lin, K.; Afzal, S.; Xu, L.; Ding, T.; Li, F.; Zhang, M. Heterogeneous photo-Fenton degradation of acid orange 7 activated by red mud biochar under visible light irradiation. *Environ. Pollut.* **2023**, *327*, 121454. [CrossRef] [PubMed]
46. Tang, W.; Zhang, H.; Yang, X.; Dai, Z.; Sun, Y.; Liu, H.; Hu, Z.; Zheng, X. Ru single atom catalyst with dual reaction sites for efficient fenton-like degradation of organic contaminants. *Appl. Catal. B-Environ.* **2023**, *320*, 121952. [CrossRef]
47. You, J.; Liu, C.; Feng, X.; Lu, B.; Xia, L.; Zhuang, X. In situ synthesis of ZnS nanoparticles onto cellulose/chitosan sponge for adsorption-photocatalytic removal of Congo red. *Carbohydr. Polym.* **2022**, *288*, 119332. [CrossRef] [PubMed]
48. Zhang, W.; Li, L.; Zhang, X.; Liu, H.; An, Y.; Zhong, Y.; Hu, Z.; Shan, X.; Wu, J.; White, M.; et al. Adsorption of Zn(ii) on amination@wood-aerogel and high-value reuse to ZnO/ZnS as an efficient photocatalyst. *J. Mater. Chem. A* **2022**, *10*, 18644–18656. [CrossRef]
49. Khan, M.E.; Mohammad, A.; Ali, W.; Khan, A.U.; Hazmi, W.; Zakri, W.; Yoon, T. Excellent visible-light photocatalytic activity towards the degradation of tetracycline antibiotic and electrochemical sensing of hydrazine by SnO_2–CdS nanostructures. *J. Clean. Prod.* **2022**, *349*, 131249. [CrossRef]
50. Qian, W.; Huang, H.; Diao, Z.; Li, H.; Liu, H.; Ye, M.; Deng, Y.; Xu, Z. Advanced treatment of dye wastewater using a novel integrative Fenton-like/MnO_2-filled upward flow biological filter bed system equipped with modified ceramsite. *Environ. Res.* **2021**, *194*, 110641. [CrossRef]
51. Zhao, L.; Zhang, J.; Zhang, Z.; Wei, T.; Zhu, Y.; Ma, J. CuO with (0 0 1)-plane exposure efficiently induces peroxymonosulfate to form Cu-$OOSO_3$-intermediates directly oxidizing organic contaminants in water. *Chem. Eng. J.* **2022**, *441*, 136100. [CrossRef]
52. Chen, L.; Wu, D.; Jiang, T.; Yin, Y.; Du, W.; Chen, X.; Sun, Y.; Wu, J.; Guo, H. A novel heterogeneous catalytic system (AC/ZVI/CaO_2) promotes simultaneous removal of phosphate and sulfamethazine: Performance, mechanism and application feasibility verification. *Water Res.* **2023**, *237*, 119977. [CrossRef] [PubMed]
53. Dong, X.; Xu, L.; Ma, J.; Li, Y.; Yin, Z.; Chen, D.; Wang, Q.; Han, J.; Qiu, J.; Yang, Z.; et al. Enhanced interfacial charge transfer and photothermal effect via in-situ construction of atom co-sharing Bi plasmonic/$Bi_4O_5Br_2$ nanosheet heterojunction towards improved full-spectrum photocatalysis. *Chem. Eng. J.* **2023**, *459*, 141557. [CrossRef]

nanomaterials

MDPI

Article

Green Synthesis of Metal Nanoparticles with Borojó (*Borojoa patinoi*) Extracts and Their Application in As Removal in Water Matrix

Erika Murgueitio Herrera [1,2,*], Gissela Jacome [2], Carina Stael [1], Geovanna Arroyo [1], Andrés Izquierdo [1], Alexis Debut [1], Patricio Delgado [1,3] and Gemma Montalvo [4,5]

1. Centro de Nanociencia y Nanotecnología, Universidad de las Fuerzas Armadas ESPE, Av. Gral. Rumiñahui s/n, Sangolqui P.O. Box 171-5-231B, Ecuador; castael@espe.edu.ec (C.S.); gvarroyo@espe.edu.ec (G.A.); padelgado2@espe.edu.ec (P.D.); arizquierdo@espe.edu.ec (A.I.); apdebut@espe.edu.ec (A.D.)
2. Departamento de Ciencias de la Tierra y de la Construcción, Universidad de las Fuerzas Armadas ESPE, Av. Gral. Rumiñahui s/n, Sangolqui P.O. Box 171-5-231B, Ecuador; gejacome2@espe.edu.ec
3. Departamento de Ciencias Exactas, Universidad de las Fuerzas Armadas ESPE, Av. Gral. Rumiñahui s/n, Sangolqui P.O. Box 171-5-231B, Ecuador
4. Departamento de Química Analítica, Química Física e Ingeniería Química, Universidad de Alcalá, Ctra. Madrid-Barcelona km 33.6, 28871 Alcalá de Henares, Madrid, Spain; gemma.montalvo@uah.es
5. Instituto Universitario de Investigación en Ciencias Policiales, Universidad de Alcalá, Libreros 27, 28801 Alcalá de Henares, Madrid, Spain

* Correspondence: esmurgueitio@espe.edu.ec

Abstract: The predominant aim of the current research was to generate a proposal for the removal of arsenic, a highly toxic pollutant, encountered within the Papallacta Lagoon in Ecuador. The average concentrations of As yielded ranges between 18 to 652 μg/L, through the use of metallic nanoparticles. Sampling was performed in the lagoon with their respective geographic locations and "in situ" parameters. Nanoparticles of Mn_3O_4 NPs, Fe_3O_4 NPs, and CuO NPs were synthesized at a 0.5 M concentration, using the precipitation method, and borojó (*Borojoa patinoi*) extract was added as an anti-caking agent as well as antioxidant. The nanoparticles were characterized by visible spectrophotometry, scanning electron microscopy (SEM), transmission electron microscopy (TEM), X-ray diffraction (XRD), and Raman spectroscopy. After arsenic removal treatment using nanoparticles, a randomized experimental design of different concentrations (5 mg/L, 10 mg/L, 25 mg/L, 50 mg/L, 100 mg/L, and 150 mg/L) was applied at laboratory level. The average diameter of Fe_3O_4NPs ranged from 9 nm to 36 nm, Mn_3O_4 NPs were 15–20 nm, and CuO NPs ranged from 25 nm to 30 nm. Arsenic removal percentages using Fe_3O_4 NPs with a concentration of 150 mg/L was 87%; with Mn_3O_4 NPs, the removal was 70% and CuO NPs of about 63.5%. Finally, these nanoparticles could be used in a water treatment plant for the Papallacta Lagoon.

Keywords: arsenic; nanoparticles; borojó (*Borojoa patinoi*); Papallacta Lagoon

1. Introduction

Arsenic is a toxic and ubiquitous metalloid that is able to act as a geogenic contaminant. It is widely distributed in nature, present in the air, water, and soil. It can be released into the environment through various natural, agricultural, and industrial processes [1–3]. This element can be found in two states, organic as well as inorganic, while in its inorganic state, it is the more dangerous version. Arsenic exists in several chemical forms, each having a different mobility, which explains its toxicity. Hereby, As^{3+} (arsenite) and As^{5+} (arsenate) are the most stable and toxic forms of arsenic [4,5]. In humans, the main route of exposure to arsenic is through ingestion of food or water, while two further routes may be through inhalation of air and through the skin. Arsenic causes abnormalities such as a variety of cancers, abnormalities in fetal development, and genetic mutation, as it accumulates in biological organisms [5–7]. Drinking water is affected in its availability and safety due to geogenic contaminants such as arsenic in groundwater. There are millions of people who can suffer serious health problems and may even die, due to high concentrations of this metalloid that exceed recommended concentrations [8]. Numerous epidemiological and scientific studies have demonstrated that ingestion of arsenic through drinking water may be a causal factor in urogenital cancers, especially bladder cancers [9].

There are several reports of arsenic exposure from drinking water in Argentina, Chile, Mexico, El Salvador, Nicaragua, Peru, Bolivia, Spain, Thailand, and China. It should be noted that the World Health Organization (WHO) categorized arsenic as a chemical element harmful to public health, indicating 10 $\mu g\ L^{-1}$ for drinking water and 100 µg/L for irrigation water [1]. The study of the Papallacta Lagoon is fundamental because it supplies one-sixth of the water to the population of Quito, Ecuador's capital, and its inhabitants. Those are mostly rural people who depend on the water and food of this area, and are exposed to a real risk of contamination due to the daily consumption of water containing arsenic [10,11]. Lagoon ecosystems are aquatic environments of critical importance for their multifunctional role in maintaining biodiversity and sustaining coastal communities [12,13]. One of these projects was realized at the CENCINAT (Center for Nanoscience and Nanotechnology of the University of the Armed Forces ESPE, in Ecuador), which identified arsenic concentrations in the Papallacta Lagoon ranging between 390 and 670 µg/L. All of these concentrations are much higher than those recommended by the WHO [14].

There are several techniques to remove arsenic from contaminated water. However, their effectiveness and costs vary widely. Among the techniques used to remove arsenic from water are coagulation, filtration, membrane separation, and ion exchange, among others. Nanotechnology offers a process that could be economically viable and innovative, applying metallic nanoparticles [15–18].

Nanoparticles of the transition metal oxides of iron, manganese, and copper are rich in oxidation states, participating in reversible redox reactions, low cost, abundant on Earth, and harmless to the environment [19]. These oxides have been used in several studies for the removal of organic contaminants and even for the improvement of biodiesel using magnetite [20–26], while hausmannite and oxides of manganese at nano and micro levels have been used in detection [27,28] and water treatment process [29,30] for the elimination of organic and inorganic contaminants such as arsenic [31]. CuO Nps have been synthesized with a green procedure and used in the removal of dyes, namely methylene blue (MB) and methyl orange (MO), from contaminated waters through adsorption on CuO nanoparticles [32–34], photodegradation of tetracycline [35], for arsenic removal [36,37], and furthermore has been combined with magnetite also for arsenic removal [38].

Most nanoparticle synthesis methods described to date rely heavily on the use of organic solvents, toxic reducing agents, high temperatures, the use of organic solvents, toxic reducing agents, high temperature, and high pressure. Therefore, nowadays there is a need to develop sustainable ecological processes, the green chemistry, which lacks to use of toxic and dangerous materials. To support the non-use of toxic reagents in the preparation of nanomaterials and to be environmentally friendly, various plant materials have been used in green nanotechnology, this means that the reaction conditions, the reducing agent, the solvent medium, and the coating agent should be economically affordable and environmentally friendly reagents. Research has been realized where capulí, mortiño, and various fruits are used for the synthesis of nanoparticles [39–41], and the fruit extract has replaced the toxic reagents, constituting an ecological synthesis of nanoparticles.

In the present study, for the synthesis of nanoparticles, the borojó plant (*Borojoa patinoi*) has been used as an anti-caking agent for the nanoparticles and as an antioxidant due to its phenolic, flavonoid, and vitamin C properties [42]. It acts as a reducing agent in the formation of magnetite, as an anti-caking agent and stabilizer for hausmannite nanoparticles, and in the formation of copper oxide NPs due to the high concentration of ascorbic acid. The Fe_3O_4 NPs, Mn_3O_4NPs, and CuONPs nanoparticles together enhance the arsenic removal process, better than doing it individually. The synthesized NPs were characterized using different analytical instruments and discussed. Furthermore, the antioxidant efficacy of borojó was evaluated against 1,1-diphenyl-2-picrylhydrazyl (DPPHradical dot). Subsequently, the NPs were used to remove arsenic first at a laboratory level and could later be used in a water treatment plant for the Papallacta Lagoon.

2. Materials and Methods

2.1. Description of the Study Area

The Papallacta Lagoon is located in the Papallacta Parish belonging to the Quijos Canton of the Napo Province. It is situated on the Quito-Papallacta-Baeza highway, just two kilometers short reaching the town of the Papallacta Parish and two hours east of the city of Quito, in north-central Ecuador. Its geographic position is 00°22′10″ latitude (south) and 78°10′06″ longitude (west). This body of water is located between the Cayambe-Coca and Antisana National Parks, having a temperature of over 4 °C, while constantly warming up in the morning and cooling down at and during the night [43].

2.2. Sampling

Seventy-five water samples were taken at the surface and subsurface in the Papallacta Lagoon, on three different dates, as in Ecuador there are only rainy and dry seasons (see Figure 1). The GPS Mobile Mapper Field Spectra 20 was used. The water samples were taken at a depth between 2 m and 15 m and with equivalent distances using a Kemmerer BTL 1.2 L SS [44]. These samples were collected according to the Ecuadorian technical standard for water quality, sampling, handling, and conservation of samples, and the random probability sampling technique was applied.

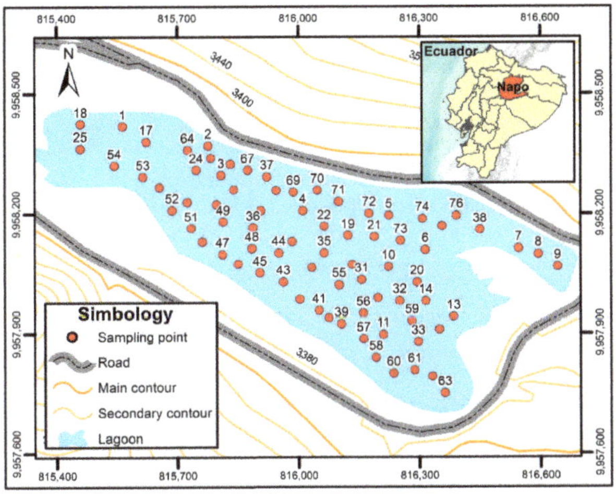

Figure 1. Location map of arsenic sampling points within the Papallacta Lagoon.

2.3. Chemicals

Reagents for water analysis were purchased from USA Hach Company P.O. Box 389 Loveland, Colorado. For the purification of the synthesis products and the preparation of the reaction solutions, ultrapure water from a Milli-Q system (Conductivity 0.28 μS/cm, Millipore, Direct Q, Darmstadt, Germany) was used, while the ethanol used was absolute grade, as with ACS (, ISO (Isopropyl Alcohol) 1728Batch 19.022.310.

2.4. Determination of Physical-Chemical Parameters

For the temperature (2550 method) and hydrogen potential (2550 method), as well as hydrogen potential (2550 method 4500-H^+ B) the HQ 30 d multiparameter equipment was used. Furthermore, for the sulfates (method 4500 $SO_4^{=}$), nitrates (method 4500 NO_3^- B), and turbidity (nephelometric method 2130 B) we used a variety of methods as described in the Manual of Standard Methods for the Examination of Water and Wastewater (APHA, 1992), realized at a spectrophotometer with UV/Visible lamp (Hach model Specord S600, Analytic Jena, Thuringia, Germany) [45]. For the total determination of arsenic, a Perkin Elmer AAnalyst 800 atomic absorption spectrometer (EAA) from Shelton, CT, USA was used, applying its FIAS injection system and Perkin Elmer WinLab32 for AA; Version 7.4.1.0730 (2014). For this process, the 3500 B methodology for arsenic in section 3114 described in the Manual of Standard Methods for the Examination of Water and Wastewater (APHA, 1992) was used. A calibration curve is performed with annual technical maintenance.

2.5. The Extract Fruit/Plant Preparation

The borojó fruit was purchased in the local market of Sangolquí, located in the Cantón of Rumiñahui, within the province of Pichincha. Afterward, the sample was cleaned and disinfected. It was then dried at 40 °C in an oven and weighed as well as labeled [39,46]. For ethanolic extraction, pulp and peel were used, and a 1:1 solution was realized (extract mass: ethanol volume 70%). The solution was concentrated by orbital shaking for 72 h. The extract was separated from the ethanol with a rotary evaporator (Buchi 850, Flawil, Switzerland). The fruit extract was filtered several times through sieves and placed in Falcon tubes to be centrifuged for 15–20 min, filtered using membranes of pore size of 0.45 µm diameter and later of 0.22 µm, while the extract was stored at 10 °C. The accuracy and precision of the equipment were not achieved as it is limited to solvent extraction, which was not performed for the given measurements.

2.6. Characterization of Fruit

First, the antioxidant activity was evaluated by the DPPH method, using Genesys UV-Vis equipment, model 10uv, Software Thermo Scientifics VISIONlite Scan version 4.0 (2002–2009) from the USA (last maintenance performed in June 2023). The experiment was realized in triplicate. Subsequently, to determine the redox potential, the cyclic voltammetry method was applied (potentiostat AUTOLAB PGSTAT128N, 2014; low isolates), following the protocol according to the user manual. The results were analyzed using Minitab version21.3, year 2024. To determine if there was a significant difference in the averages, a multifactorial analysis of variance was applied, and significant differences between averages ($p < 0.05$) were identified using Fisher's LSD test.

2.7. Synthesis of Nanoparticles of Fe_3O_4NPs, Mn_3O_4NPs, CuONPs

The NPs were prepared by the precipitation method and at room temperature of 20 °C, Fe_3O_4NPs were synthesized according to the existing protocol (Murgueitio's et al., 2018) [18]. A 0.5 M solution of $FeCl_3.7H_2O$ (Batch SG34191112; Wodehouse Road, Mumbai 400005, India) was prepared. Borojum extract previously alkalinized with 1M sodium hydroxide was added to a pH between 9–10, with a volume ratio of (1:1). The pH was controlled with a multiparameter HQ equipment. Subsequently, they were dried on a heating plate at 80 °C. Hausmannite (Mn_3O_4) NPs were synthesized using a modified version of the Stähli procedure in alkali, based on the protocol in [46,47]. Two solutions of 0.5 M $MnCl_2$, each prepared from $MnCl_2.H_2O$ (Sigma-Aldrich Laborchemikalien GmbH D-30926 Seelze lot 4265A, Steinheim, Germany), one at 20 °C and the other at <5 °C, were used. Alkalinized borojum solutions were added to each solution at pH 9–10. Subsequently, they were mixed in an oxygen line for 30 min and 20 °C. The solid obtained was centrifuged and washed several times with milliQ water.

The synthesis of the NPs-CuO was based on the existing protocol [46]. A 0.5 M solution of $CuCl_2$ ($CuCl_2.2H_2O$ Sigma Aldrich, Saint Louis, MO, USA, CAS 10125-13-0) was sIG an aeration line was added for 30 min. The solid obtained was centrifuged and washed several times with milliQ water.

All the solids obtained were washed with abundant milliQ water, centrifuged (Hermle centrifuge, model Z206A, Gosheim, Germany), and filtered through filter paper and several filters of 0.45 μm and 0.22 μm, consecutively. They were subsequently stored in an airtight bottle for subsequent characterization, at room temperature (25 °C) and away from any exposure to light.

2.8. Materials and Equipment Used in the Characterization of NPs

The nanoparticles are characterized to determine their average size and morphology. In addition, data on absorption bands at specific wavelengths are provided. Particle size characterization was performed using TEM. The protocol consists of letting a drop of colloidal suspension of nanoparticles evaporate on a carbon-coated copper grid and then analyzed on a transmission electron microscope (model Fei Tecnai G2 Spirit Twin, Eindhoven, The Netherlands) operating at 80 kV [48]. Calibration involved the diffraction of the gold particles and the measurement of the distance between two gold nanoparticles (0.24 nm apart). The use of the pre-calibrated web for reference has been at https://www.agarscientific.com/tem/calibration-standards (accessed on 1 April 2020). The ImageJ-win64 software was used to analyze the size of the nanoparticles. In addition, a Genesys 10uv UV-Vis spectrophotometer equipped with the Thermo Scientifics VISIONlite Scan version 4.0 (2002–2009) Software (Waltham, MA, USA) was used, while maintenance is performed annually. The baseline is obtained using deionized water as reference and the method of analysis is performed according to the equipment manual Thermo Fisher Scientific (Waltham, MA, USA). Automatic self-calibration was performed to ensure optimal wavelength accuracy and reproducibility.

Finally, semi-quantitative analysis of the elemental composition of each sample was performed in the chamber of a TESCAN (Brno, Czech Republic) brand scanning electron microscope (SEM,), model MIRA 3, using the Bruker X-Flash 6 | 30 energy dispersive X-ray detector (Bruker, Berlin, Germany) with 123 eV resolution for MnKα [48]. Maintenance, alignment verification, and calibration are performed annually. Calibration was performed with the help of https://www.tedpella.com/calibration_html/SEM_Calibration.aspx (accessed on 1 April 2020) networks [48].

The Raman spectra were collected with a Thermo Scientific confocal DXR Raman spectrometer (Waltham, MA, USA), a 780 nm excitation laser was used for the iron oxides and copper oxide, for the manganese oxides a Raman Cora 5001 equipment was used, with a 785 nm excitation laser.

2.9. Arsenic Removal with Mn_3O_4 NPs, Fe_3O_4 NPs and CuO NPs

We worked with the arsenic concentrations of 5 mg/L, 10 mg/L, 25 mg/L, 50 mg/L, 100 mg/L, and 150 mg/L, in synthetic water in the laboratory, simulating the concentration of As total in the Papallacta Lagoon. Each arsenic dilution had three replicates to perform the arsenic removal tests, while to each was added a 1:1 ratio (v/v) of nanoparticles. They were then shaken in a Wise Shake SHO-2D digital orbital shaker for 24 h at 150 rpm. Subsequently, they were filtered several times to a 0.22 μm filter. Finally, the concentration of arsenic in the final solutions was determined. The analyses were realized individually for each type of NPs.

3. Results and Discussion

3.1. Physical-Chemical Analysis of Water Samples from the Papallacta Lagoon

Samples were taken during the rainy season, at a maximum depth of 15 m and a minimum depth of 2 m. The sampling points are indicated in Figure 1. The average water temperature was 12.76 °C ± 0.52, similar to that obtained in previous research [49] and the average pH was 7.43 ± 0.32. These pH values were similar to the pH values obtained previously [50], where pH values between 7 and 9 were obtained from the same lagoon. The average sulfate concentration was 42.5 mg/L, nitrates 2.2 mg/L, and turbidity 70.6 UTN (Nephelometric Turbidity Unit), which is within the established limits (Table 1 of Book

VI of TULSMA Annex 1 [51]). The arsenic concentration data from the Papallacta lake obtained in this study range from 18 to 652 μg/L. Figure 2 demonstrates the distribution of As concentration in the lake, where the concentrations are lower than the concentrations previously obtained [52], where arsenic concentrations ranged from 677 to 1203 μg/L of As within a rainy season. This decrease in arsenic concentrations in the lagoon is attributed to the dilution conducted during the winter season in this body of water. During the winter season, the lagoon increases its volume from 8,000,000 m^3 to a volume of approximately 13,500,000 m^3.

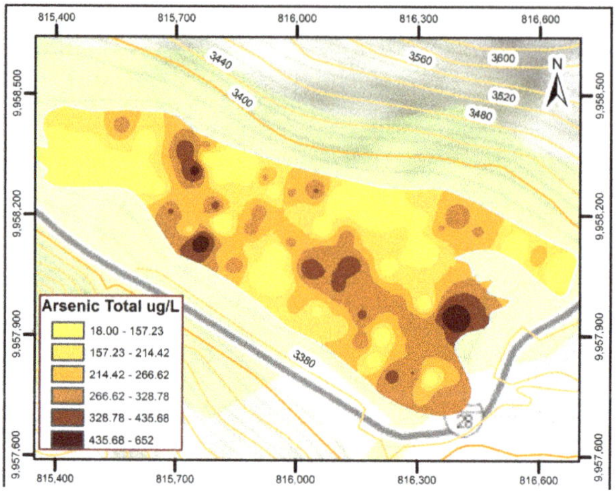

Figure 2. As total concentration in the Papallacta Lagoon.

3.2. Characterization of the Fruit Extract

The hydrogen donating capacity to DPPH radicals of 342.8 ± 31.0 μm Trolox equivalents/100 g fresh fruit, which is similar to that published by [53]. Compared to other fruits, the level is intermediate.

3.3. Characterization of Mn_3O_4 NPs, Fe_3O_4 NPs, and CuO NPs

The particles were sintered for this investigation according to the protocols described in Section 2.5. In the UV-Vis study, the Mn_3O_4 NPs had an absorption maximum of 284 nm, which agrees with the given data of [54]. CuO NPs are evidenced by their absorption maximum at 250 nm which is similar to that published by Prabu and Losetty (2024) [55] and Renuga et al., 2020 [56]. Fe_3O_4 NPs can be observed to peak in the region of 295–301 nm Fe_3O_4 NPs and wavelength in the band centered at 278 nm (See Figure 3a).

In the TEM analyses, the "Fiji is just" program was used to process and analyze the digital images. Figure 4a shows the highest percentage is nanoparticles Fe_3O_4 with diameters between 9 to 36 nm while the lowest percentage of nanoparticles is about 117 to 144 nm. With respect to Mn_3O_4 the NPs predominant diameters are between 15 to 20 nm (Figure 4b). For CuO NPS the highest percentage of nanoparticles is found in diameters between 25 and 30 nm (Figure 4c).

XRD studies are observed in Figure 5. The peak of the diffraction peaks of (101), (112), (200), (103), (211), (004), (220), (105), (312), (303), (321), (224), and (400) are coincident with ICDD n. No. 080-0382 and can be attributed to the tetragon in the tetragonal hausmannite phase, of the Mn_3O_4 spinel, which is similar to that published by [57–59] (Figure 5a). According to the quantification report 100% belong to tenorite (Figure 5b). The diffraction peaks (110) (002) (111) (202) (020) (202) (113) (311) (004) are indicative of a typical structure with the monoclinic topology of CuO or tenorite which is in agreement with studies by [60,61] (Figure 5c).

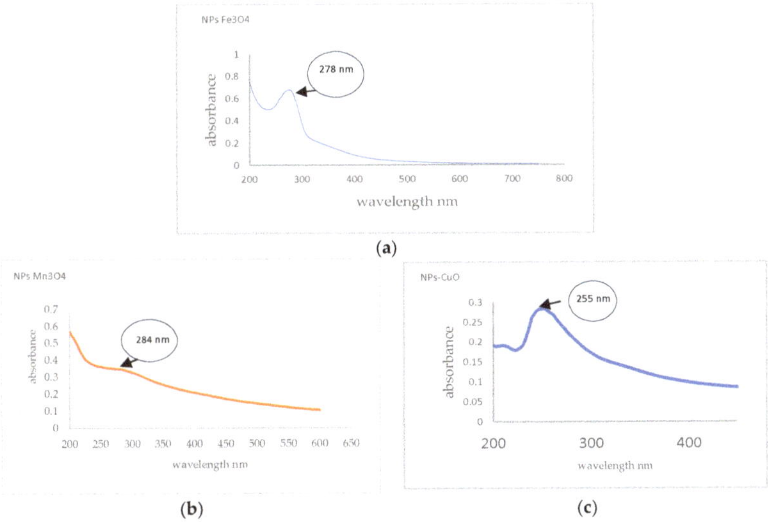

Figure 3. (**a**) Spectrum of UV-Vis Fe_3O_4 NPs, (**b**) Mn_3O_4 NPs, (**c**) CuO NPs.

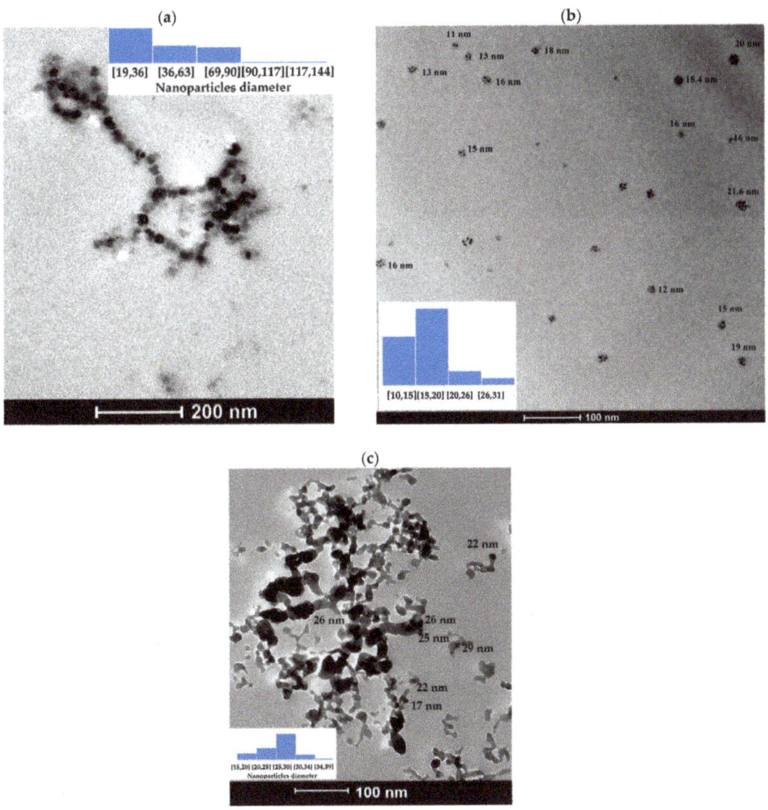

Figure 4. (**a**) Fe_3O_4 NPs, (**b**) Mn_3O_4 NPs, (**c**) CuO NPs.

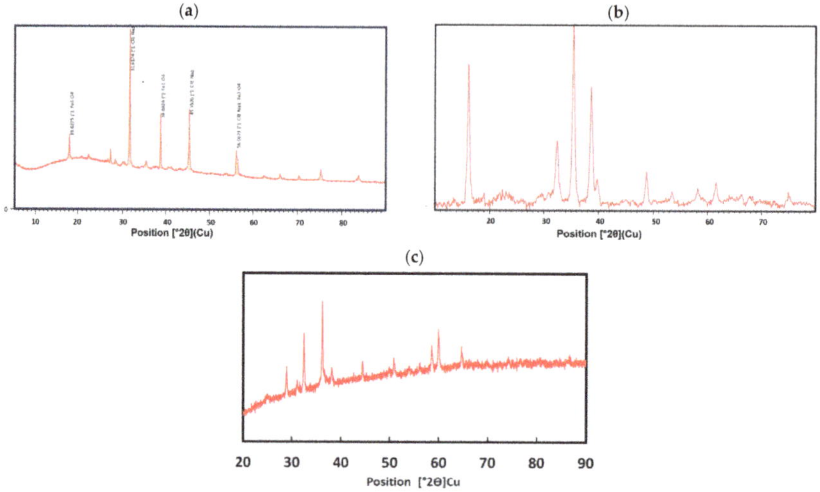

Figure 5. Powder X-ray diffraction (XRD) pattern of the synthesized: (**a**) NPs-Fe_3O_4, (**b**) NPs-Mn_3O_4, (**c**) NPs-CuO.

For NPs-Fe_3O_4, hematite bands are evidenced, similar to the studies of [62,63], which would indicate that oxidation to ferrous to ferric oxides occurred in the synthesis process (Figure 6a). The Raman spectrum of iron shows the peaks corresponding to the magnetite peak at 660 cm^{-1} [62,63] and 298 cm^{-1} based on [63]. Furthermore, we performed a Raman analysis. In the spectrum of the CuO nanoparticles, a band appears at 616 cm^{-1}, identified as the Cu-O stretching vibration band, which coincides with the studies carried out by [62–66] (Figure 6b). Additionally, the characteristic peak of copper hydroxide of 500 cm^{-1} can be observed, which coincides with the research of Mayer in 1992, which has been formed in the synthesis process of copper oxides.

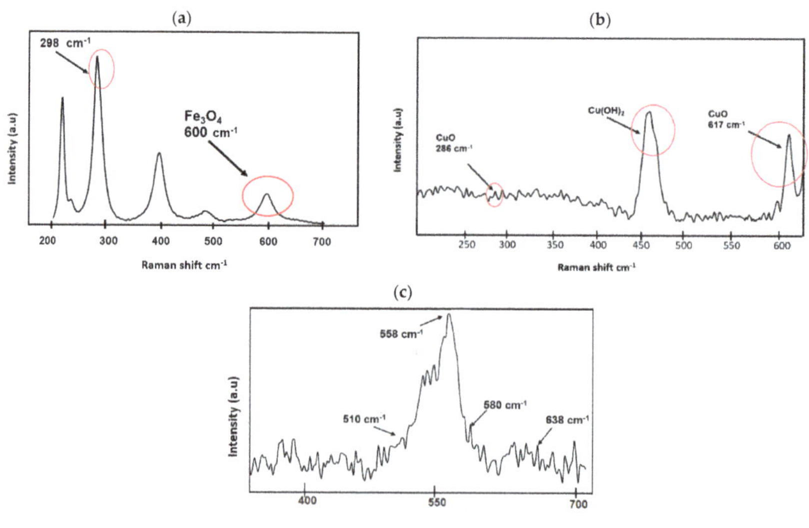

Figure 6. Raman spectra (**a**) CuO NPs, (**b**) Fe_3O_4 NPs, (**c**) manganese oxides NPs.

The Raman technique is used to analyze the local structure of manganese dioxides, especially in the case of samples with poor crystallinity [66]. Figure 6c illustrates the Raman spectrum of manganese oxide with a birnessite-type structure. The vibrational peculiarity of birnessite-type MnO_2 is its low Raman activity. Three main characteristics can be recognized between 500–510, 575–585, and 625–650 cm^{-1}. The two high wavenumber bands dominate all the spectra, while the bands in the low-frequency region appear with rather weak intensity. In the Raman spectrum of MnO_2 (Figure 6c), three Mn-O lattice vibrations characteristic of birnessite-type MnO_2 were observed at 510, 580, and 638 cm^{-1} [30].

Figure 7a illustrates the SEM images of NPs-Fe_3O_4, which indicates the crystal structure, similar to that previously described [66–68]. The Mn_3O_4 NPs are observed to be found within the organic matter of borojó, or similar to that reported by [69,70]. That is the reason why the RX diffraction of the CuO NPs, is similar to the structure of [71]. Crystalline structures are clearly visible. In the samples of magnetite, copper oxides, and manganese oxides, the surface morphology cannot be seen very clearly, because there is the borojó extract, which is the organic matter, causing the sample to exhibit low conductivity.

Figure 7. SEM images of (**a**) NPs-Fe_3O_4, (**b**) NPs-Mn_3O_4, (**c**) NPs-CuO.

In the studies of energy dispersive X-ray spectroscopy, we can observe in Figure 8a, for the NPs-Fe_3O_4, the presence of precursor reagents of the magnetite synthesis. For the hausmannite nanoparticles (Figure 8b) there is additionally the sodium peak whose origin is due to the alkalinization with sodium hydroxide in the synthesis process, which has not been completely eliminated in the successive washes. Finally, in Figure 8c, the chloride peak is visible, which originates from the copper chloride used in the preparation of the tenorite.

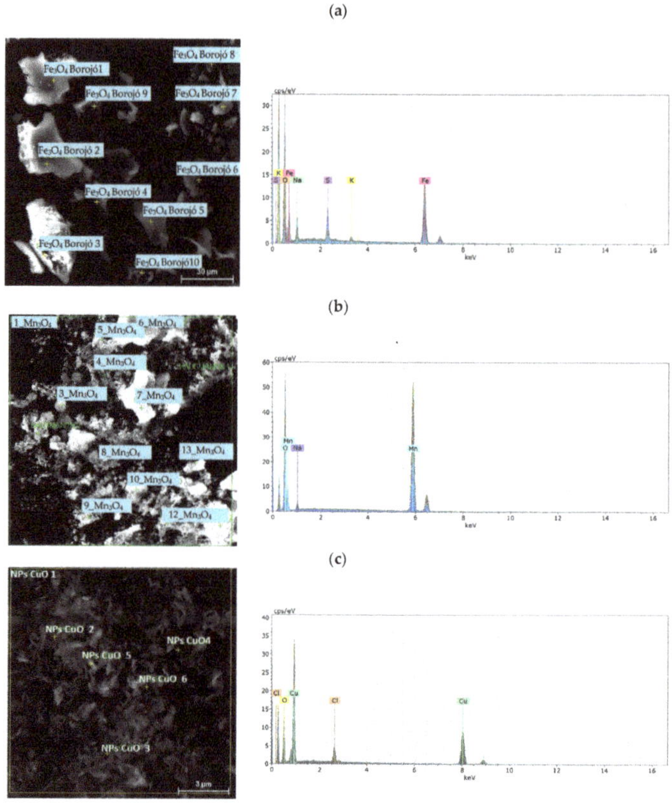

Figure 8. Energy dispersive X-ray spectroscopy of nanoparticles of (**a**) NPs-Fe_3O_4, (**b**) NPs-Mn_3O_4, (**c**) NPs-CuO.

3.4. *As Removal at the Laboratory Level*

The average removal percentages (%) of arsenic using NPs-Fe_3O_4 were the highest with a removal of 87.60% for initial concentration (Ci) 100 mg/L, and the least effective was with CuO-NPs, where a removal of 65.05% was achieved for Ci 150 mg/L (see Figure 9a,b).

(a)

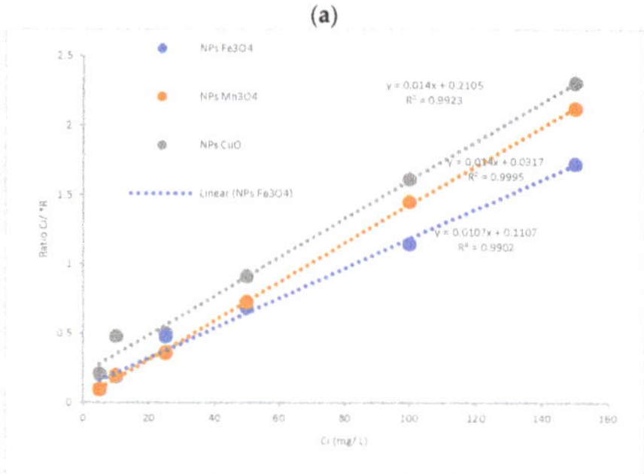

(b)

Ci As mg/L	*R (%) NPs Fe_3O_4	*R(%) NPs Mn_3O_4	*R(%) NPs CuO
150	87.02	70.70	65.05
100	87.60	69.07	62.03
50	72.90	69.02	55.04
25	52.28	69.79	49.94
10	52.89	51.04	20.95
5	52.68	51.95	24.13

*R % removal of As

Figure 9. Arsenic removal: (**a**) initial and final concentrations of As using NPs-Fe_3O_4, NPs-Mn_3O_4, NPs-CuO. (**b**) Table of removal percentages.

The removal mechanism of Fe_3O_4 NPs and CuO NPs with arsenic (Figure 10) is due to the effective affinity between arsenic and iron, which is due to an adsorption process [72,73], which involves chemical and physical adsorption. Additionally, the removal process may be due to the surface interaction between As(V) and ferric (hydric) oxides, which is mainly an internal sphere complexation [31,74], with respect to competing ions, sulfates, chlorides, and bicarbonates do not enter the first coordination sphere.

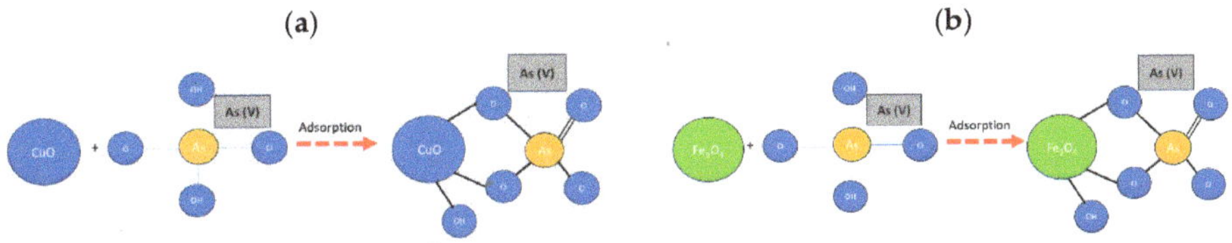

Figure 10. Arsenic adsorption removal process: (**a**) with CuO nanoparticles, (**b**) with Fe_3O_4 nanoparticles.

As removal with Mn_3O_4 NPs was 67.6% (at pH 2.0–8.0), monodentate and bidentate surface complexes are suggested for the adsorption of arsenic in the composite (5–16 mg/g).

Hausmannite is composed of $MnO{\cdot}Mn_2O_3$ NPs, as most of these works indicate that arsenate adsorbs on the edge of MnO_2, joining the Mn(IV) octahedra in a bidentate binuclear form [75]. More recent work has yielded that arsenate can bind to Mn(IV) octahedra in a mononuclear monodentate form and to Mn(III) octahedra in a mononuclear bidentate form.

With respect to the structure of borojó polyphenols, they enhance the stability of nanoparticles, which causes an increase in electrostatic and/or steric repulsion between particles and a decrease in van der Waals interactions and magnetic force, leading to a reduction in agglomeration and oxidation of nanoparticles [76] (Equation (1)).

$$-nFe^{2+} + +2Ar\,(OH)\,n \rightarrow nFe^0 + 2nAr = O + 2nH^+ \tag{1}$$

where Ar is the phenyl group and n is the number of hydroxyl groups oxidized by Fe^{2+}. The position of the hydroxyl groups on the phenolic. In addition, iron nanoparticles are highly efficient for arsenic removal, and these characteristics are based, in general, on the higher affinity of this oxyanion for active sites of the type ≡Fe–OH and ≡Fe–O–OH, which predominate on the surface of these substrates. However, due to the core-shell type structure of these nanoparticles, multiple removal processes occur, such as adsorption, absorption, precipitation, co-precipitation, oxidation, and reduction, which will predominate depending on the redox nature of the contaminant and the conditions of the medium (pH, ionic strength and redox potential) [76].

4. Conclusions

Nanoparticles of NPs-Mn_3O_4, NPs-Fe_3O_4, and NPs-CuO with borojum were used for arsenic removal within a water matrix.

Mn_3O_4 NPs, Fe_3O_4 NPs, and CuO NPs were synthesized by green chemistry method using borojum as an anti-caking agent and as a low-cost approach. In addition to the arsenic removal process performed by the Fe_3O_4, Mn_3O_4, and CuO nanoparticles, there is also an adsorption process by the borojum organic matter.

The properties, such as wavelength, size, distribution, morphology, and mineralogy were determined by UV-Vis, TEM, SEM, EDS, and XRD techniques, in which we could verify that the size of the NPs-Fe_3O_4 is between 9 nm to 46 nm, NPs-Mn_3O_4 average 66 nm, and NPs-CuO 5 nm to 47 nm. XRD studies confirm the mineralogy of the nanoparticles, as well as EDS and SEM studies, ratify the chemical elements present in the nanoparticles.

In addition, it was observed that the application of NPs-Mn_3O_4, NPs-Fe_3O_4, and NPs-CuO removed an average of 66% of arsenic.

Finally, it has been demonstrated that NPs could be used to remove arsenic in the waters of the Papallacta Lagoon, as part of a water treatment plant.

Author Contributions: Conceptualization, E.M.H.; Methodology, C.S. and G.A.; Formal analysis, A.I. and A.D.; Investigation, E.M.H., P.D. and G.J.; Writing—original draft, A.D.; Writing—review & editing, E.M.H. and G.M. All authors have read and agreed to the published version of the manuscript.

Funding: This research received no external funding.

Data Availability Statement: The original contributions presented in the study are included in the article, further inquiries can be directed to the corresponding author.

Acknowledgments: The authors thank the Center for Nanoscience and Nanotechnology for its accessibility to the collaborating students and use of equipment, materials, and reagents. We also acknowledge Karla Vizuete for the reading of TEM, SEM, and DRX images. We are also grateful to the Universidad de Alcalá, Department of Analytical Chemistry, Physical Chemistry, and Chemical Engineering with the "PROGRAMA GINER DE LOS RIOS".

Conflicts of Interest: The authors declare no conflict of interest.

References

1. Hussain, M.M.; Bibi, I.; Niazi, N.K.; Shahid, M.; Iqbal, J.; Shakoor, M.B.; Ahmad, A.; Shah, N.S.; Bhattacharya, P.; Mao, K.; et al. Arsenic biogeochemical cycling in paddy soil-rice system: Interaction with various factors, amendments and mineral nutrients. *Sci. Total Environ.* **2021**, *773*, 145040. [CrossRef] [PubMed]
2. Usese, A.; Chukwu, O.; Mahmudur, M.; Naidu, R.; Islam, S.; Olusegun, E. Enrichment, contamination and geo-accumulation factors for assessing arsenic contamination in sediment of a Tropical Open Lagoon, Southwest Nigeria. *Tecnol. E Innovación Ambient.* **2017**, *8*, 126–131. [CrossRef]
3. Jiang, Y.; Zhong, W.; Yan, W.; Yan, L. Arsenic mobilization from soils in the presence of herbicides. *J. Environ. Sci.* **2019**, *85*, 66–73. [CrossRef] [PubMed]
4. Sevak, P.; Pushkar, B. Arsenic pollution cycle, toxicity and sustainable remediation technologies. A comprehensive review and bibliometric analysis. *J. Environ. Manag.* **2024**, *349*, 119504. [CrossRef]
5. Jimenez, P.A.J.; Díaz, X.; Silva, M.L.N.; Vega, A.; Curi, N. Assessing and Understanding Arsenic Contamination in Agricultural Soils and Lake Sediments from Papallacta Rural Parish, Northeastern Ecuador, via Ecotoxicology Factors, for Environmental Embasement. *Sustainability* **2023**, *15*, 3951. [CrossRef]
6. Pizzali, M.M.; Robles, P.; Mendoza, M.; Torres, C. Ingesta de arsénico: El impacto en la alimentación y la salud humana. *Rev. Peru. De Med. Exp. Y Salud Publica* **2018**, *35*, 93–102. [CrossRef]
7. Sun, H.; Cheng, H.; Lin, L.; Deng, K.; Cui, X. Bioaccumulation and sources of metal(loid)s in lilies and their potential health risks. *Ecotoxicol. Environ. Saf.* **2018**, *151*, 228–235. [CrossRef]
8. Mukherjee, A.; Coomar, P.; Sarkar, S.; Johannesson, K.H.; Fryar, A.E.; Schreiber, M.E.; Ahmed, K.M.; Alam, M.A.; Bhattacharya, P.; Bundschuh, J.; et al. Arsenic and other geogenic contaminants in global groundwater. *Rev. Earth Environ.* **2024**, *5*, 312–328. [CrossRef]
9. Fu, H.; Murali, A.; Damodaran, C. Ingestion of Inorganic Arsenic in Drinking Water Induced Urogenital Carcinogenesis in Mice Models. *Urol. Oncol. Semin. Orig. Investig.* **2024**, *2*, s35. [CrossRef]
10. Cumbal, L.; Aguirre, V.; Murgueitio, E.; Tipán, I.; Chávez, C. *El Origen del Arsénico en las Aguas y Sedimentos de la Laguna Papallacta*; Litter, M., Ed.; Sangolqui, Ecuador Libro de Resúmenes del Taller de Distribución del As en Iberoamérica: Sangolqui, Ecuador, 2006; pp. 80–81.
11. Jiménez, P. Diagnóstico de la Presencia y Contaminación por Arsénico en el Suelo y Agua de Consumo en la Parroquia Papallacta. Bachelor's Thesis, Facultad de Ingenieria en Geologia y Petróleos, Quito, Ecuador, 2018.
12. Mosley, L.M.; Priestley, S.; Brookes, J.; Dittmann, S.; Farkaš, J.; Farrell, M.; Ferguson, A.J.; Gibbs, M.; Hipsey, M.; Huang, J.; et al. Extreme eutrophication and salinisation in the Coorong estuarine-lagoon ecosystem of Australia's largest river basin (Murray-Darling). *Mar. Pollut. Bull.* **2023**, *188*, 114648. [CrossRef]
13. Stocco, A.; Dupré, L.; Pranovi, F. Exploring the interplay of landscape changes and ecosystem services maximization in man-managed lagoon areas. *Estuar. Coast. Shelf Sci.* **2024**, *296*, 108597. [CrossRef]
14. World Health Organization. Arsenic and Arsenic Compounds. Environmental Health Criteria 224. 2001. Available online: http://www.inchem.org/documents/ehc/ehc/ehc224.htm (accessed on 15 May 2024).
15. Martinson, C.; Reddy, K. Adsorption of arsenic(III) and arsenic(V) by cupric oxide nanoparticles. *J. Colloid Interface Sci.* **2009**, *336*, 406–411. [CrossRef] [PubMed]
16. Habuda-Stanić, M.; Nujić, M. Arsenic removal by nanoparticles: A review. *Environ. Sci. Pollut.* **2015**, *22*, 8094–8123. [CrossRef] [PubMed]

17. Babaee, Y.; Mulligan, C. Removal of arsenic (III) and arsenic (V) from aqueous solutions through adsorption by Fe/Cu nanoparticles. *Chem. Technol. Biotechnol.* **2017**, *93*, 63–71. [CrossRef]
18. Murgueitio, E.; Cumbal Flores, L.; Toulkeridis, T. Removal of arsenic and heavy metals from contaminated water with emerging sorbents. In *International Conference on Applied Technologies*; Springer Nature: Cham, Switzerland, 2022; pp. 196–211.
19. Prasad, K.; Sreekanth, T.; Kim, J. Facile synthesis of Mn_3O_4 nanoparticles towards high performance asymmetric supercapacitors. *Vacuum* **2024**, *221*, 112930. [CrossRef]
20. Zhao, Z.; Zhang, Y.; Li, Y.; Quan, X.; Zhao, Z. Comparing the mechanisms of ZVI and Fe_3O_4 for promoting waste-activated sludge digestion. *Water Res.* **2018**, *144*, 126–133. [CrossRef]
21. El-Sawaf, A.; Hemdan, M.; Selim, H.; Nassar, A. Revolutionizing water treatment: Enhanced flux and selectivity in polyethersulfone mixed matrix membrane through magnetic CuO-functionalized Fe_3O_4 nanoparticles for synthetic oily produced water remediation. *Surf. Interfaces* **2024**, *46*, 104142. [CrossRef]
22. Uslu, S.; Maki, D.; Al-Gburi, A.S.K. Investigation of nanoparticle (Fe_3O_4) addition to 3rd generation biodiesel (spirulina microalgae)/diesel mixture as an innovative fuel according to different engine variables: An RSM optimization. *Energy Convers. Manag.* **2024**, *310*, 118481. [CrossRef]
23. Payami, E.; Mohammadzadeh, A.; Safa, K.; Teimuri-Mofrad, R. Ferrocene surface-modified Fe_3O_4 nanoparticles as prominent electrode material for supercapacitor application. *J. Energy Storage* **2024**, *88*, 111624. [CrossRef]
24. Payami, E.; Aghaiepour, A.; Mohamma, R.; Teimuri-Mofrad, R. Design and synthesis of ternary GO-Fc/Mn_3O_4/PANI nanocomposite for energy storage applications. *J. Alloys Compd.* **2020**, *829*, 154485. [CrossRef]
25. Teimuri-Mofrad, R.; Payami, E.; Ahadzadeh, I. Synthesis, characterization and electrochemical evaluation of a novel high performance GO-Fc/PANI nanocomposite for supercapacitor applications. *Electrochim. Acta* **2019**, *321*, 134706. [CrossRef]
26. Payami, E.; Ahadzadeh, I.; Mohammadi, R.; Teimuri-Mofrad, R. Design and synthesis of novel binuclear ferrocenyl-intercalated graphene oxide and polyaniline nanocomposite for supercapacitor applications. *Electrochim. Acta* **2020**, *342*, 136078. [CrossRef]
27. Chang, Y.; Yu, D.; Long, Y.; Xu, J.; Luo, X.; Ye, R. Large-scale fabrication of single-crystalline Mn_3O_4 nanowires via vapor phase growth. *J. Cryst. Growth* **2005**, *279*, 88–92. [CrossRef]
28. Murugadoss, G.; Ma, J.; Ning, X.; Kumar, M. Selective metal ions doped CeO_2 nanoparticles for excellent photocatalytic activity under sun light and supercapacitor application. *Inorg. Chem. Commun.* **2019**, *109*, 107577. [CrossRef]
29. Sophia, P.J.; Aravindkumar, S.; Chander, D.S.; Kumar, K.D.A.; Kumar, M.R. Cost-effective synthesis and studies of novel hybrid ternary Cd_{1-x} Ag_xS nanoparticles. *J. Nanophotonics* **2020**, *14*, 046016.
30. Kumar, M.R.; Murugadoss, G.; Pirogov, A.; Thangamuthu, R. A facile one step synthesis of SnO_2/CuO and CuO/SnO_2 nanocomposites: Photocatalytic application. *J. Mater. Sci. Mater. Electron.* **2018**, *29*, 13508–13515. [CrossRef]
31. Song, J.; Huang, G.; Han, D.; Hou, Q.; Liu, R. Alpha-MnO_2 nanoneedle embedded in MgO-chitosan biochar for higher removal of arsenic from groundwater: Co-effects of oxidation and adsorption. *J. Alloys Compd.* **2023**, *947*, 169643. [CrossRef]
32. Atri, A.; Dhaouadi, F.; Mechi, N.; Sellaoui, L.; Echabaane, M.; Chaabane, R.B.; Erto, A.; Badawi, M.; Lamine, A.B. Removal of textile pollutants from aqueous medium using biosynthesized CuO nanoparticles: Theoretical comparative investigation via analytical model. *Heliyon* **2024**, *10*, 5. [CrossRef]
33. Gautam, P.; De, A.; Sinha, I.; Behera, C.; Singh, K. Waste to wastewater treatment: Synthesis of CuO nanoparticles from discarded printed circuit boards for the removal of Methylene Blue under visible light. *Mater. Today Proc.* **2024**, *in press*. [CrossRef]
34. Inamdar, A.; Rajenimbalkar, R.; Thabet, A.; Shelke, S.; Inamdar, S. Environmental applications of flame synthesized CuO nanoparticles through removal of Congo Red dye. *Mater. Today Proc.* **2023**, *92*, 515–521. [CrossRef]
35. Moradi, A.; Khamforoush, M.; Rahmani, F.; Ajamei, H. Synthesis of 0D/1D electrospun titania nanofibers incorporating CuO nanoparticles for tetracycline photodegradation and modeling and optimization of the removal process. *Mater. Sci. Eng. B* **2023**, *297*, 116711. [CrossRef]
36. Khan, A.; Tahir, K.; Khan, Q.; Albalawi, K.; Jevtovic, V.; Almarhoon, Z.; Khan, M. Scaled-up development of recyclable Pd@ ZnO/CuO nanostructure for efficient removal of arsenic from wastewater. *J. Mol. Struct.* **2022**, *1260*, 132828. [CrossRef]
37. Liu, H.; Li, P.; Qiu, F.; Zhang, T.; Xu, J. Controllable preparation of FeOOH/CuO@ WBC composite based on water bamboo cellulose applied for enhanced arsenic removal. *Food Bioprod. Process.* **2020**, *123*, 177–187. [CrossRef]
38. Sun, T.; Zhao, Z.; Liang, Z.; Cui, F. Efficient degradation of p-arsanilic acid with arsenic adsorption by magnetic CuO-Fe_3O_4 nanoparticles under visible light irradiation. *Chem. Eng. J.* **2018**, *334*, 1527–1536. [CrossRef]
39. Murgueitio, E.; Cumbal, L.; Debut, A.; Landivar, J. Synthesis of Iron Nanoparticles using Extracts of Native Fruits of Ecuador, as Capuli (*Prunus serotina*) and Mortiño (*Vaccinium floribundum*). *Biol. Med.* **2016**, *8*, 282. [CrossRef]
40. Vasco, C.; Ruales, J.; Kamal-Eldin, A. Total phenolic compounds and antioxidant capacities of major fruits from Ecuador. *Food Chem.* **2008**, *111*, 816–823. [CrossRef]
41. Kumar, B.; Smita, K.; Cumbal, L.; Debut, A. Green synthesis of silver nanoparticles using Andean blackberry fruit extract. *Saudi J. Biol. Sci.* **2017**, *24*, 45–50. [CrossRef]
42. Sotelo, I.; Casas, N.; Camelo, G. Borojó (*Borojoa Patinoi*): Fuente de Polifenoles. Revista de la Facultad de Química Farmacéutica. *Vitae* **2010**, *17*, 2145–2660.
43. Molina Santos, M.G. *Efectos del Derrame de Petróleo Sobre la Comunidad Fitoplactónica de la Laguna de Papallacta y sus Principales Afluentes*; Universidad Internacional del Ecuador: Quito, Ecuador, 2013; pp. 24–25.

44. INEN. NTE INEN-ISO 5667-1. Enero 2014. Available online: https://www.insistec.ec/images/insistec/02-cliente/07-descargas/NTE%20INEN%201108%20-%20AGUA%20POTABLE.%20REQUISITOS.pdf (accessed on 20 March 2022).
45. APHA. *Standard Methods for the Examination of Water and Wastewater*; American Public Health Association: Washington, DC, USA, 1992; Volume 18.
46. Portehault, D.; Cassaignon, S.; Nassif, N.; Baudrin, E.; Jolivet, J. A core–corona hierarchical manganese oxide and its formation by an aqueous soft chemistry mechanism. *Angew. Chem. Int. Ed.* **2008**, *47*, 6441–6444. [CrossRef]
47. Stähli, E. Über Manganate (IV) Mit Schichten-Strucktur. Ph.D. Thesis, University of Bern, Bern, Switzerland, 1968.
48. Murgueitio-Herrera, E.; Falconí, C.E.; Cumbal, L.; Gómez, J.; Yanchatipán, K.; Tapia, A.; Martínez, K.; Sinde-Gonzalez, I.; Toulkeridis, T. Synthesis of Iron, Zinc, and Manganese Nanofertilizers, Using Andean Blueberry Extract, and Their Effect in the Growth of Cabbage and Lupin Plants. *Nanomaterials* **2022**, *12*, 1921. [CrossRef]
49. Villacreces, L. Brief Analysis of Chemical monitoring Results of Lake Papallacta to Determine Control Parameters for an Antique oil Spill. 2015. Available online: https://www.researchgate.net/publication/274567038_Brief_analysis_of_chemical_monitoring_results_of_Lake_Papallacta_to_determine_control_parameters_for_an_antique_oil_spill (accessed on 15 April 2021).
50. Molina-Santos, M.; Terneus-Jácome, E.; Yánez-Moretta, P.; Cueva-Sánchez, M. Resiliencia de la comunidad fitoplanctónica en la laguna andina de Papallacta y sus afluentes, ocho años después de un derrame de petrolero. *La Granja Rev. De Cienc. De La Vida* **2018**, *28*, 67–83. [CrossRef]
51. Ministerio del Ambiente de la República del Ecuador. *Texto Unificado de la Legislación Secundaria del Ministerio del Ambiente*; Ministerio del Ambiente de la República del Ecuador: Quito, Ecuador, 2015; pp. 295–332.
52. Zuñiga Salazar, M.M. *Síntesis de Sorbentes Híbridos Para la Remoción Selectiva de Arsénico con Aplicación en la Descontaminación de Aguas Superficiales*; Universidad de la Fuerzas Armadas "ESPE": Sangolqui, Ecuador, 2009; pp. 111–113.
53. Rincon-Velandia, J.; Garzón, A. Actividad Antioxidante, Contenido de Vitamina C y fenoles totales en el borojó (*Borojoa Patinoi* Cuatrec). *Vitae* **2014**, *19*, S442–S444.
54. Souri, M.; Hoseinpour, V.; Shakeri, A.; Ghaemi, N. Optimisation of green synthesis of MnO nanoparticles via utilising response surface methodology. *IET Nanobiotechnol.* **2018**, *12*, 822–827. [CrossRef] [PubMed]
55. Prabu, P.; Losetty, V. Green synthesis of copper oxide nanoparticles using *Macroptilium lathyroides* (L) leaf extract and their spectroscopic characterization, biological activity and photocatalytic dye degradation study. *J. Mol. Struct.* **2024**, *1301*, 137404. [CrossRef]
56. Renuga, D.; Jeyasundari, J.; Athithan, A.; Jacob, Y. Synthesis and characterization of copper oxide nanoparticles using Brassica oleracea var. italic extract for its antifungal application. *Mater. Res. Express* **2020**, *7*, 045007. [CrossRef]
57. Tholkappiyan, R.; Naveen, A.; Vishista, K.; Hamed, F. Investigation on the electrochemical performance of hausmannite Mn_3O_4 nanoparticles by ultrasonic irradiation assisted co-precipitation method for supercapacitor electrodes. *J. Taibah Univ. Sci.* **2018**, *12*, 669–677. [CrossRef]
58. Suhaimi, L.; Bahtiar, S.; Sarina, A.; Khairunnisya, K. Recovery of Manganese from Manganese Ore Reductive Acid Leaching Process Using Reeds (Imperata Cylindrica) as Reducing Agent. *Metalurgi* **2023**, *38*, 1–24. [CrossRef]
59. Said, M.; Harbrecht, B. Size-controlled synthesis of Mn_3O_4 nanoparticles: Characterization and defect chemistry. *J. Nanoparticle Res.* **2019**, *21*, 68. [CrossRef]
60. Neto, M.E.; Britt, D.; Jackson, K.; Braccini, A.; Inoue, T.; Batista, M. Early development of corn seedlings primed with synthetic tenorite nanofertilizer. *J. Seed Sci.* **2020**, *42*, e202042040. [CrossRef]
61. Gopal, V.; Chellapandian, K. Synthesis of hybrid framework of tenorite and octahedrally coordinated aluminosilicate for the robust adsorption of cationic and anionic dyes. *Environ. Res.* **2023**, *220*, 115111. [CrossRef]
62. Gunawardana, B.; Singhal, N.; Swedlund, P. Dechlorination of pentachlorophenol by zero valent iron and bimetals: Effect of surface characteristics and bimetal preparation procedure. *Proc. Annu. Int. Conf. Soils Sediments Water Energy* **2012**, *17*, 8.
63. Hanesch, M. Raman spectroscopy of iron oxides and (oxy) hydroxides at low laser power and possible applications in environmental magnetic studies. *Geophys. J. Int.* **2009**, *177*, 941–948. [CrossRef]
64. Deng, Y.; Handoko, A.D.; Du, Y.; Xi, S.; Yeo, B.S. In situ Raman spectroscopy of copper and copper oxide surfaces during electrochemical oxygen evolution reaction: Identification of CuIII oxides as catalytically active species. *ACS Catal.* **2016**, *6*, 2473–2481. [CrossRef]
65. Murthy, P.; Venugopalan, V.; Arunya, D.; Dhara, S.; Pandiyan, R.; Tyagi, A. Antibiofilm activity of nano sized CuO. In Proceedings of the International conference on nanoscience, engineering and technology, ICONSET 2011, Chennai, India, 28–30 November 2011; IEEE: New York City, NY, USA, 2011; pp. 580–583.
66. Robles Ardila, D.; Rodríguez Pardo, N.; Pataquiva-Mateus, A. Síntesis de nanopartículas de magnetita a partir del extracto de cáscara de papaya para la degradación de colorantes azoicos en soluciones acuosas. *Ingeniare Rev. Chil. De Ing.* **2019**, *27*, 431–442. [CrossRef]
67. Corredor Acuña, J.; Echeverría Echeverría, F. Preparación y evaluación mediante oxidación cíclica a 700 °C de recubrimientos electroless ni-p y ni-p-magnetita. *Rev. Latinoam. De Metal. Y Mater.* **2012**, *32*, 107–120.
68. Li, Q.; Kartikowati, C.W.; Horie, S.; Ogi, T.; Iwaki, T.; Okuyama, K. Correlation between particle size/domain structure and magnetic properties of highly crystalline Fe_3O_4 nanoparticles. *Sci. Rep.* **2017**, *7*, 9894. [CrossRef]
69. Rosaiah, P.; Zhu, J.; Hussain, O.; Qiu, Y. Synthesis of flower-like reduced graphene oxide–Mn_3O_4 nanocomposite electrodes for supercapacitors. *Appl. Phys. A* **2018**, *124*, 1–9. [CrossRef]

70. Park, Y.; Lee, S.W.; Kim, K.; Min, B.; Kumar Nayak, A.; Pradhan, D.; Sohn, Y. Understanding hydrothermal transformation from Mn_2O_3 particles to $Na_{0.55}Mn_2O_4 \cdot 1.5\ H_2O$ nanosheets, nanobelts and single crystalline ultra-long $Na_4Mn_9O_{18}$ nanowires. *Sci. Rep.* **2015**, *5*, 18275. [CrossRef]
71. Khan, M.; Nayan, N.; Ahmad, M.; Soon, C. Surface study of CuO nanopetals by advanced nanocharacterization techniques with enhanced optical and catalytic properties. *Nanomaterials* **2020**, *10*, 1298. [CrossRef]
72. Sun, T.; Zhao, Z.; Liang, Z.; Liu, J.; Shi, W.; Cui, F. Efficient As (III) removal by magnetic CuO-Fe_3O_4 nanoparticles through photo-oxidation and adsorption under light irradiation. *J. Colloid Interface Sci.* **2017**, *495*, 168–177. [CrossRef]
73. Villafañe, G.; Bazán, V.; Brandaleze, E.; López, A.; Maratta, A. Solid phase extraction of arsenic on modified MWCNT/Fe3O4 magnetic hybrid nanoparticles from copper ores samples with ETAAS determination. *Talanta Open* **2022**, *6*, 100149. [CrossRef]
74. Feng, L.; Cao, M.; Ma, X.; Zhu, Y.; Hu, C. Superparamagnetic high-surface-area Fe_3O_4 nanoparticles as adsorbents for arsenic removal. *J. Hazard. Mater.* **2012**, *217*, 439–446. [CrossRef] [PubMed]
75. Babaeivelni, K.; Khodadoust, A.; Bogdan, D. Adsorption and removal of arsenic (V) using crystalline manganese (II,III) oxide: Kinetics, equilibrium, effect of pH and ionic strength. *J. Enviromental Sci. Health* **2014**, *49*, 1462–1473. [CrossRef] [PubMed]
76. Manquián-Cerda, K.; Cruces, E.; Rubio, M.A.; Reyes, C.; Arancibia-Miranda, N. Preparation of nanoscale iron (oxide, oxyhydroxides and zero-valent) particles derived from blueberries: Reactivity, characterization and removal mechanism of arsenate. *J. Ecotoxicol. Environ. Saf.* **2017**, *145*, 69–77. [CrossRef]

nanomaterials

Article

Aluminum-Crosslinked Nanocellulose Scaffolds for Fluoride Removal

Ken I. Johnson [1], Sunil K. Sharma [1], Priyanka R. Sharma [1,2], Abdulrahman G. Alhamzani [3] and Benjamin S. Hsiao [1,*]

1 Department of Chemistry, Stony Brook University, Stony Brook, NY 11790, USA; priyanka.sharma@wmich.edu (P.R.S.)
2 Department of Chemical and Paper Engineering, Western Michigan University, Kalamazoo, MI 49008, USA
3 Chemistry Department, College of Science, Imam Mohammad Ibn Saud Islamic University (IMSIU), Riyadh 11623, Saudi Arabia; agalhamzani@imamu.edu.sa
* Correspondence: benjamin.hsiao@stonybrook.edu; Tel.: +1-631-632-7793

Abstract: Anionic carboxylated cellulose nanofibers (CNF) are effective media to remove cationic contaminants from water. In this study, sustainable cationic CNF-based adsorbents capable of removing anionic contaminants were demonstrated using a simple approach. Specifically, the zero-waste nitro-oxidization process was used to produce carboxylated CNF (NOCNF), which was subsequently converted into a cationic scaffold by crosslinking with aluminum ions. The system, termed Al-CNF, is found to be effective for the removal of fluoride ions from water. Using the Langmuir isotherm model, the fluoride adsorption study indicates that Al-CNF has a maximum adsorption capacity of 43.3 mg/g, which is significantly higher than that of alumina-based adsorbents such as activated alumina (16.3 mg/g). The selectivity of fluoride adsorption in the presence of other anionic species (nitrate or sulfate) by Al-CNF at different pH values was also evaluated. The results indicate that Al-CNF can maintain a relatively high selectivity towards the adsorption of fluoride. Finally, the sequential applicability of using spent Al-CNF after the fluoride adsorption to further remove cationic contaminant such as Basic Red 2 dye was demonstrated. The low cost and relatively high adsorption capacity of Al-CNF make it suitable for practical applications in fluoride removal from water.

Keywords: nanocellulose; aluminum; ionic crosslinking; cationic adsorbent; fluoride removal

1. Introduction

The world's supply of fresh water is under immense pressure as the global population rapidly expands. Today, approximately 2 billion people live without access to clean potable water [1]. Among varying water pollution issues, the problem of fluoride contamination in drinking water has become a notable challenge in some regions. While low doses of fluoride (around 0.7 ppm) can improve dental health by reducing the likelihood of cavities and tooth decay, high concentrations of fluoride in drinking water can lead to negative health effects such as dental and skeletal fluorosis, arthritis, bone damage, osteoporosis, and chronic issues [2,3]. The sources of fluoride pollution vary widely in different regions but are typically attributed to natural pollution and/or anthropogenic causes. Natural pollution is usually caused by fluoride salts that are held inside rocks and sediment, which can be released into the environment over time through weathering and other geological processes. In this case, regions dependent on ground water for drinking are particularly susceptible to the harmful effects of high fluoride concentrations. In contrast, fluoride pollution due to anthropogenic causes can range from the inappropriate use of agricultural chemicals to the mismanagement of industrial waste [4].

Recently, nanocelluloses, such as cellulose nanofibers (CNF) and cellulose nanocrystals (CNC), have been recognized as effective and sustainable nature-based water remediation materials, and they can be prepared from any lignocellulose biomass feedstock [5,6]. These materials not only retain environmentally friendly cellulosic properties, but also possess characteristics such as large surface area, abundant functionalities and scaffolding stability, thus offering new opportunities for a broad spectrum of water purification applications. The typical production of nanocellulose consists of combined chemical (e.g., from popular TEMPO-mediated oxidation [7] to the recent nitro-oxidation process (NOP) [8]) and mechanical (e.g., high pressure homogenization and ultrasonic) [6] treatments to fibrillate microscale fibers into nanoscale fibers. The conventional chemical treatments involve the conversion of primary hydroxyl groups in the anhydroglucose units of the cellulose chain into negatively charged groups such as carboxylate groups (by TEMPO and NOP treatments) that are capable of adsorbing cationic contaminants such as toxic metal and ammonium ions. However, the anionic nanocellulose scaffold is not effective at removing anionic contaminants, such as fluoride and nitrate ions.

In this study, we used a simple approach to convert anionic carboxylate nanocellulose into a cationic scaffold by crosslinking the nanocellulose with aluminum ions. The chosen nanocellulose was based on CNF produced from raw jute fibers using the zero-waste NOP approach [8]. The NOP approach can be understood by a simple analogy borrowed from the kitchen. To make vegetable soup, one starts with vegetables and water. When cooked, the nutrients are fused with the water and the vegetables are turned into insoluble pulp. In NOP, the essential ingredient, nitric acid, is the same basic chemical used to produce synthetic nitrogen fertilizers for farming. NOP works because nitric acid is one of the most widely used digestion reagents for the decomposition of organic matter, as well as a primary oxidant for the oxidation of cellulose. In the NOP treatment, the lignocellulosic component is simultaneously pulped (delignified) and oxidized, resulting in a carboxylated cellulose scaffold that can be electrostatically swollen and mechanically defibrillated in the nanoscale with low energy. As the processed effluent from NOP contains nutrients released from the feedstock, it can be neutralized into effective fertilizers to augment synthetic fertilizers. Consequently, the NOP approach represents a zero-waste process, where the production cost of CNF by NOP can be a fraction of that by conventional methods.

In this study, raw jute was chosen as a model non-woody biomass feedstock to prepare NOCNF. It possesses a looser structure and relatively lower lignin content, similar to many agricultural residues and natural fibers, which need less energy (and chemical treatments) to defibrillate. It has recently been pointed out that non-woody biomasses, although abundant, are relatively underutilized around the globe when compared to woods [6]. To convert carboxylated CNF into a cationic scaffold, we used aluminum ions as an ionic crosslinking agent. Aluminum is the most abundant metallic element in the Earth's crust, and its +3 charge makes the ionic species an effective crosslinking agent to bind the cellulose nanofiber scaffold (the resulting composite scaffold is termed Al-CNF). The presence of aluminum species renders the scaffold a positively charged medium which can adsorb negatively charged contaminants such as fluoride from water.

We note that the removal of fluoride contaminants can be carried out through various techniques, including ion-exchange, adsorption, coagulation–precipitation/electro coagulation, and filtration techniques such as reverse osmosis [9–11]. The advantage of adsorption over other techniques includes its high removal efficiency and potentially low cost. One disadvantage over other techniques is a sensitivity towards pH and other competing salts which may be in solution [12–14]. Currently, the common adsorption materials include aluminum oxide and activated carbon, which generally offer low-cost and high adsorption efficiency features to remove fluoride. We argue that the approach of using Al-CNF as an adsorptive medium for fluoride removal may offer another cost-effective pathway that can complement the existing adsorbents, especially when CNFs are prepared using the zero-waste NOP approach. Furthermore, the use of Al-CNF can mitigate the risk of secondary contamination from the use of typical aluminum-based adsorbents (such as activated alumina nanoparticles) at neutral pH levels.

2. Materials and Methods

2.1. Materials

Untreated jute fibers were provided by Toptrans Bangladesh Ltd. (Dhaka, Bangladesh), Nitric acid (65 wt%), sodium nitrite (97%), aluminum sulfate hydrate ($\geq$97 wt%), hydrochloric acid (1.0 N), and sodium fluoride ($\geq$99%) were purchased from Sigma Aldrich, St. Louis, MO, USA. Sodium hydroxide (99%) was purchased from Macron Fine Chemicals, Thermo Fisher Scientific, Waltham, MA, USA. Sodium bicarbonate (99%) was purchased from Fisher Scientific, Waltham, MA, USA. All chemicals were used without further purification.

2.2. Preparation of Nitro-Oxidized Cellulose Nanofibers (NOCNF)

NOCNF were prepared using the following procedures. In brief, 1 g of untreated ground jute was placed in a 1 L round-bottom flask, to which 14 mL of nitric acid was subsequently added. When the samples became completely soaked in the acid, 0.96 g of sodium nitrite was added to the reacting mixture under continuous stirring. Upon the addition of sodium nitrite, red/brown fumes evolved inside the flask, which was sealed with stoppers. The reaction was performed at 50 °C for 12 h, and then subsequently quenched by adding 250 mL of deionized water to the flask. The resulting mixture was transferred to a beaker for decantation. After the solids settled, the supernatant liquid was removed and neutralized for fertilization study, which will be described elsewhere. The slurry was further washed with deionized water and the decantation step was repeated until the pH level of the upper supernatant was above 2.5. The fiber slurry was then transferred to a dialysis bag (Spectral/Por, Thermo Fisher Scientific, Waltham, MA, USA, MWCO: 6–8 kD) and equilibrated for 4–5 days until the conductivity of the wash water reached below 5 µS/cm. As the pH value of the resulting cellulose nanofibers was between 2.8 and 3.0 due to the presence of carboxylic acid groups, the agglomeration of fibers was seen during the dialysis. To convert the carboxylic acid groups into carboxylate groups, the resulting fibrous suspension was treated with 4 wt% sodium bicarbonate solution, until the pH level reached 7.5. The treatment produced carboxylate groups on cellulose chains and negative charges on the cellulose surface. To create nanofibers, the slurry was passed through a homogenizer at 250 bar for 1 pass to fibrillate the microfibers into NOCNF.

2.3. Preparation of Al-CNF Composite Suspension

The preparation of aluminum-crosslinked NOCNF suspensions was carried out as follows: 100 g of 0.294 wt% suspension of NOCNF was added to a 250 mL beaker under vigorous stirring using a one-inch Teflon coated stir bar. In this preparation, one mole of aluminum ion was added to one mole of the glucose unit in nanocellulose to ensure an excess amount of aluminum ions for crosslinking. In this scenario, with the 1.00 molar solution of aluminum ions from aluminum sulfate hydrate, 3630 µL was added dropwise to the NOCNF suspension. It was found that upon adding approximately half of the aluminum solution, a rigid gel formed immediately. Stirring was continued to break down the gel to allow for the continuing addition of aluminum solution. Upon the complete aluminum solution addition, the suspension was continuously stirred for an additional 3 h. Subsequently, the final suspension was centrifuged, then decanted, and the supernatant was removed. Deionized water was then added to the sample to resuspend the final product. This procedure was repeated until the supernatant of the suspension maintained a conductivity below 10 µS/cm. The final suspension sample was kept at 2–4 °C until further study.

2.4. Characterization of NOCNF and Al-CNF

2.4.1. Conductometric Titration

In this test, 40 g of NOCNF suspension was diluted to 0.2 wt% using 160 g of water under stirring at a rate of 300 rotations per minute (rpm). The dilution process allowed the suspension to be stirred easily while avoiding gelation during the addition of salt solutions. The pH level of the suspension was then adjusted to 2.5 using dilute hydrochloric acid, and subsequently titrated with 0.05 M of sodium hydroxide in 1 mL increments until the pH value of 10 was reached (a more dilute sodium hydroxide solution was used in the late stage of the dilution).

The conductivity of the suspension was measured during titration and plotted against the volume of the titrant. The resulting graph usually exhibited three linear regions. From volume 0 to V_1, the increase in titrant should decrease the conductivity until a low plateau volume (V_1) is reached. This is the result of neutralizing the excess amount of free acid in the suspension. The further addition of titrant does not change the conductivity of the suspension because the process neutralizes protons bound to the carboxylic acid groups on CNF. This is true to a certain volume (V_2) of titrant used, above which additional titrant introduces free hydroxide and increases the conductivity. The carboxylic acid concentration can thus be calculated using Equation (1) [8].

$$\frac{\text{mmol carboxylate}}{\text{g NOCNF(s)}} = \frac{(V_2 - V_1) \times C_t}{m} \tag{1}$$

where C_t represents the titrant concentration. The numerator is typically converted into the unit of mmols. The symbol m represents the mass of solid NOCNF, which can be calculated by first converting the weight percentage (wt%) of the suspension into a fraction (i.e., the mass of solid NOCNF over the mass of suspension). The solid mass is then quantified by multiplying by the mass of original suspension used before dilution. The final carboxylic acid concentration represents the carboxylic acid content in mmols of carboxylic acid per gram of solid NOCNF.

2.4.2. Fourier-Transform Infrared Spectrometry

Fourier-transform infrared (FTIR) spectroscopy was recorded on a Nicolet iS10 FT-IR Spectrometer by Thermo Scientific using the attenuated total reflectance (ATR) mode. Sixteen scans were averaged at a resolution of 4. The instrument was equipped with a DTGS KBr detector (Thermo Fisher Scientific, Waltham, MA, USA), KBr beam splitter (Waltham, MA, USA), IR source, Smart iTR accessory, and diamond window. Samples were generally measured from 400 to 4000 cm^{-1}.

2.4.3. X-ray Photoelectron Spectroscopy

X-ray photoelectron spectroscopy (XPS) was performed on a custom-built instrument. The instrument contained an X-ray source from PHI Electronics (Chanhassen, MN, USA), a spectrometer from V.G. Scientific CLAM 100 (VG Scientific, Waltham, MA, USA), and a VGX900I controller (Waltham, MA, USA) data collection system. The chosen X-ray source was Al $K\alpha_{1,2}$, and the measurement was made under an ultrahigh vacuum of 1.0×10^{-8} torr. The measurement was collected at a 90° take-off angle relative to the sample film. The sample charging was corrected using the C1s peak of adventitious carbon (a thin layer of carbonaceous material on the surface of the air-exposed sample) at 284.8 mV. The Na KLL Auger peak was typically observed at 497 eV from the COONa group.

2.4.4. Zeta Potential

The average zeta potential of the suspension samples was measured using a ZetaProbe Analyzer (Colloidal Dynamics, Palm Valley, FL, USA). The value was determined based on the assumption that the CNF has a dielectric constant of 5.0 and a density of 1.5 g/mL. In the measurement, aqueous suspensions were stirred at 300 rpm and titrated with an auto-burette. The instrument was equipped with a niobium electrokinetic sonic amplitude (ESA) electrode and was calibrated using a KSiW suspension. Additionally, samples were treated with either hydrochloric acid or sodium hydroxide using an auto-burette to achieve the desired pH value before the analysis.

2.4.5. UV–VIS Analysis

A Genesys 10S UV–Vis Spectrophotometer (Thermo Scientific, Waltham, MA, USA) was used to measure the transmittance (%) of the sample in a quartz cuvette from 190 to 1100 nm using a 1 nm increment. The instrument was equipped with a Shimadzu SPD-M20A UV/VIS (Shimadzu, Kyoto, Japan) photodiode array detector.

2.4.6. Scanning Electron Microscopy

Scanning electron microscopy (SEM) was carried out on a ZEISS Crossbeam 340 instrument (ZEISS, Oberkochen, Germany) using an electron high tension (EHT) of 3 kV and an accelerating voltage of 30 kV. Samples were prepared by casting a 0.1 wt% suspension of NOCNF on a silicon wafer, and then coating by gold sputtering. The instrument was also equipped with energy-dispersive X-ray spectroscopy (EDS) capability for elemental mapping. The EDS analysis was conducted in conjunction with SEM imaging.

2.4.7. Transmission Electron Microscopy

Transmission electron microscopy (TEM) was conducted on a JEOL JEM 1400 instrument (JEOL, Peabody, MA, USA) using an accelerating voltage of 120 kV. The samples were prepared on 300 mesh copper grids (Ted Pella Inc., Redding, CA, USA) by casting 10 μL of a 0.01 wt% sample onto the grid. After removing the excess fluid, the sample was stained with 10 μL of 2 wt% aqueous uranyl acetate solution. The excess solution was removed, and the grid was left to air dry.

2.4.8. Atomic Force Microscopy

Atomic Force Microscopy (AFM) measurements were performed using a Bruker Dimension ICON instrument (Bruker, Billerica, MA, USA). This instrument was equipped with a Bruker OTESPA tip (the radius was 10 nm). In sample preparation, 10 μL of a 0.005 wt% suspension was deposited on the surface of a silica plate and air-dried, and the measurement was carried out in the tapping mode.

2.4.9. Wide-Angle X-ray Diffraction

A MiniFlex instrument from Rigaku, Japan, was used to measure the wide-angle X-ray diffraction (WAXD) patterns. The samples were measured from 5 degrees to 85 degrees in steps of 0.02 degrees, at a speed of 5 degrees per minute. The scan axis was set to theta/2-theta in a continuous mode, where the intensity was measured in counts per second. In this measurement, the voltage and current were set to 40 kV and 15 mA, respectively, using Cu Kα radiation. Additionally, an incident side and receiving side Soller slit of 5.0 degrees along with an incident-beam divergence-limiting slit of 1.250 degrees were used in the measurement.

2.4.10. ^{13}C and ^{27}Al Cross-Polarization Magic-Angle Spinning Nuclear Magnetic Resonance

^{13}C and ^{27}Al cross-polarization magic-angle spinning nuclear magnetic resonance (CP-MAS-NMR) of Al-CNF were carried out using a Bruker Utrashield 500 WB plus (500 MHz) instrument (Bruker, Billerica, MA, USA). The instrument was equipped with a 2.5 mm triple resonance MAS NMR probe, capable of spinning the sample up to 35 KHz. The chosen resonance frequency was 10,000 Hz, and the samples were spun at the magic angle at a speed of 10 KHz.

2.4.11. Thermogravimetric Analysis

Thermogravimetric analysis (TGA) was taken on a TA Q50 instrument (TA Instrument, New Castle, DE, USA). In this measurement, the sample was heated from 25 °C to 800 °C at a rate of 10 °C per minute under a nitrogen atmosphere. The sample was loaded on a platinum pan, where its mass was monitored as a function of temperature. The derivate weight, shown as %/min, was also plotted as a function of temperature, and its profile is typically referred to as derivative thermogravimetry (DTG). The DTG profile usually exhibits a clearer change in the weight loss.

2.4.12. The Fluoride Adsorption Study

The fluoride adsorption capacity of Al-CNF was determined using different concentrations of sodium fluoride solutions (5 ppm to 2500 ppm), where the concentration was determined by a fluoride ion-selective electrode (ISE, San Diego, CA, USA, model number ISEF121, HACH, USA). In this study, 3 mL of Al-CNF suspension of known weight percent (wt%) was mixed with 3 mL of sodium fluoride of known concentration. After equilibrating for an hour, the samples were centrifuged at 10,000 rcf for 5 min to separate the solid and liquid phases. The supernatant portion was analyzed using the ISE and compared against the control samples (i.e., 3 mL of water and 3 mL of sodium fluoride solution with known concentration). To study the pH effect on the fluoride adsorption capacity, the pH value of the sample was systematically adjusted before the incubation period.

The methodology to determine the fluoride removal efficiency is as follows. The initial concentration of fluoride is the concentration of fluoride solution to be treated with Al-CNF (the fluoride solution was mixed with the Al-CNF suspension in a 1:1 volume ratio). The equilibrium fluoride concentration thus equals half of the initial fluoride concentration. All samples are then tested against the control solution (i.e., the 1:1 mixture of the same fluoride solution and water). The efficiency is defined as the percentage removal of fluoride between the control and the sample. Difference in concentration is the difference in the fluoride concentration between the sample and its respective control. As a result, the amount of fluoride removed represents the conversion of fluoride concentration into solid fluoride. The quantity of Al-CNF is derived from the weight percent (wt%) and the volume of Al-CNF suspension used. Q_e represents the ratio between the fluoride removed and the quantity of Al-CNF used (mg/g).

Furthermore, mixed solutions of fluoride, sulfate, and nitrate were prepared to evaluate the selectivity of Al-CNF. For this purpose, a solution containing equal moles of three salts, sodium fluoride, sodium sulfate, and sodium nitrate, was prepared and then mixed with Al-CNF (1:1 by volume) and then left to equilibrate overnight.

3. Results and Discussion

3.1. Structure and Functionality of NOCNF and Al-CNF

It is interesting to observe that at the same concentration of 0.5 wt%, the Al-CNF suspension is much more viscous than the NOCNF suspension. With a slight increase in concentration, the Al-CNF system can readily form a stable gel with solid-like property. Figure 1a illustrates the transmittance curves of two suspensions based on NOCNF (0.5 wt%) and Al-CNF (0.5 wt%; to prepare this suspension, 250 mL of 0.5 wt% NOCNF suspension was mixed with 15.4 mL of 1.0 M aluminum solution using aluminum sulfate), respectively, as a function of the wavelength. Visually, the NOCNF suspension is relatively transparent, but the Al-CNF suspension is near-opaque. For example, at the wavelength of 800 nm, the % transmittance of the NOCNF suspension is 79%, but that of the Al-CNF suspension is only 13% (i.e., a six-fold reduction). This indicates that the scaffolds in Al-CNF are much more tightly bound, and some clusters may be formed. The presence of Al-CNF cluster can scatter the light and greatly decrease the transmittance. In contrast, the NOCNFs are relatively more homogeneously dispersed in the suspension, leading to a higher transmittance value. Figure 1b displays the zeta potential values of the NOCNF and Al-CNF suspensions as a function of the pH value, which was adjusted using dilute concentrations of hydrochloric acid or sodium hydroxide. This study confirms the anionic characteristics of NOCNF at the pH level below 7.0, where its zeta potential is in the range of −50 mV. At a high pH value (e.g., pH = 9.4), the zeta potential of the NOCNF is zero, indicating that it is not charged. In contrast, the zeta potential values of the Al-CNF suspension are seen to be around zero in the pH range tested (pH from 4.2 to 9.2, Figure 1b), indicating that the Al-CNF sample is in the neutral state. As the ionic bonding between the carboxylate group of NOCNF and the aluminum ion is very strong, it is not affected by the change in pH value.

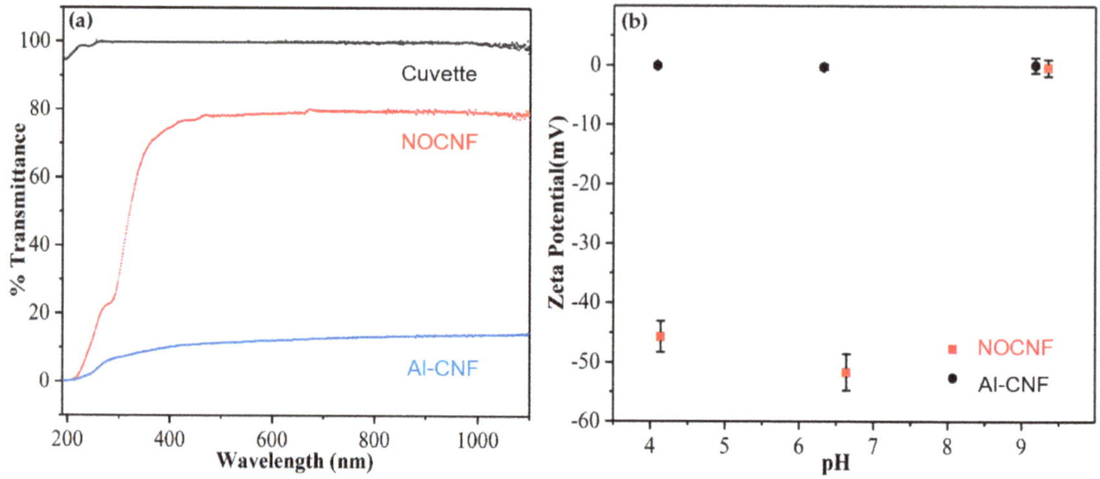

Figure 1. (**a**) The % transmittance curves of nanocellulose suspensions (0.5 wt%) containing NOCNF and Al-CNF as a function of the wavelength. (**b**) Zeta potential of corresponding NOCNF and Al-CNF suspensions as a function of pH.

Figure 2a shows the FTIR spectra of dried NOCNF, Al-CNF, and the Al-CNF–sodium fluoride (NaF) mixture (from the 1:1 mixture of Al-CNF suspension and 500 ppm of sodium fluoride solution). All three samples exhibit the signature cellulose vibrational peaks, such as the carbonyl bond stretching at 1595 cm^{-1}, C-OH stretching at 3311 cm^{-1}, C-H stretching at 2901 cm^{-1} (all three characteristic peaks were indicated by yellow lines in Figure 2a), and C-O stretching at 1035 cm^{-1} [15]. It is interesting to note that the 1595 cm^{-1} becomes slightly broader with the addition of aluminum, possibly due to the formation of aluminum–carboxylate ionic interaction, where this peak becomes relatively more intense due to the presence of sodium fluoride, which possesses a vibrational peak at approximately the same location.

Regarding the crystalline structure of varying samples, Figure 2b shows the WAXD profiles of dried NOCNF, Al-CNF, and Al-CNF/NaF mixture samples, and pure NaF crystals. Both NOCNF and Al-CNF samples exhibit the typical WAXD pattern of the cellulose I structure, in which the three major diffraction peaks at 2θ positions of 16°, 18°, and 24° can be indexed by the (101), $(10\overline{1})$, and (002) planes, respectively [16]. This confirms that the cellulose I crystal structure is maintained after NOP treatment. In the Al-CNF/NaF mixture, several new peaks appear (2θ positions at 20°, 24°, 40°, 48°, and 57°). The peaks at 2θ positions of 40° and 57° can be indexed by the (200) and (220) planes of the NaF crystal, respectively. The formation of the NaF crystal is due to the presence of excess NaF solution in the Al-CNF scaffold after being dried. However, the 20°, 24°, and 48° peaks that appear are not from the NaF crystal structure. This indicates the formation of a new crystal structure, probably from the interactions between aluminum and fluoride ions, such as the formation of aluminum fluoride hydrate crystals (e.g., β-AlF_3•$3H_2O$).

Al-CNF was analyzed using ^{27}Al CPMAS-NMR, and the result is shown in Figure 2c. In this figure, a distinct chemical shift in the aluminum species at 18.6 ppm can be observed, which is consistent with the known range for AlO_6 [17]. This is expected, as the aluminum ion can coordinate with the oxygen atoms in molecules of water, hydroxide, alcohol, and carboxylic acid. Specifically, Al^{3+} can form up to six coordinate bonds with these oxygen-containing groups in an octahedral arrangement. The ^{13}C CPMAS-NMR spectra for NOCNF and Al-CNF are shown in Figure 2d. Both spectra exhibit a similar structure: the peaks at 63–65 ppm are due to the C6 primary alcohol, peaks at 72–75 ppm are due to the C5, C3, and C2 carbons, peaks at 84–89 ppm represent the C4 carbon, and the peak at 105 ppm represents the C1 carbon. These features have been explained in our previous publication [8]. Perhaps the only feature difference between the two spectra is the peak at 175 ppm due to the C6 carbonyl carbon. The slight broadening of the 175 ppm in Al-CNF is due to the bonding between the aluminum and carboxylate groups.

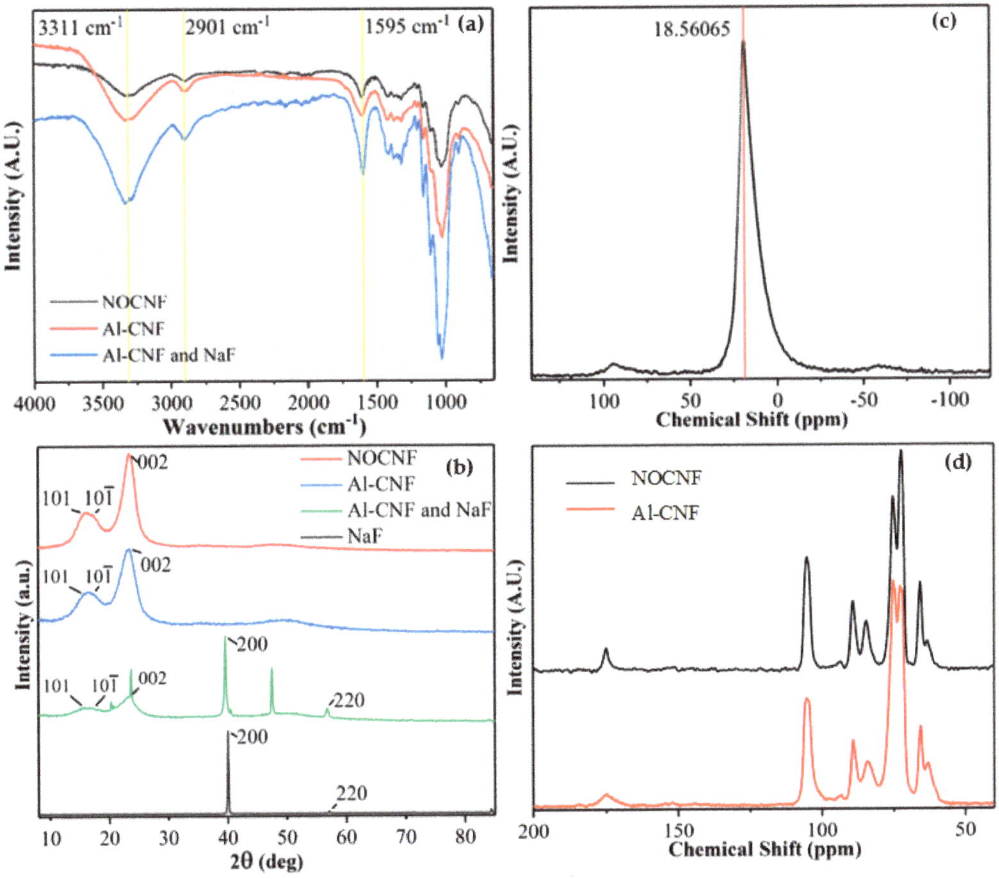

Figure 2. (**a**) The FTIR spectra of NOCNF, Al-CNF, and 1:1 Al-CNF mixed with 500 ppm of fluoride solution. (**b**) WAXD of NOCNF, Al-CNF, and 1:1 Al-CNF mixed with 500 ppm fluoride solution and sodium fluoride. The new peaks (24° and 48°) that appear in the Al-CNF/NaF sample indicate that crystallization occurs between aluminum and fluoride (e.g., the formation of aluminum fluoride hydrate crystals such as β-$AlF_3{\bullet}3H_2O$). (**c**) The ^{27}Al CPMAS-NMR spectra of Al-CNF (a major peak appears at 18.561 ppm). Additionally, (**d**) ^{13}C CPMAS-NMR spectra of NOCNF and Al-CNF samples.

XPS scans of NOCNF and Al-CNF samples are shown in Figure 3a, which displays the elemental composition as well as the chemical and electronic states of the atoms within these materials. Using the sensitivity factor for C1s, O1s, and Al2p of 1.00, 2.93, and 0.753, respectively, the elemental composition of COONa was determined to be 73.0% carbon and 27.0% oxygen. Similarly, the elemental composition of COOAl was calculated to be 58.7% carbon, 34.1% oxygen, and 7.2% aluminum. The C1s peaks in Figure 3a associated with the COONa and COOAl groups were further deconvoluted, and the results are illustrated in Figure 3b. The C1s peak of COONa was deconvoluted into four peaks using a procedure including the application of a Shirley background, and fitting with four Gaussian functions constrained by peak position, peak area, and FWHM. The fitting results indicate the presence of C–C and C–H bonds (65.5%) corresponding to the cellulose backbone and hydrocarbon components, C–OH bonds (21.7%) corresponding to the carbon and hydroxyl bonds in cellulose, O–C–O bonds (8.7%) corresponding to the acetal groups in cellulose, and O=C–O bonds (3.9%) corresponding to the carboxylate groups. In contrast, the C1s peak of COOAl was deconvoluted into five peaks using the above procedures. The results indicate the presence of C–C and C–H bonds (70.2%) from the cellulose backbone and hydrocarbon component, C–OH bonds (17.4%) from the carbon and hydroxyl bonds in cellulose, O–C–O bonds (8.2%) corresponding to the acetal groups in cellulose, O=C–O bonds (2.1%) corresponding to the carboxylate groups, and a small peak due to the O=C–OAl bonds (2.1%) at around 288 nm (the inset in Figure 3b shows the deconvolution of the COOAl peak from 293 to 287 nm, where the results indicate the bonding between aluminum and carboxylate groups).

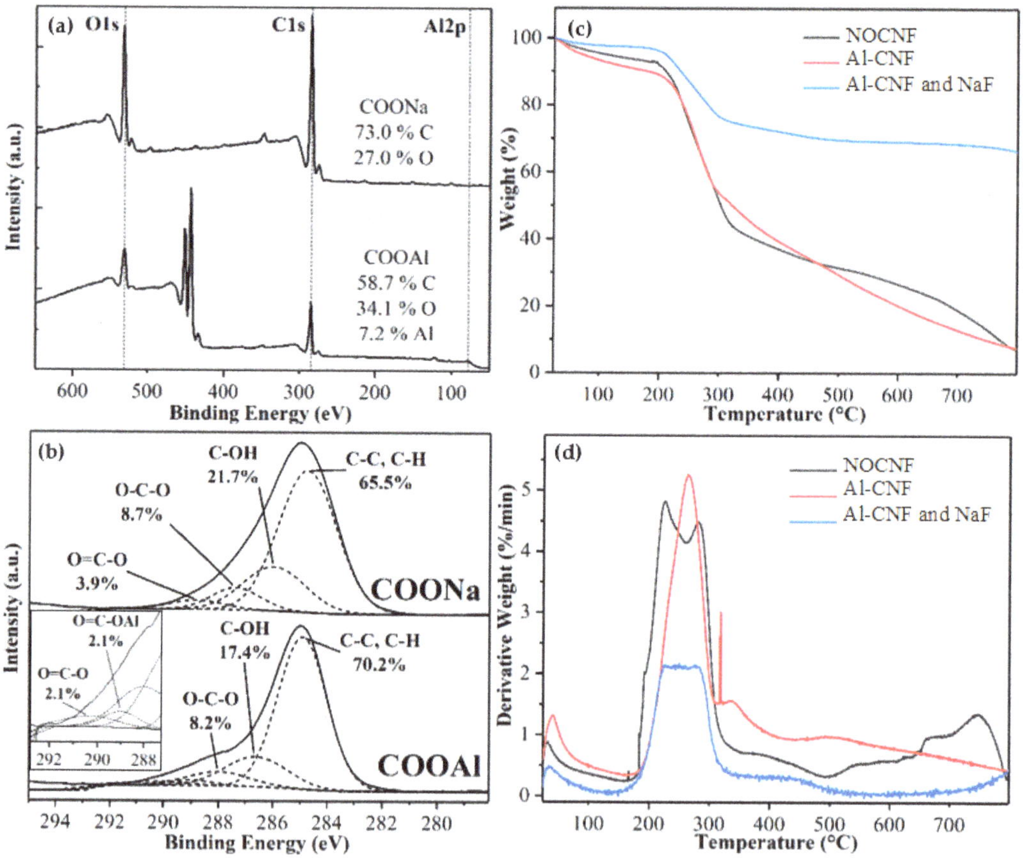

Figure 3. (**a**) XPS scans of NOCNF (indicating the COONa group) and Al-CNF (indicating the COOAl group). (**b**) The C1s peak in XPS associated with the COONa and COOAl groups, and the corresponding peak deconvolution showing relative concentrations of different bonds. (**c**) TGA profiles of NOCNF, Al-CNF, and Al-CNF/NaF mixture. (**d**) Corresponding DTG of NOCNF, Al-CNF, and Al-CNF/NaF mixture.

Figure 3c,d shows the TGA and DTG curves of NOCNF, Al-CNF, and Al-CNF/NaF mixture, respectively. All samples show that the onset temperatures of degradation begin at about 160 °C. The major peaks in the DTG curves for three samples occur in the range between 160 and 370 °C, representing the degradation of glucuronic acid and D-glucose in the cellulose chains. However, the TGA curves show that the associated weight losses for NOCNF and Al-CNF are similar and relatively large (about 60%), while that for Al-CNF/NaF is relatively small (about 30%). An additional peak at about 350 °C is seen in the DTG curve of Al-CNF, which may be a result of the CNF network crosslinked by aluminum with a slight increase in thermal stability. It is interesting to observe that the residual weight of Al-CNF/NaF is 68.9%. This may be largely due to the unabsorbed sodium fluoride crystals in the Al-CNF scaffold, which is consistent with the WAXD results in Figure 2b.

The morphology of the NOCNF, Al-CNF, and Al-CNF/NaF samples was investigated using several microscopic techniques. The AFM images of (i) NOCNF and (ii) Al-CNF samples cast on the silica wafer substrate are shown in Figure 4. It can be seen that the fiber length of NOCNF is relatively short, rendering its morphology more similar to that of CNC. A noticeable distinction is seen between the NOCNF and Al-CNF samples. In NOCNF, the fibers are relatively dispersed, and a higher degree of aggregation is observed in Al-CNF. Fiber aggregation behavior induced by the aluminum crosslinking interaction is expected in Al-CNF.

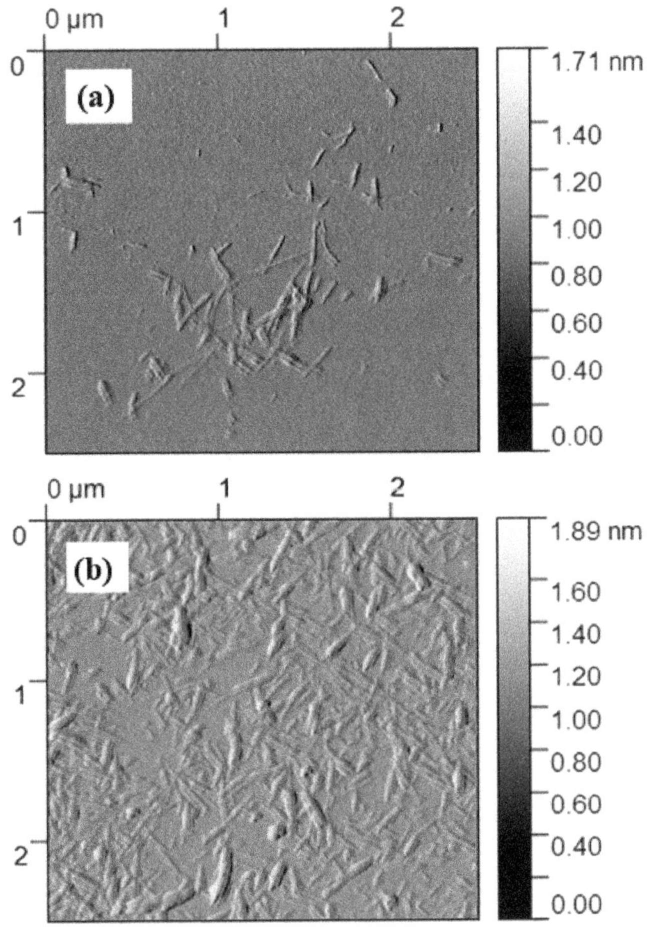

Figure 4. AFM images of (**a**) NOCNF and (**b**) Al-CNF samples cast and dried on the silica wafer substrate.

Figure 5 shows the SEM images and corresponding EDS profiles of Al-CNF and Al-CNF/NaF samples. In Figure 5a, the SEM image of Al-CNF exhibits a relatively porous structure due to the crosslinking effect of aluminum, causing a high degree of local CNF aggregation. The corresponding EDS profile confirms the presence of aluminum with no signs of sodium or sulfur, corroborating the completion of ion exchange by the adopted NOP approach. The presence of silicon comes from the substrate and gold from the sputter coating. In contrast, the SEM image in Figure 5b of Al-CNF/NaF exhibits a relatively smooth surface structure with low porosity. This is consistent with the adsorption of fluoride in the Al-CNF scaffold, and the formation of free NaF crystals. The corresponding EDS profile in Figure 5b confirms the presence of sodium and fluoride elements, where the relative low fluoride concentration (as compared to the sodium concentration) indicates that some fluoride was adsorbed and hidden in the Al-CNF scaffold. This hypothesis is verified by the EDS elemental colored SEM image of the Al-CNF/NaF sample, with (1) discrete NaF nanocrystals with a cubic structure (~20 nm) resulted from a cubic motif where both Na^+ and F^- occupy octahedral coordination sites, and (2) sodium-rich Na and F nanostructures (probably not in the crystal form) which can be seen in Figure 6. The TEM images showing the fine morphological structures of NOCNF, Al-CNF, and Al-CNF/NaF samples are illustrated in Figure 7. The TEM images are consistent with the AFM images, which indicates the relative length of nanofibers, and the aggregation of nanofibers in Al-CNF. The TEM image of the Al-CNF/NaF mixture shows an even more aggregated structure, containing discrete black nanoparticles (NaF nanocrystals) and dark thin films covering the nanofiber texture. This thin-film structure appears to contain Al-CNF scaffolds which adsorb fluoride ions and are then further covered by sodium ions (as in Figure 6).

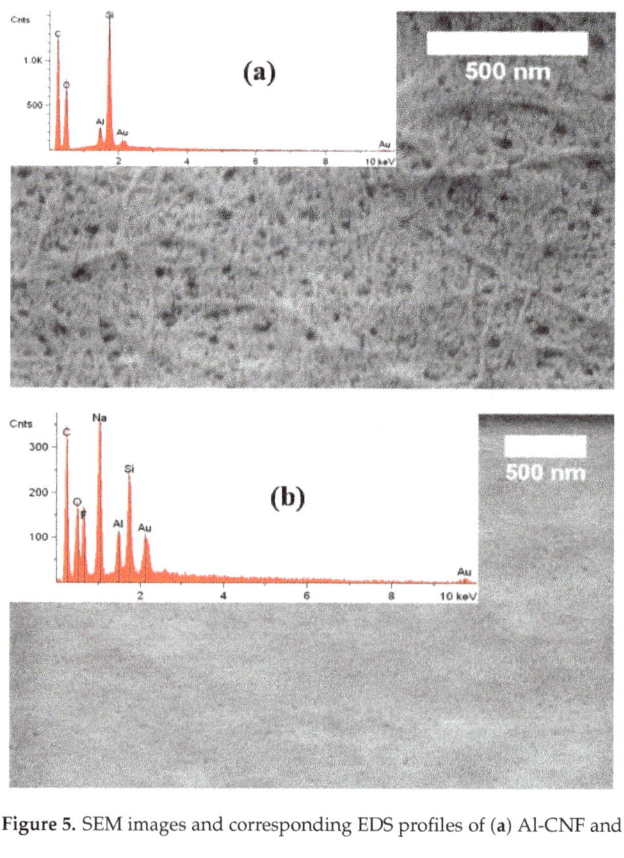

Figure 5. SEM images and corresponding EDS profiles of (**a**) Al-CNF and (**b**) Al-CNF/NaF samples.

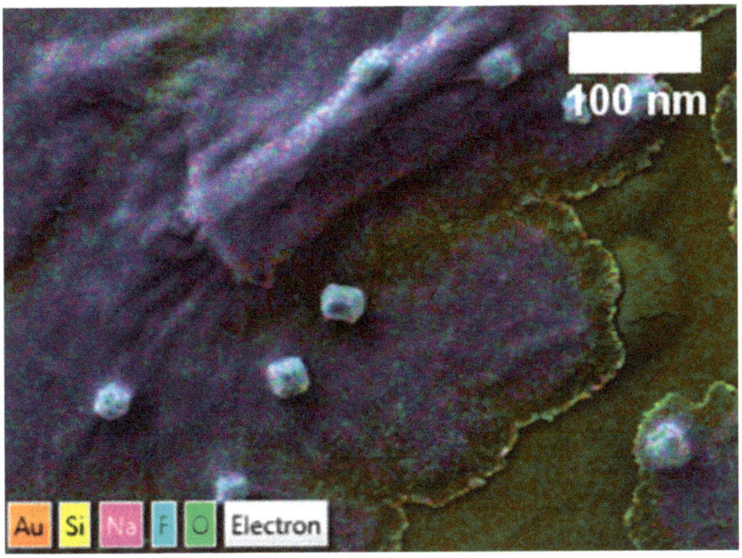

Figure 6. EDS color layered with the SEM image of the Al-CNF/NaF sample.

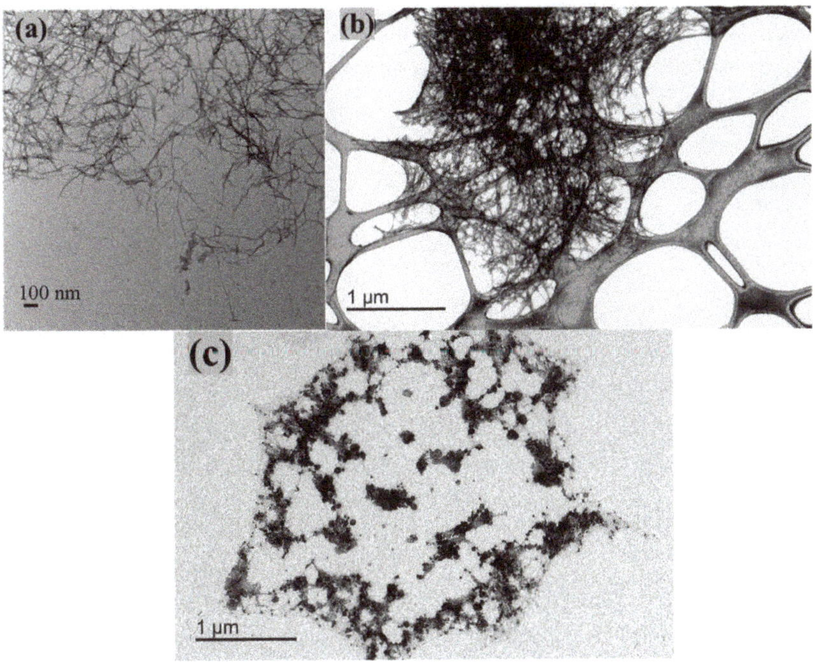

Figure 7. TEM images of (**a**) NOCNF, (**b**) Al-CNF, and (**c**) Al-CNF/NaF samples.

3.2. Fluoride Adsorption Study Using Al-CNF

The fluoride adsorption study of Al-CNF was carried out using the sodium fluoride solutions of different concentrations. The experimental results are summarized in Table 1. Based on these results, Figure 8a shows the experimental adsorption capacity (Q_e) in the mg/g of Al-CNF plotted against the equilibrium fluoride concentration (C_e) in mg/L. In this figure, it can be seen that the fluoride adsorption capacity increases with the equilibrium fluoride concentration until the adsorption process reaches a plateau value (or asymptote). This plateau value (or the maximum experimental adsorption capacity) is 42.1 mg/g. Figure 8a indicates that with an increase in equilibrium fluoride concentration, the aluminum ions (cross-linkers) in the Al-CNF scaffold can further bond to fluoride ions, resulting in rapid uptake at the initial adsorption stage. As the active aluminum sites in Al-CNF become fully occupied, the scaffold becomes saturated and can no longer adsorb additional fluoride ions, resulting in the plateau region. We hypothesize that fluoride adsorption follows the monolayer adsorption process, which can be described by the Langmuir isotherm model with the following expression [18,19].

$$\frac{C_e}{Q_e} = \frac{C_e}{Q_m} + \frac{1}{Q_m b} \tag{2}$$

where Q_e represents the equilibrium adsorption capacity, C_e is the equilibrium concentration of fluoride, Q_m is a constant representing the maximum adsorption capacity, and b is a constant related to free adsorption energy. According to Equation (2), Q_m can be determined from the inverse slope of the Langmuir plot (i.e., the plot of C_e/Q_e against C_e), which is shown in Figure 8b. This figure exhibits a good linear relationship with an R^2 value (0.999), which indicates the validity of the Langmuir isotherm model. We also used the Freundlich isotherm model, assuming the adsorption follows the multilayer deposition pathway to analyze the isotherm results, and the analysis results were not as good (the R^2 value was around 0.98) as those from the Langmuir isotherm model. This indicates that the adsorption mechanism follows the monolayer deposition pathway, which is consistent with the prior hypothesis that the mechanism of fluoride removal is a chemisorptive ligand exchange reaction involving the formation of the internal complexation of fluoride with aluminum [20].

Table 1. The results of the fluoride adsorption study using Al-CNF (the calculated Q_m is 43.3 mg).

Calc. Eqlbrm. Con. of Fluoride, C_e (ppm F^-)	Con. of Water + Fluoride Sol. 1:1 (ppm F^-)	Con. of Al-CNF (0.0760 wt%) + Fluoride Solution 1:1 (3 mL each) (ppm F^-)	Diff. in Con. (ppm)	Percent Removal (%)	Fluoride Removed (mg)	Quantity of Al-CNF (g)	Q_e (mg/g)	C_e/Q_e
0.625	0.4072	0.04239	0.3648	89.6	0.002189	0.00228	0.960	0.651
6.25	6.040	4.954	1.086	18.0	0.006516	0.00228	2.86	2.19
62.5	65.61	59.09	6.52	9.9	0.0391	0.00228	17.2	3.64
500	481.6	467.1	14.5	3.0	0.0870	0.00228	38.2	13.1
2500	2243	2227	16	0.7	0.096	0.00228	42.1	59.3

Based on the Langmuir fitting analysis, the maximum adsorption capacity (Q_m) of Al-CNF is 43.6 mg/g. Compared to other published adsorbent materials used for fluoride removal, Al-CNF performs relatively well; the comparison is shown in Table 2. It can be seen that Al-CNF outperforms most naturally occurring inorganic minerals, such as alumina. Additionally, Al-CNF tends to do better than graphite, carbon nanotube, or chitosan-based materials. While some synthetic organic adsorbents (e.g., glutaraldehyde-crosslinked calcium alginate) can outperform Al-CNF, Al-CNF possesses the characteristic of better sustainability.

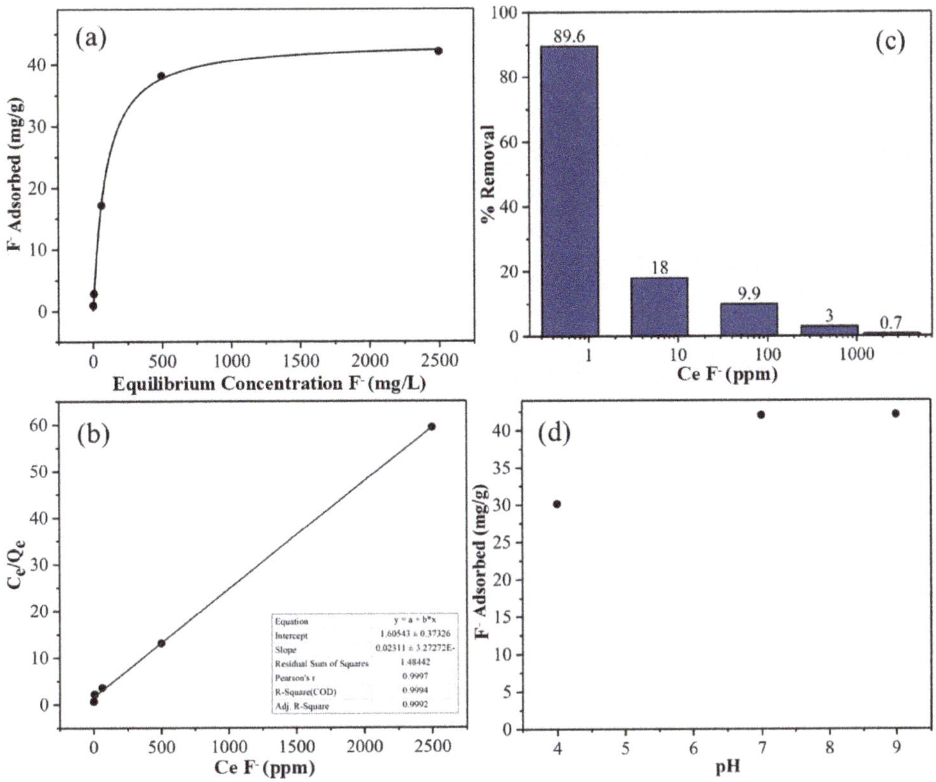

Figure 8. (**a**) Adsorption capacity of Al-CNF as a function of the fluoride concentration with the Langmuir isotherm fitting. The maximum experimental adsorption capacity is 42.1 mg/g measured at an equilibrium concentration of 2500 ppm fluoride solution. (**b**) The Langmuir isotherm fitting curve, where C_e/Q_e is plotted over C_e with an R^2 value of 0.999. A maximum theoretical adsorption capacity (Q_m) can be calculated to be 43.3 mg/g. (**c**) Percentage removal of fluoride using Al-CNF with varied equilibrium fluoride concentrations. (**d**) The adsorption capacity of Al-CNF at pH = 4, 7, and 9.

Table 2. The maximum adsorption capacity values (Q_m) of varying adsorbents and Al-CNF.

Adsorbent	Maximum Adsorption Capacity (mg/g)	Reference
Alum sludge	5.4	[21]
Activated alumina (γ-Al_2O_3)	16.3	[12]
Alum-impregnated activated alumina	40.7	[22]
Quick lime	16.7	[23]
Schwertmannite	55.3	[24]
Chitosan-based mesoporous alumina	8.3	[25]
Glutaraldehyde-crosslinked calcium alginate	73.6	[26]
Amorphous alumina supported on carbon nanotubes	28.7	[27]
Al-CNF	43.6	This study

In Figure 8c, the percentage removal is seen to be inversely related to the equilibrium fluoride concentration. At low fluoride concentrations, Al-CNF offers plenty of aluminum binding sites where most of the fluoride can be removed from water (e.g., at an equilibrium fluoride concentration of 0.625 ppm, the percent removal is 89.6%, but the adsorption capacity of Al-CNF is only 0.96 mg/g). With the increasing fluoride concentration, the total amount of fluoride that is removed becomes greater, but the percentage removal is less because the Al-CNF scaffold becomes saturated.

Figure 8d depicts the adsorption capacity of Al-CNF at pH = 4, 7, and 9. It can be seen that the adsorption capacity remains constant at 43 mg/g for pH = 7 and 9, but decreases to 30 mg/g at pH = 4. Table 3 shows the concentration of aluminum in solution after the fluoride removal study at pH = 4, 7, and 9. It can be seen that at pH = 7, the smallest aluminum concentration of 2.7 ppb is detected, followed by 11.8 ppb at pH = 9 and 560 ppb at pH = 4. This study indicates that Al-CNF is not effective in acidic conditions. In other words, at pH = 4, a significant amount of aluminum can be dissolved, and the fluoride adsorption capacity of Al-CNF then decreases rapidly. This is because the carboxylate groups are more likely to bind with protons than aluminum ions at low pH values. This will be a limitation for the application of Al-CNF in acid conditions. However, the system of Al-CNF is quite effective for fluoride removal in neutral and basic conditions.

Table 3. The pH effect on the solvation of aluminum from Al-CNF after fluoride removal.

pH	Dissolved Al^{3+} (ppb)
Acidic	560
Basic	11.8
Neutral	2.7

The zeta potential value of Al-CNF (Figure 1b) in the range of pH 4–9 is about zero, which indicates that Al-CNF is not charged in the suspension. The most plausible mechanism for the F^- sorption process is ion exchange, i.e., some of the aluminum–carboxylate bonds (on NOCNF) are replaced with alumina–fluoride bonds. Some researchers have reported that anions such as $SO_3{}^{2-}$, $CO_3{}^{2-}$, and $NO_3{}^-$ can be replaced with fluoride ions [28], which should also be the case for COO^-. This hypothesis is certainly consistent with the formation of aluminum fluoride hydrate crystals (e.g., β-AlF_3•$3H_2O$).

3.3. Sequential Applicability and Comparative Studies of Al-CNF

It does not appear that Al-CNF can be regenerated and reused like other adsorbent materials. This is because, in the conventional regeneration process using acid, Al-CNF can release aluminum, which would make the system inefficient. To enhance the sustainability of Al-CNF, we have investigated the sequential applicability of fluoride-adsorbed Al-CNF for the possible removal of other cationic pollutants. To test this hypothesis, a cationic Basic Red 2 dye was chosen to test the removal capacity of flocculated fluoride-adsorbed Al-CNF (treated with a 500 ppm fluoride solution). Figure 9a shows the adsorption capacity of Basic Red 2 using fluoride-adsorbed Al-CNF. It is interesting to see that even at a 2500 ppm equilibrium concentration, the adsorption capacity of fluoride-adsorbed Al-CNF against the Basic Red 2 dye does not reach the plateau value, indicating that the removal efficiency can be even higher. Using 0.126 wt% of fluoride-adsorbed Al-CNF at a 1:1 ratio to a 5000 ppm initial concentration of Basic Red 2, the maximum experimental removal efficiency was found to be 1923 mg/g (i.e., C_e at 2500 ppm in Figure 9a). We hypothesize that the isotherm adsorption data (Table 4) cannot be described by the Langmuir isotherm model, i.e., the adsorption mechanism does not follow a monolayer deposition process. In this case, we consider the use of the Freundlich isotherm model, assuming that the adsorption follows the multilayer deposition pathway. The expression for the Freundlich isotherm model can be described as follows [29,30].

$$\log Q_e = logK_F + \frac{1}{n}logC_e \quad (3)$$

where Q_e represents the adsorption capacity at equilibrium, C_e represents the equilibrium concentration of the adsorbate, and K_f and n are constants where K_f is calculated as the y-intercept when plotting a linear curve of logQ_e over logC_e. The results of the Freundlich isotherm fitting curve for the Basic Red 2 removal using the fluctuated fluoride-adsorbed Al-CNF adsorbent are shown in Figure 9b. It can be seen that a good linear fitting of the experimental results using the Freundlich model (R^2 is greater than 0.99) is found, again validating the selection of this analysis. In Figure 9c, the percentage removal is seen to increase with the equilibrium Basic Red 2 concentration. The results indicate that the adsorption of the Basic Red 2 molecules by the fluoride-adsorbed Al-CNF may not only be through electrostatic interaction, and that hydrophobic interaction and hydrogen bonding between the CNF scaffold and the dye molecules can also play a role. These effects will be explored in detail later. It is imperative to note that the concentration of aluminum after the dye adsorption has not been determined. This is important, considering the possibility of aluminum and fluoride losses. Nevertheless, the sequential applicability of the fluoride-adsorbed Al-CNF adsorbent for cationic contaminant removal can improve the sustainability of this absorbent system.

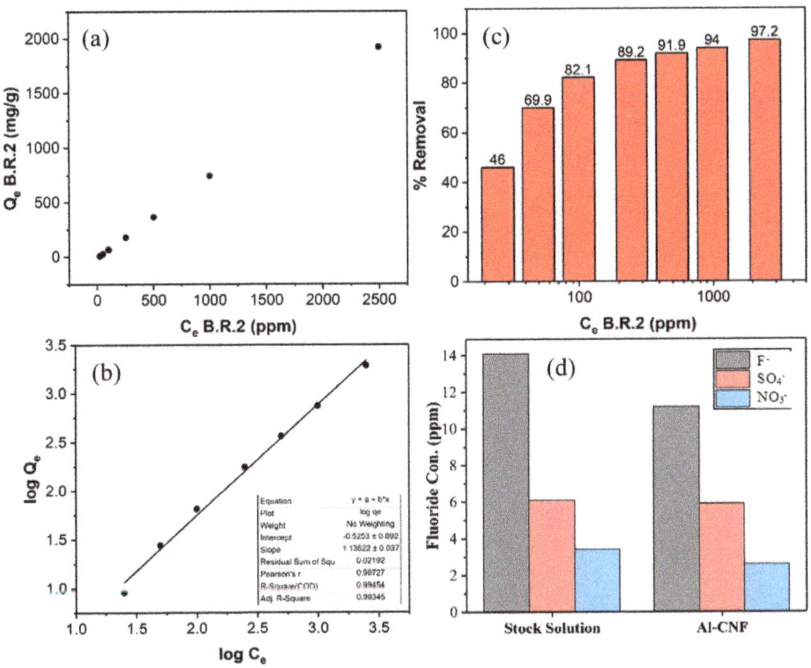

Figure 9. (**a**) The adsorption capacity of Basic Red 2 using the flocculated fluoride-adsorbed Al-CNF adsorbent. The maximum experimental adsorption capacity is 1923 mg/g using 0.126 wt% flocculated Al-CNF and 1:1 5000 ppm (initial) Basic Red 2. (**b**) The Freundlich isotherm fitting curve of Basic Red 2 removal using the fluctuated fluoride-adsorbed Al-CNF adsorbent. (**c**) Percentage removal of Basic Red 2 using flocculated fluoride-adsorbed Al-CNF. (**d**) The adsorption study using mixed ions of fluoride, sulfate, and nitrate. The stock solution is compared with the solution treated with Al-CNF.

Finally, the selectivity of Al-CNF for anionic contaminant removal in the mixed solutions of fluoride, sulfate, and nitrate was investigated. The selection of nitrate and sulfate was made because both are common and enforceable regulated anionic contaminants, and nitrate is on the National Primary Drinking Water Regulations (NPDWR) standard list and sulfate is on the secondary regulation list. In this study, a solution containing three salts, namely sodium fluoride, sodium sulfate, and sodium nitrate, was mixed with Al-CNF (1:1 by volume) and then left to equilibrate overnight. The results are illustrated in Figure 9d, which shows that the fluoride concentration decreased from 14.1 ppm to 11.2 ppm, the sulfate concentration decreases from 6.1 ppm to 5.9 ppm, and the nitrate concentration decreased from 3.4 ppm to 2.6 ppm. The decrease in fluoride is very comparable to the results in Figure 8, indicating that the presence of other anions, such as sulfate and nitrate, does not significantly reduce the removal of fluoride. Such behavior has been reported in other aluminum-based adsorbents such as aluminum sulfate, alumina, and kaolinite [31,32]. This behavior can be explained because of how the different anions can bind with aluminum. It is known that fluoride binds with aluminum through an inner-sphere complex, while sulfate binds through both an inner- and outer-sphere complex, and nitrate binds through an outer-sphere complex [33]. Furthermore, non-ionic hydroxyls on the CNF surface have also been shown to form complexes with aluminum, such as polyphenols [34].

Table 4. Experimental data for Basic Red 2 removal using flocculated fluoride-adsorbed Al-CNF.

Calc. Equilibrium Con. of B.R.2, C_e (ppm)	Absorbance Measured at 518 nm of Water + B.R.2 Solution 1:1 (3 mL Each)	Absorbance Measured at 518 nm of Flocculated Al-CNF (0.1264 wt%) + B.R.2 Sol. 1:1 (3 mL Each)	Con. of Flocculated Al-CNF (0.126 wt%) + B.R.2 Sol. 1:1 (3 mL each)	Conc. Diff. (ppm)	Eff. (%)	B.R.2 Removed (mg)	Quantity of Flocculated Al-CNF (g)	Q_e (mg/g)	Log C_e	Log Q_e
25	0.1220	0.0147	13.51	11.49	46.0	0.035	0.00379	9.090	1.398	0.959
50	0.2447	0.0220	15.06	34.94	69.9	0.105	0.00379	27.64	1.699	1.442
100	0.5039	0.0354	17.91	82.09	82.1	0.246	0.00379	64.94	2.000	1.813
250	1.1850	0.0786	27.11	222.89	89.2	0.669	0.00379	176.3	2.398	2.246
500	2.2398 *	0.1411	40.40	459.60	91.9	1.379	0.00379	363.6	2.699	2.561
1000	4.6762 *	0.2340	60.17	939.83	94.0	2.819	0.00379	743.5	3.000	2.871
2500	10.9905 *	0.2787	69.68	2430.32	97.2	7.291	0.00379	1923	3.398	3.284

* Corrected for Dilution.

4. Conclusions

A new class of adsorbent, Al-CNF, prepared by the ionic crosslinking of an anionic nanocellulose scaffold produced from the zero-waste nitro-oxidization treatment of lignocellulose biomass, was demonstrated. The use of aluminum crosslinkers converts the anionic scaffold into a cationic adsorbent, which is effective for fluoride removal. The structure, functionality, and fluoride removal capability of Al-CNF were characterized by varying microscopic, spectroscopic, X-ray diffraction, and thermal analysis techniques. Using Langmuir isotherm modeling, the fluoride adsorption results from Al-CNF yielded the maximum adsorption capacity (Q_m) of 43.6 mg/g, which is comparable or exceeds the fluoride removal performance of existing alumina adsorbents. However, Al-CNF exhibited some limitations in usage under the acidic conditions. The selectivity of fluoride removal in the presence of other anionic contaminants (e.g., sulfate and nitrate ions) remained high in neutral conditions. Finally, the sequential applicability of using spent Al-CNF after fluoride adsorption (i.e., fluoride-adsorbed Al-CNF) to further remove cationic molecules such as Basic Red 2 dye was demonstrated. The practice can further enhance the sustainability aspect of this already promising adsorbent system for the removal of multiple contaminants with opposite surface charge.

Author Contributions: Methodology, K.I.J., S.K.S., P.R.S. and B.S.H.; Validation, K.I.J., S.K.S., P.R.S. and A.G.A.; Formal analysis, K.I.J.; Investigation, K.I.J. and P.R.S.; Resources, B.S.H.; Data curation, K.I.J.; Writing—original draft, K.I.J.; Writing—review & editing, A.G.A. and B.S.H.; Supervision, B.S.H.; Project administration, B.S.H.; Funding acquisition, B.S.H. All authors have read and agreed to the published version of the manuscript.

Funding: The authors thank the National Science Foundation (NSF-DMR-2216585) for their financial support of this work.

Data Availability Statement: The original contributions presented in the study are included in the article, further inquiries can be directed to the corresponding author.

Acknowledgments: The authors thank Hui Chen, Hao-Yen Chang, Bernice Pham, and Min Liu for experimental assistance in this study. The authors also extend their appreciation to the Deanship of Scientific Research at Imam Mohammad Ibn Saud Islamic University (IMSIU, Research Partnership No. IMSIU-RP23052).

Conflicts of Interest: The authors declare no conflict of interest.

References

1. WHO. Drinking Water. Available online: https://www.who.int/news-room/fact-sheets/detail/drinking-water (accessed on 20 May 2024).
2. NIH. Fluoridated Water. Available online: https://www.cancer.gov/about-cancer/causes-prevention/risk/myths/fluoridated-water-fact-sheet (accessed on 20 May 2024).
3. Guissouma, W.; Hakami, O.; Al-Rajab, A.J.; Tarhouni, J. Risk assessment of fluoride exposure in drinking water of Tunisia. *Chemosphere* **2017**, *177*, 102–108. [CrossRef] [PubMed]
4. Ali, S.; Thakur, S.K.; Sarkar, A.; Shekhar, S. Worldwide contamination of water by fluoride. *Environ. Chem. Lett.* **2016**, *14*, 291–315. [CrossRef]
5. Das, R.; Lindström, T.; Sharma, P.R.; Chi, K.; Hsiao, B.S. Nanocellulose for sustainable water purification. *Chem. Rev.* **2022**, *122*, 8936–9031. [CrossRef] [PubMed]
6. Das, R.; Lindström, T.; Khan, M.; Rezaei, M.; Hsiao, B.S. Nanocellulose preparation from diverse plant feedstocks, processes, and chemical treatments: A review emphasizing non-woods. *BioResources* **2024**, *19*, 1865–1924. [CrossRef]
7. Isogai, A.; Saito, T.; Fukuzumi, H. TEMPO-oxidized cellulose nanofibers. *Nanoscale* **2011**, *3*, 71–85. [CrossRef] [PubMed]
8. Sharma, P.R.; Joshi, R.; Sharma, S.K.; Hsiao, B.S. A Simple Approach to Prepare Carboxycellulose Nanofibers from Untreated Biomass. *Biomacromolecules* **2017**, *18*, 2333–2342. [CrossRef] [PubMed]
9. Onyango, M.S.; Kojima, Y.; Aoyi, O.; Bernardo, E.C.; Matsuda, H. Adsorption equilibrium modeling and solution chemistry dependence of fluoride removal from water by trivalent-cation-exchanged zeolite F-9. *J. Colloid Interface Sci.* **2004**, *279*, 341–350. [CrossRef] [PubMed]
10. Owusu-Agyeman, I.; Reinwald, M.; Jeihanipour, A.; Schäfer, A.I. Removal of fluoride and natural organic matter from natural tropical brackish waters by nanofiltration/reverse osmosis with varying water chemistry. *Chemosphere* **2019**, *217*, 47–58. [CrossRef] [PubMed]
11. Moussa, D.T.; El-Naas, M.H.; Nasser, M.; Al-Marri, M.J. A comprehensive review of electrocoagulation for water treatment: Potentials and challenges. *J. Environ. Manag.* **2017**, *186*, 24–41. [CrossRef]
12. Ku, Y.; Chiou, H.-M. The Adsorption of Fluoride Ion from Aqueous Solution by Activated Alumina. *Water Air Soil Pollut.* **2002**, *133*, 349–361. [CrossRef]
13. Ghorai, S.; Pant, K.K. Investigations on the column performance of fluoride adsorption by activated alumina in a fixed-bed. *Chem. Eng. J.* **2004**, *98*, 165–173. [CrossRef]
14. López Valdivieso, A.; Reyes Bahena, J.L.; Song, S.; Herrera Urbina, R. Temperature effect on the zeta potential and fluoride adsorption at the α-Al_2O_3/aqueous solution interface. *J. Colloid Interface Sci.* **2006**, *298*, 1–5. [CrossRef] [PubMed]
15. Xu, F.; Yu, J.; Tesso, T.; Dowell, F.; Wang, D. Qualitative and quantitative analysis of lignocellulosic biomass using infrared techniques: A mini-review. *Appl. Energy* **2013**, *104*, 801–809. [CrossRef]
16. Garvey, C.J.; Parker, I.H.; Simon, G.P. On the Interpretation of X-ray diffraction powder patterns in terms of the nanostructure of Cellulose I fibres. *Macromol. Chem. Phys.* **2005**, *206*, 1568–1575. [CrossRef]
17. Haouas, M.; Taulelle, F.; Martineau, C. Recent advances in application of 27Al NMR spectroscopy to materials science. *Prog. Nucl. Magn. Reson. Spectrosc.* **2016**, *94*, 11–36. [CrossRef] [PubMed]
18. Langmuir, I. The Adsorption of Gases on Plane Surface of Glass, Mica and Platinum. *J. Am. Chem. Soc.* **1918**, *40*, 1361–1402. [CrossRef]
19. Van Der Kamp, K.A.; Qiang, D.M.; Aburub, A.; Wurster, D.E. Modified Langmuir-like Model for Modeling the Adsorption from Aqueous Solutions by Activated Carbons. *Langmuir* **2005**, *21*, 217–224. [CrossRef] [PubMed]
20. Ayoob, S.; Gupta, A.K.; Bhakat, P.B.; Bhat, V.T. Investigations on the kinetics and mechanisms of sorptive removal of fluoride from water using alumina cement granules. *Chem. Eng. J.* **2008**, *140*, 6–14. [CrossRef]
21. Sujana, M.G.; Thakur, R.S.; Rao, S.B. Removal of Fluoride from Aqueous Solution by Using Alum Sludge. *J. Colloid Interface Sci.* **1998**, *206*, 94–101. [CrossRef]
22. Tripathy, S.S.; Bersillon, J.-L.; Gopal, K. Removal of fluoride from drinking water by adsorption onto alum-impregnated activated alumina. *Sep. Purif. Technol.* **2006**, *50*, 310–317. [CrossRef]
23. Islam, M.; Patel, R.K. Evaluation of removal efficiency of fluoride from aqueous solution using quick lime. *J. Hazard. Mater.* **2007**, *143*, 303–310. [CrossRef]
24. Eskandarpour, A.; Onyango, M.S.; Ochieng, A.; Asai, S. Removal of fluoride ions from aqueous solution at low pH using schwertmannite. *J. Hazard. Mater.* **2008**, *152*, 571–579. [CrossRef]
25. Jagtap, S.; Yenkie, M.K.N.; Labhsetwar, N.; Rayalu, S. Defluoridation of drinking water using chitosan based mesoporous alumina. *Microporous Mesoporous Mater.* **2011**, *142*, 454–463. [CrossRef]

26. Vijaya, Y.; Popuri, S.R.; Reddy, A.S.; Krishnaiah, A. Synthesis and characterization of glutaraldehyde-crosslinked calcium alginate for fluoride removal from aqueous solutions. *J. Appl. Polym. Sci.* **2011**, *120*, 3443–3452. [CrossRef]
27. Li, Y.-H.; Wang, S.; Cao, A.; Zhao, D.; Zhang, X.; Xu, C.; Luan, Z.; Ruan, D.; Liang, J.; Wu, D.; et al. Adsorption of fluoride from water by amorphous alumina supported on carbon nanotubes. *Chem. Phys. Lett.* **2001**, *350*, 412–416. [CrossRef]
28. Gai, W.Z.; Deng, Z.Y. A comprehensive review of adsorbents for fluoride removal from water: Performance, water quality assessment and mechanism. *Environ. Sci. Water Res. Technol.* **2021**, *7*, 1362–1386. [CrossRef]
29. Jaroniec, M. Adsorption on heterogeneous surfaces: The exponential equation for the overall adsorption isotherm. *Surf. Sci.* **1975**, *50*, 553–564. [CrossRef]
30. Douglas, L.M.; Vermeulen, T. Binary Langmuir and Freundlich isotherms for ideal adsorbed solutions. *J. Phys. Chem.* **1981**, *85*, 3247–3250.
31. Meenakshi, S.; Sundaram, C.S.; Sukumar, R. Enhanced fluoride sorption by mechanochemically activated kaolinites. *J. Hazard. Mater.* **2008**, *153*, 164–172. [CrossRef]
32. Aoudj, S.; Drouiche, N.; Hecini, M.; Ouslimane, T.; Palaouane, B. Coagulation as a post-treatment method for the defluoridation of photovoltaic cell manufacturing wastewater. *Procedia Eng.* **2012**, *33*, 111–120. [CrossRef]
33. Sivasankar, V.; Rajkumar, S.; Murugesh, S.; Darchen, A. Influence of shaking or stirring dynamic methods in the defluoridation behavior of activated tamarind fruit shell carbon. *Chem. Eng. J.* **2012**, *197*, 162–172. [CrossRef]
34. Aziz, S.B.; Nofal, M.M.; Ghareeb, H.O.; Dannoun, E.M.A.; Hussen, S.A.; Hadi, J.M.; Ahmed, K.K.; Hussein, A.M. Characteristics of Poly(vinyl alcohol) (PVA) based composites integrated with green synthesized Al^{3+}-metal complex: Structural, optical, and localized density of state analysis. *Polymers* **2021**, *13*, 1316. [CrossRef]

nanomaterials

MDPI

Article

Cellulose Sulfate Nanofibers for Enhanced Ammonium Removal

Ken I. Johnson [1], William Borges [2], Priyanka R. Sharma [1,3], Sunil K. Sharma [1], Hao-Yen Chang [1], Mortaga M. Abou-Krisha [4], Abdulrahman G. Alhamzani [4] and Benjamin S. Hsiao [1,*]

1 Department of Chemistry, Stony Brook University, Stony Brook, New York, NY 11790, USA; priyanka.sharma@wmich.edu (P.R.S.)
2 Department of Biochemistry and Molecular Biology, Brown University, Providence, RI 02906, USA
3 Department of Chemical and Paper Engineering, Western Michigan University, Kalamazoo, MI 49008, USA
4 Department of Chemistry, College of Science, Imam Mohammad Ibn Saud Islamic University (IMSIU), Riyadh 11623, Saudi Arabia; mmaboukrisha@imamu.edu.sa (M.M.A.-K.); agalhamzani@imamu.edu.sa (A.G.A.)
* Correspondence: benjamin.hsiao@stonybrook.edu; Tel.: +1-631-6327793

Abstract: In this study, a sulfonation approach using chlorosulfonic acid (CSA) to prepare cellulose sulfate nanofibers (CSNFs) from raw jute fibers is demonstrated. Both elemental sulfur content and zeta potential in the CSNFs are found to increase with increasing CSA content used. However, the corresponding crystallinity in the CSNFs decreases with the increasing amount of CSA used due to degradation of cellulose chains under harsh acidic conditions. The ammonium adsorption results from the CSNFs with varying degrees of sulfonation were analyzed using the Langmuir isotherm model, and the analysis showed a very high maximum ammonium adsorption capacity (41.1 mg/g) under neutral pH, comparable to the best value from a synthetic hydrogel in the literature. The high ammonium adsorption capacity of the CSNFs was found to be maintained in a broad acidic range (pH = 2.5 to 6.5).

Keywords: nanocellulose; ammonium removal; cellulose sulfate; jute

1. Introduction

Recent research activities in developing new sustainable materials from natural resources for water purification have gained notable interest beyond the subjects of biochar- and zeolite-based materials. Specifically, nanomaterials based on carboxylated cellulose nanofibers have attracted a great deal of attention due to several newly developed processes that can effectively oxidize the cellulose surface (converting the hydroxyl group into a carboxyl group on the C6 position of the glucose repeating unit), bring negative charges on the cellulose surface, facilitate the defibrillation process, and produce no waste [1,2]. However, there is one limitation of using nanocellulose with carboxylic acid or carboxylate functional groups, that is, its efficacy at different pH values.

Similar limitations have also been noted in ion-exchanging materials containing carboxyl or carboxylate groups, which are termed weak acid cation exchangers (WACEs) [3]. In WACEs, the carboxylic acid (carboxyl) groups can be protonated, and the carboxylate groups are present in the form of a salt with a counter ion such as sodium. The hydrogen in the carboxyl group of WACEs cannot be exchanged by neutral salts, such as sodium chloride. Instead, they are highly reactive towards alkaline species, such as sodium hydroxide and calcium bicarbonate [4]. The hardness of water can be used to determine whether water has a relatively high concentration of dissolved multivalent cations (e.g., calcium and magnesium). If the alkalinity is greater than the hardness, the ions contributing to the hardness are removed (some sodium ions are also removed). The amount of ions (sodium and hardness) removed will be equal to the alkalinity neutralized via the carboxylic acid groups. If the hardness is greater than the alkalinity, only ions contributing to the hardness are removed. All ion-exchange systems have a degree of selectivity, and typically have greater affinity for ions with higher molecular weights and higher ionic charges [5]. A distinct exception to this trend exists with WACEs. The carboxylate ion, being a conjugate base and also an alkaline salt, reacts with acid to form carboxylic acid, and has a higher affinity for hydrogen than any other cation [6]. This is also observed in the regeneration processes of WACEs, which require only a dilute and stoichiometric solution of acid [7]. The implications of this property mean that bound impurities may be easily lost due to the interaction with acid, but also that the regeneration of WACEs is particularly efficient and cost-effective.

The hydrogen WACEs can be converted to an ionic WACE using alkaline salts, such as sodium bicarbonate [8]. Ionic WACEs can be used to exchange both alkaline and neutral salts, and generally follow the selectivity preference for ions with higher molecular weights and higher ionic charges. As a result, ionic WACEs cannot be regenerated using concentrated sodium chloride, but can be regenerated to the hydrogen WACE using a dilute and stoichiometric solution of acid, and subsequently converted to an ionic WACE using an alkaline salt.

Strong acid cation-exchange (SACE) materials usually have a sulfonic acid functional group and follow the selectivity for ions with higher molecular weights and higher ionic charges. SACEs can be regenerated using concentrated sodium chloride, unlike WACEs, which must first be converted into a hydrogen WACE using an acid and then to an ionic WACE using a base [8]. This gives SACEs an advantage in that they require one less step in the regeneration process. Sulfur, being more electronegative than carbon, gains relatively more electron density from oxygen and takes the electron density away from hydrogen. SACEs act as strong acids; their affinity for hydrogen is lower than other monovalent cations, such as sodium and potassium. Compared with WACEs, SACEs have the advantage of operating in a broader pH range, faster kinetics, and a lack of swelling behaviors due to the relatively higher solubility of WACEs.

In this study, cellulose sulfate nanofibers (CSNFs) were prepared from raw jute feedstock using a heterogenous sulfonation approach, involving the use of chlorosulfonic acid (CSA). This chemical process has been reported in the literature but has not been conducted on raw lignocellulosic biomass [9–13]. The structure, morphology, and chemical modification of the resulting CSNFs prepared using different CSA amounts were characterized by transmission electron microscopy (TEM), Fourier transform infrared spectroscopy (FTIR), nuclear magnetic resonance (NMR), wide-angle X-ray diffraction (WAXD), thermogravimetric analysis (TGA), elemental sulfur content, and zeta potential. The ammonium adsorption efficiency of the CSNFs in the suspension state was also tested. The results were analyzed using the Langmuir isotherm model to yield the maximum adsorption capacity, and the adsorption performance was evaluated at various pH conditions.

2. Materials and Methods

2.1. Materials

Jute fibers were obtained from Toptrans Bangladesh Ltd. in Dhaka, Bangladesh. Anhydrous N,N-dimethylformamide (DMF) was purchased from Fisher Scientific, Waltham, MA, USA. Chlorosulfonic acid (CSA, 99%) was purchased from VWR, Radnor, PA, USA. Ammonium chloride (98%) and hydrochloric acid (1.0 N) were purchased from Sigma Aldrich, St. Louis, MO, USA. Sodium hydroxide (99%) was purchased from Macron Fine Chemicals, Center Valley, PA, USA. All chemicals were used without further purification. Cellulose dialysis tubing with a molecular weight cutoff of 12–14 kDa made by Spectra/Por was also purchased from VWR, Radnor, PA, USA.

2.2. Preparation of Cellulose Sulfate Nanofibers

Reaction conditions for preparation of cellulose sulfate nanofibers (CSNFs) are given in Table 1. In all reactions, 2.0 g of ground raw jute was suspended in 35 mL of anhydrous

DMF for 1 h. An appropriate amount of CSA ranging from 1.0 to 2.5 mL was added (by dropwise addition) to the suspension using a serological pipet. The suspension was then stirred at room temperature for 6 h. The reaction was then quenched by adding 100 mL of deionized (DI) water, and allowed to stir for 30 min. The resulting suspension was vacuum-filtered until the fibers began to swell. Once the fibers swelled to a certain point, vacuum filtration became inefficient, at which point the sample was suspended in a dialysis tube. The dialysis water was changed twice a day until the conductivity of the wash water did not change over a 24 h period.

Table 1. Reaction conditions for preparation of various CSNF samples.

CSNF	Raw Jute (g)	DMF (mL)	Time (Hours)	CSA (mL)	Yield (%)
1	2.0	35	6	1.0	54.1
2	2.0	35	6	1.5	50.3
3	2.0	35	6	2.0	47.5
4	2.0	35	6	2.5	44.2

After dialysis, samples were placed in 500 mL media bottles and diluted to 500 mL using DI water. The suspensions were passed through a high-pressure homogenizer at 400 bars, which noticeably changed the viscosity to a viscous but flowable liquid. After letting the suspension equilibrate over 24 h, a small portion was weighed and dried to determine the weight percent (wt%) and product yield of the samples.

2.3. *Cellulose Sulfate Nanofiber Characterization*

2.3.1. Transmission Electron Microscopy

Transmission electron microscopy (TEM) was conducted on a JEOL JEM 1400 instrument (Tokyo, Japan) at an accelerating voltage of 120 kV. The samples were prepared on 300-mesh copper grids (Ted Pella Inc., Redding, CA, USA) by casting 10 µL of a 0.01 wt% sample onto a grid. After removing the excess fluid, the sample was stained with 10 µL of 2 wt% aqueous uranyl acetate solution. The excess solution was subsequently removed, and the coated grid was left to air-dry.

2.3.2. Fourier Transform Infrared Spectroscopy

Fourier transform infrared spectroscopy (FTIR) was recorded on a Nicolet iS10 FT-IR Spectrometer by Thermo Scientific, Waltham, MA, USA, using the attenuated total reflectance (ATR) mode. Each spectrum was averaged from the measurements of 4 scans. The chosen instrument was equipped with a DTGS KBr detector, a KBr beam splitter, Smart iTR accessory, and a diamond window. Each sample was scanned from 400 to 4000 cm^{-1}.

2.3.3. ^{13}C Cross Polarization Magic-Angle Spinning Nuclear Magnetic Resonance

^{13}C Cross polarization magic-angle spinning nuclear magnetic resonance (CPMAS-NMR) measurements of the chosen samples were carried out by a Bruker Utrashield 500 WB plus (500 MHz) instrumen (Bruker Corporation, Billerica, MA, USA), which was equipped with a 2.5 mm triple resonance MAS NMR probe, capable of spinning samples up to 35 KHz. The chosen resonance frequency was 10,000 Hz, and the samples were spun at the magic angle with a speed of 10 KHz.

2.3.4. Sulfur Elemental Analysis

A FlashEA 1112 instrument (Thermo Scientific, Waltham, MA, USA) was used to determine the elemental sulfur content of the samples. This instrument was equipped with a flash dynamic combustion chamber, capable of reaching a temperature of 1800 °C. A gas chromatography and thermal conductivity detector were used to quantify the sulfur content.

2.3.5. Wide-Angle X-ray Diffraction

A MiniFlex instrument from Rigaku Japan was used to measure and record wide-angle X-ray diffraction (WAXD) patterns. The samples were measured from 5 degrees to 100 degrees (2 theta (2θ) diffraction angle), in steps of 0.02 degrees and at a rate of 5 degrees per minute. The scan axis was set to a theta/2-theta continuous mode, where the intensity measured was in counts per second. The voltage and current were set to 40 kV and 15 mA, respectively, using Cu Kα radiation. An incident-side and receiving-side soller slit of 5.0 degrees along with an incident beam divergence-limiting slit of 1.25 degrees were used in this measurement.

The WAXD patterns were analyzed as follows. The measured patterns were first baselined using a straight line of constant value equal to that of the lowest value in the range of 5–45° 2θ. Six Gaussian curves were then used to fit the diffraction pattern, five for each lattice plane of cellulose crystals and one for the amorphous background. In the fitting procedure, the centers for 101, $10\overline{1}$, amorphous background, 021, 002, and 040 peaks were constrained to 14–16, 16–18, 19.5–22.5, 19.5–21.5, 22.5–23.5, and 34.5–35.5° 2θ, respectively. Constraints for the peak centers were disabled after several fitting iterations, so that constraints did not heavily influence the final peak positions. After several iterations, the full width at half maximum (FWHM) for the strongest crystalline peak, 002, was determined. This FWHM (002) was used to set the constraints for all other crystalline peaks in the pattern, specifically $\pm$ 0.4. In other words, the 002 peak in each sample was used as a reference to guide the FWHM determination for other crystalline peaks. In contrast, the FWHM for the amorphous cellulose (peak 3) was not constrained. The typical FWHM for amorphous cellulose was greater than 9° because in some instances the FWHM had to be initially set to a high value, such as 9, before fittings would converge. In general, the peak area was not constrained, except for constraining the area to positive values only. Fittings were iterated until converged, with a tolerance of 1.0×10^{-6}.

2.3.6. Thermogravimetric Analysis

Thermogravimetric analysis (TGA) was performed on a TA Q50 instrument (TA Instruments, New Castle, DE, USA). In this measurement, a sample was heated from 25 °C to 850 °C at 10 °C per minute, under a nitrogen atmosphere, where its mass was monitored over a temperature range. Sample was loaded and measured on a platinum pan. The derivative weight, shown as %/min, was also plotted against temperature to obtain derivative thermogravimetry (DTG) profiles, which showed more clearly the small changes in weight during the heating scan.

2.3.7. Zeta Potential

The averaged zeta potential values of the samples were taken on a ZetaProbe Analyzer by Colloidal Dynamics, Ponte Vedra Beach, FL, USA. Samples were measured with the assumption of a dielectric constant of 5.0 and a density ratio of 1.5 g/mL. In this test, samples were stirred at 300 rpm and titrated with an auto-burette. DI water was used as a solvent for all samples. The instrument was equipped with a niobium electrokinetic sonic amplitude electrode and calibrated with a KSiW suspension. Samples were treated with either hydrochloric acid or sodium hydroxide using an auto-burette to achieve the desired pH before the analysis.

2.3.8. Ammonium Ion-Selective Electrode

Ammonium concentration was determined using an ammonium ion-selective electrode (ISE). The IntelliCAL ISENH4181 probe by Hach (Hach Company, Loveland, CO, USA) was equipped with a non-refillable Dritek gel reference and double junction encased in an epoxy body. An Ag/AgCl reference electrode and solid-state PVC membrane sensor were also included in this instrument.

2.3.9. Ammonium Adsorption Data Analysis Using Langmuir Isotherm Model

The ammonium adsorption data analysis using the Langmuir isotherm model [14] was carried out as follows. The Langmuir isotherm model can be expressed as

$$\text{Langmuir equation } Q_e = \frac{Q_m b C_e}{1 + bC_e} \quad (1)$$

where Q_e represents the equilibrium adsorption capacity, C_e is the equilibrium concentration of ammonium, Q_m is the monolayer maximum adsorption capacity, and b is the Langmuir adsorption constant. Q_m can be determined as the inverse slope calculated from the linear fit when plotting C_e/Q_e against C_e.

3. Results and Discussion

The objective of this research is to develop cellulose sulfate nanofibers (CSNFs) that are not soluble in water and that have a cellulose surface with a high degree of sulfonation. During the sample preparation, the use of chlorosulfonic acid (CSA) can degrade the lignin component and form water-soluble compounds (pulping) and form sulfate groups on the cellulose surface (cellulose sulfonation) [9–13]. The strong charge of the sulfate group on the cellulose surface can introduce repulsive forces between adjacent cellulose fibrils and defibrillate the macroscale fibers into nanoscale fibers or nanofibers. It has been shown that using too much acid could cause increased cleaving of glycosidic linkages in cellulose chains, resulting in a lower degree of polymerization. It has also been shown that the sulfonation reaction of polysaccharides using an ionic liquid medium can dissolve the cellulose chains and form a homogenous reaction [15]. However, in this study, as DMF is a poor solvent for cellulose but a polar protic solvent for CSA, the sulfonation reaction was carried out through a heterogenous substitution with DMF. However, we note that although water is also a polar protic solvent, it cannot be used for sulfonation because CSA can react rapidly with water. It has been reported that sulfonation occurs primarily on the C6 primary alcohol group of cellulose and forms cellulose sulfate, according to spectroscopic analysis [11].

3.1. Product Yield, Crystallinity, and Morphology of CSNFs

The product yield, crystallinity, and morphology of CSNF products prepared under different reaction conditions were investigated, and the results are shown in Table 1. It is seen that as the amount of CSA increases, the yield decreases. This is expected because CSA is a harsh acid and it can degrade the cellulose component at higher concentrations. In our study, the lowest amount of CSA used of 1.0 mL leads to a product yield of 54.1%. In this case, the molar ratio of cellulose (anhydroglucose) to CSA is 1:1.2, where CSA is in a slight excess to cellulose.

At first glance, one may assume that the yield of 54.1% indicates that a significant fraction of biomass components (e.g., lignin and hemicellulose) is removed, leaving behind mostly semi-crystalline cellulose nanofibers. To explore the relationship between the product yield and cellulose crystallinity, WAXD patterns of raw jute and four CSNF samples (as illustrated in Table 1) with peak deconvolution analysis results are illustrated in Figure 1. In this figure, peak areas corresponding to each lattice plane and the amorphous component are determined by the peak deconvolution method, and the results are summarized in Table 2. As shown in this table, the crystallinity of raw jute is around 53.8%, and that of the CSNF 1, CSNF 2, CSNF 3, and CSNF 4 samples is 47.8%, 42.2%, 23.9%, and 19.5%, respectively. The results indicate that as the CSA content increases, the product yield decreases, and the crystallinity decreases. The decrease in product yield exhibits a linear relation, but the decrease in crystallinity shows a non-linear behavior with the CSA content consumed.

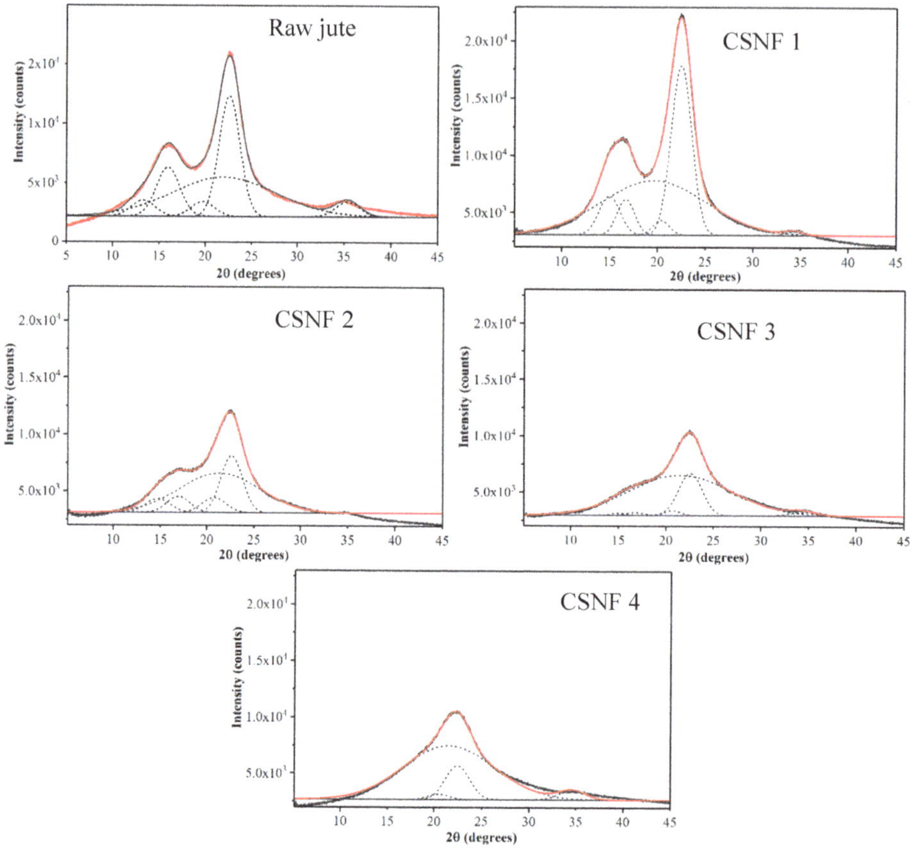

Figure 1. Wide-angle X-ray diffraction (WAXD) patterns of raw jute and cellulose sulfate nanofibers (CSNFs) with peak deconvolutions (dotted lines) to estimate the crystallinity. The red line represents the fit of the experimental data.

However, upon a closer examination of Figure 1, it becomes apparent that the lattice planes corresponding to 101 and $10\bar{1}$ are depleted with the higher amount of CSA used (these peaks are virtually non-existent in the CSNF 4 sample). In native cellulose I_β structure, the $10\bar{1}$ and 002 lattice planes are described as hydrophilic and hydrophobic planes, respectively [16]. This description loosely describes the hydrophilic plane as having exposed hydroxyl groups, while the hydrophobic plane as having buried hydroxyl groups. However, it is generally agreed that both can contribute to the combination of hydrophilic and hydrophobic surfaces, with differences being only marginal [17]. We speculate that because DMF is a hydrophilic solvent, it may be less accessible to the 002 plane and as a result degrade the 002 plane more slowly relative to the other hydrophilic crystalline planes (e.g., $10\bar{1}$). However, this hypothesis needs further verification. Another explanation is that the relative peak areas between 101, $10\bar{1}$, and 002 are similar to those found in raw jute, but as the relative intensity of the amorphous peak increases, deconvolution of the 101 and $10\bar{1}$ planes may become unreliable. As for why the 18.3% decrease in crystallinity occurs in CSNF 3 (when compared to raw jute), it can be attributed to the fact that the molar ratio of cellulose (anhydroglucose) to CSA in CSNF 3 is 1:2.4, where CSA is in greater excess to cellulose. As CSA sulfonates all the accessible primary alcohols on the cellulose surface, it may then start to degrade the cellulose component.

Table 2. Peak deconvolution results of WAXD patterns of raw jute and CSNF samples in Table 1.

		Raw Jute		
Peak (Index)	Area	FWHM	Peak position	Area (%)
1 (101)	4,313.1	3.0	13.2	4.3
2 ($10\bar{1}$)	13,371.7	3.0	15.9	13.2
3 (amorphous)	46,812.4	13.2	22.0	46.2
4 (021)	3,958.2	3.0	19.7	3.9
5 (002)	28,842.5	2.6	22.6	28.5
6 (040)	3981.1	3.0	35.4	3.9
		CSNF 1		
Peak (Index)	Area	FWHM	Peak position	Area (%)
1 (101)	9,317.6	2.6	15.0	7.5
2 ($10\bar{1}$)	7,116.5	2.2	16.7	5.7
3 (amorphous)	65,030.8	12.8	19.6	52.2
4 (021)	2,718.1	2.0	20.4	2.2
5 (002)	39,594.5	2.5	22.5	31.8
6 (040)	717.9	2.0	34.3	0.6
		CSNF 2		
Peak (Index)	Area	FWHM	Peak position	Area (%)
1 (101)	4,004.2	3.1	14.7	6.2
2 ($10\bar{1}$)	4,463.5	2.9	16.9	6.9
3 (amorphous)	37,201.5	10.0	21.3	57.8
4 (021)	4,207.8	3.0	20.8	6.5
5 (002)	14,452.1	2.7	22.6	22.5
6 (040)	~0	2.0	34.0	~0
		CSNF 3		
Peak (Index)	Area	FWHM	Peak position	Area (%)
1 (101)	452.9	2.0	15.0	0.7
2 ($10\bar{1}$)	720.1	2.6	16.8	1.1
3 (amorphous)	47,997.6	12.5	21.5	76.1
4 (021)	1,021.1	2.2	20.6	1.6
5 (002)	11,760.9	2.9	22.5	18.6
6 (040)	1,155.0	3.3	34.4	1.8
		CSNF 4		
Peak (Index)	Area	FWHM	Peak position	Area (%)
1 (101)	~0	2.6	14.4	~0
2 ($10\bar{1}$)	~0	2.6	16.7	~0
3 (amorphous)	4,802.1	11.2	21.5	80.5
4 (021)	482.9	2.6	20.5	1.9
5 (002)	3,000.9	3.0	22.5	13.4
6 (040)	821.9	3.4	34.6	4.2

Figure 2 displays TEM images of CSNF 1 at different magnifications. Nanoscale fibers can be observed, although they are not well dispersed, an indication of only partial fibrillation. There is also presence of dark blob-shaped masses mixed with the fibrillar structure. Considering there is no presence of free salts in the cellulose sulfate after dialysis, this may be concluded to be some organic high-molecular-weight material, such as lignin aggregation. It should be noted that TEM images could only be captured for CSNF 1, indicating that the fibrillar structure in other reactions must be more degraded.

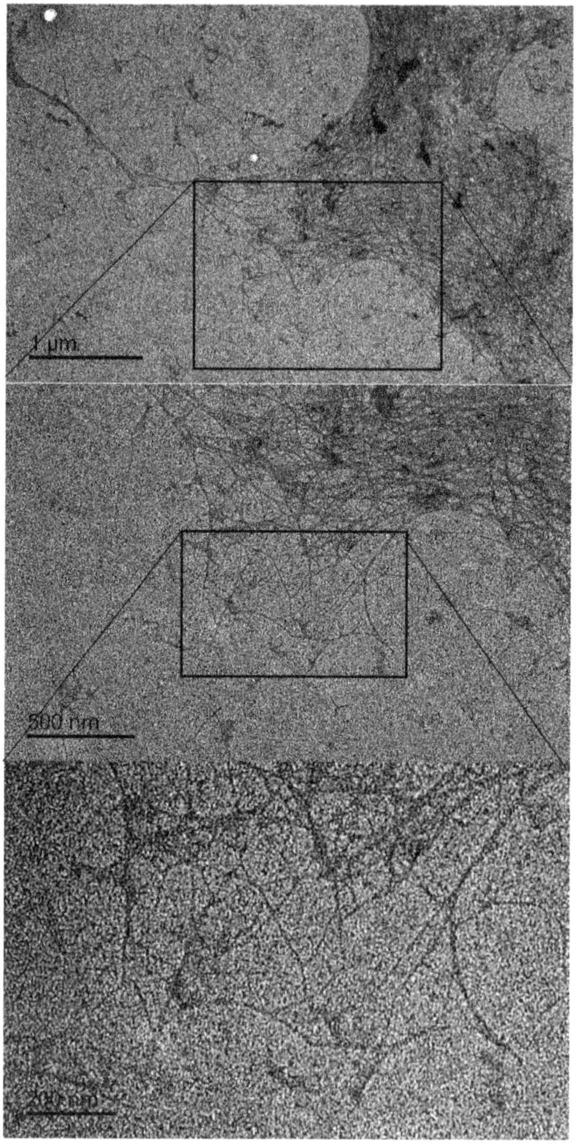

Figure 2. TEM images of CSNF 1 sample at different magnifications.

3.2. Structure and Composition Characterization of Cellulose Sulfate Nanofibers

The FTIR spectra of raw jute and various CSNF samples are shown in Figure 3a. Typical bonds associated with cellulose or cellulose derivatives are seen. For example, O-H stretching at 3330 cm^{-1} from the hydroxyl group, C-H stretching at 2900 cm^{-1}, and C-O stretching at 1035 cm^{-1} from the glycosidic linkage are seen [18]. Additionally, stretches corresponding to the sulfate ester group are present in the CSNF samples: the S=O stretch is detected at 1204 cm^{-1} and S-O stretch detected at 813 cm^{-1} [13].

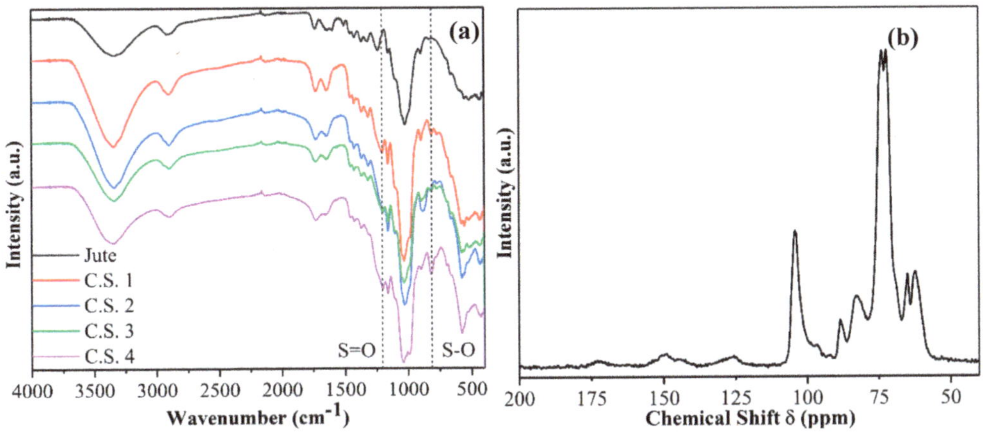

Figure 3. (**a**) Fourier transform infrared spectroscopy (FTIR) of various CSNF samples, and (**b**) ^{13}C cross polarization magic-angle spinning nuclear magnetic resonance (CPMAS-NMR) of CSNF 1.

The ^{13}C CPMAS-NMR spectrum of sample CSNF 1 is shown in Figure 3b, which also exhibits expected peaks typically associated with cellulose and cellulose derivatives. For example, peaks in the region of 63–65 ppm represent the C6 primary alcohol; peaks in the region of 72–75 ppm represent C5, C3, and C2 carbons; peaks in the region of 84–89 ppm represent the C4 carbon; and the peak at 105 ppm represents the C1 carbon. From the NMR spectrum, there is also a very clear indication of the presence of hemicelluloses and lignin. For example, the peak found at 125 ppm is associated with the H-substituted aromatic carbon, typically found on the lignin component [19]. The peak at 150 ppm can be ascribed to C3 and C4 carbons in the guaiacyl units or C3 and C5 carbons in the syringyl units. The peak at 175 ppm is ascribed to carbonyl carbon found in the acetate group of xylan.

The elemental sulfur contents of CSNF samples prepared under various sulfonation reactions are shown in Figure 4a. The figure displays an increasing percentage of elemental sulfur detected in reactions using more CSA. Reaction series of the CSNF 1, CSNF 2, CSNF 3, and CSNF 4 samples have elemental sulfur contents of 2.2, 4.0, 4.9, and 6.2 wt%, respectively. There is a slightly greater disparity in sulfur content between the CSNF 1 and CSNF 2 samples, in what otherwise is a linear-like increase in elemental sulfur. It is noted that the increase in sulfur content follows an expected trend, which is also consistent with the notion that the increase in CSA can produce samples with increasing ammonium adsorption capacity that will be discussed later. The elemental sulfur analysis clearly corroborates the presence of sulfonic acid functional groups in the CSNF samples.

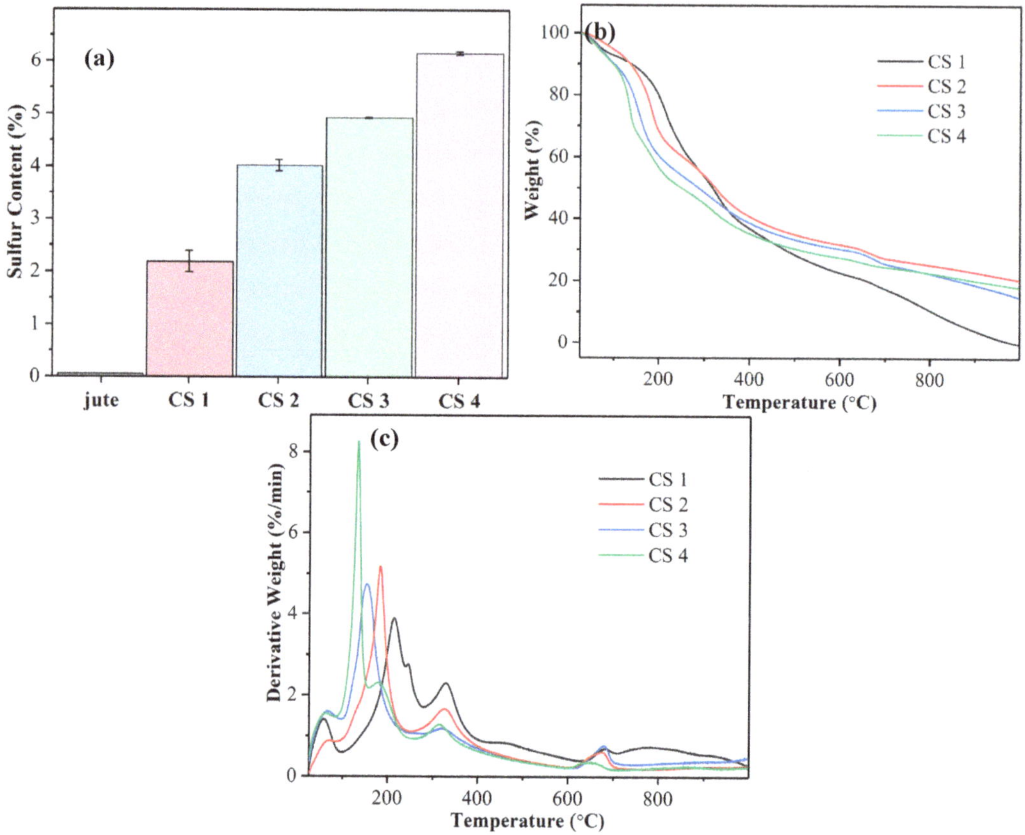

Figure 4. (**a**) Elemental sulfur analysis of raw jute and CSNF samples. (**b**) Thermogravimetric analysis (TGA) and (**c**) derivative thermogravimetry (DTG) curves of CSNF samples.

The TGA and DTG analysis results of the CSNF samples are shown in Figure 4b,c, respectively. These profiles display a significant lowering of the onset temperatures for degradation in CSNFs as compared to the TGA and DTG profiles of raw jute [2]. In these figures, the most intense derivative weight peak shifts to a lower degradation temperature as the elemental sulfur content increases. It is assumed that the change does not belong to the degradation of lignin, which is known to be around 300–400 °C [20]. Because the degree of sulfonation on the cellulose surface is systematically varied, it is conceivable that the observed shift in degradation temperature is caused by an increasingly substituted cellulose derivative. As such, a possible explanation is acid-catalyzed dehydration of the alcohol group in cellulose. Research on the effects of acid-impregnated cellulose on pyrolysis has shown a decrease in the degradation temperature [21]. The peak at 330 °C is present for all samples at the same temperature and is ascribed to lignin. The peak at 680 °C is ascribed to the degradation of the high-molecular-weight lignin component.

3.3. Evaluation of Ammonium Removal Efficiency

To evaluate the ammonium removal efficiency of the CSNFs, the following analyses were carried out. Firstly, the equilibrium adsorption capacity for ammonium removal (Q_e) was plotted as a function of the equilibrium ammonium concentration (C_e) for various CSNF samples (Figure 5a). In this figure, all adsorption curves follow an expected trend, i.e., the adsorption capacity increases rapidly initially, and then the increase tapers off, reaching an asymptote. The plateau values of Q_e for CSNF 1, CSNF 2, CSNF 3, and CSNF 4 are 19.3, 30.0, 32.1, and 35.1 mg of ammonium per gram of CSNF, respectively. A higher adsorption capacity is observed in the CSNFs prepared with a greater amount of CSA.

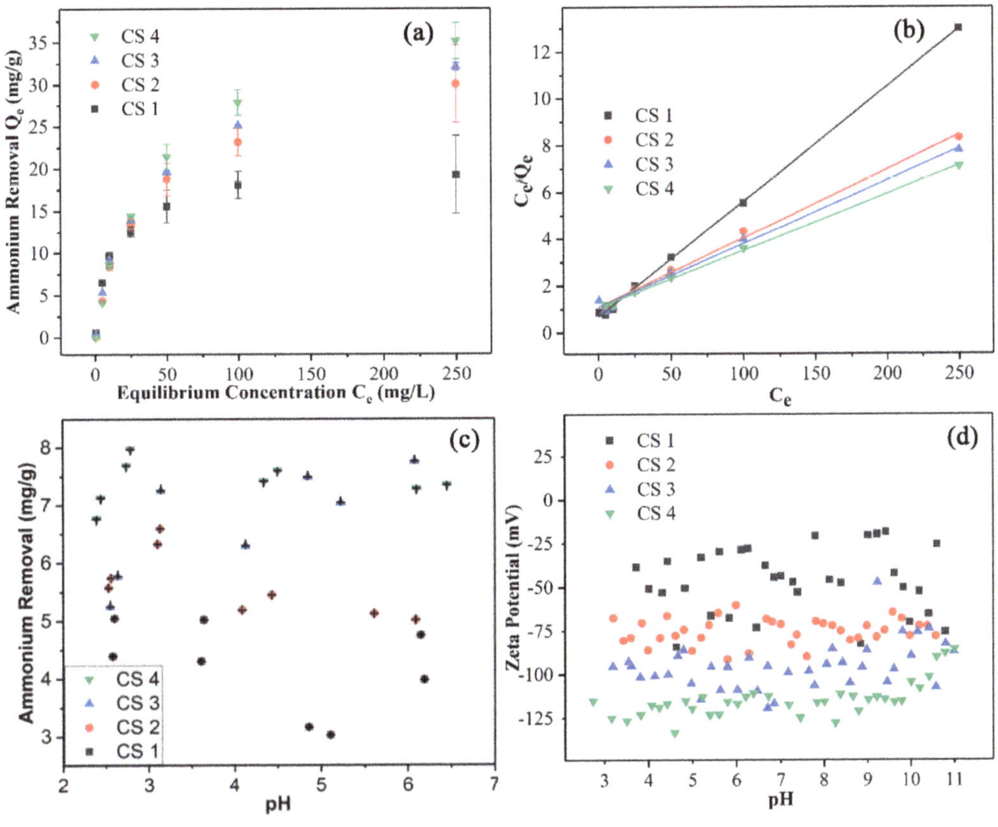

Figure 5. (**a**) The equilibrium adsorption capacity of CSNFs as a function of the equilibrium ammonium concentration. (**b**) Langmuir isotherm fittings for various CSNF samples. (**c**) Ammonium adsorption (removal) capacities of CSNFs as a function of the pH value. (**d**) Zeta potential measurements of CSNFs at various pH levels.

The ammonium removal results in Figure 5a were analyzed using the Langmuir isotherm model, which assumes that the adsorption process follows the monolayer mechanism. In this analysis, the C_e/Q_e value was plotted against C_e (Figure 5b). The figure indicates that the data exhibit a linear relationship for each CSNF sample. The fitting results are summarized in Table 3, showing that all fits have a R^2 value larger than 0.99. These results support the use of the Langmuir isotherm model. The isotherm data were also analyzed with the Freundlich isotherm model, assuming that the adsorption follows the multilayer deposition pathway. But this analysis yielded a significantly lower R^2 value, indicating that the ammonium adsorption by the CSNFs follows the monolayer process.

Table 3. Langmuir isotherm fitting parameters for cellulose sulfate (CS).

CSNF	Intercept	Slope	R^2	Q_m (mg/g)
1	0.676	0.0492	0.999	20.329
2	1.096	0.0294	0.996	33.990
3	1.079	0.0272	0.992	36.791
4	1.071	0.0243	0.998	41.084

Using Equation (1), the maximum adsorption capacity (Q_m) can be calculated as the inverse of the slope in Figure 5b. All the Q_m values for the CSNFs (CSNF 1: 20.3 mg/g, CSNF 2: 34.0 mg/g, CSNF 3: 36.8 mg/g, CSNF 4: 41.1 mg/g) are higher than those of carboxylated cellulose nanofibers produced by the nitro-oxidization or TEMPO-mediated oxidation approach (Q_m = 22.7 mg/g for nitro-oxidized CNF, and Q_m = 18.2 mg/g for TEMPO-oxidized CNF) [22]. However, we note that although the sulfonated nanocellulose system exhibits better ammonium adsorption performance than the carboxylated nanocellulose system, the current sulfonation approach involves the use of DMF that requires the recycling of spent waste. We further note that the Q_m value of the best performing CSNF 4 sample (41.1 mg/g) is significantly better than that of zeolite (6.3 mg/g) [23], mordenite (9.5 mg/g) [24], clinoptilolite (11.2 mg/g) [25], and biochar (2.8 mg/g) [26] in the literature, and it is approaching the best reported Q_m value of a synthetic polymeric hydrogel (42.7 mg/g) [27].

The effects of pH level on the adsorption capacity and zeta potential of the CSNF samples were also investigated, and the results are illustrated in Figure 5c,d, respectively. Figure 5c displays the effective ammonium removal capacity in the pH range of 2.5–6.5. This pH range was chosen to ensure that the ammonium would not be converted to ammonia, since ammonium, with a pKa of 9.3, can convert to ammonia at higher pH values. Based on the Henderson–Hasselbalch equation, it can be calculated that at pH = 7.3, there is a 1% conversion of ammonium to ammonia. While ammonia is soluble in water, it exists as a gas and can evaporate. To avoid these issues, the chosen pH value of the adsorption study was below 7.0. In Figure 5c, it is seen that the adsorption capacity of the CSNFs generally increases with the content of CSA used. In fact, a higher ammonium adsorption capacity can be correlated with the increase in negative zeta potential of the sample (Figure 5d). These results confirm that that the surface charge due to the sulfonic acid group of the CSNFs cannot be effectively screened with the increase in acid, while the sulfonic acid groups remain highly selective for adsorbing higher-molecular-weight cations, such as ammonium ions. The zeta potentials of the various CSNF samples were investigated with a wider pH range between 3 and 11 (Figure 5d). It is seen that the zeta potentials remain about constant in each sample, which can be attributed to the strong acidic characteristics of the sulfonic acid group, i.e., it is not protonated under the chosen acidic conditions, unlike carboxylic acid. The above properties of the CSNFs are in contrast to those of carboxylated nanocellulose, whose zeta potential and ammonium adsorption capacity both approach 0 as the acidity is increased. This study indicates that cellulose sulfate nanofibers can be used in a more robust manner than carboxylated cellulose nanofibers.

4. Conclusions

Sulfonation of raw lignocellulosic biomass can create cellulose sulfate nanofibers directly, where the presence of sulfonic acid groups on the nanofiber surface are more effective to adsorb ammonium cations in a relatively broad pH range (~2.5–neutral) than carboxylated cellulose nanofibers. In this study, the sulfonation approach involved the use of chlorosulfonic acid and N,N-dimethylformamide. An increase in chlorosulfonic acid content can severely decrease the yield and crystallinity of cellulose sulfate nanofibers. The presence of nanofibers was confirmed when using a nearly stoichiometric quantity of chlorosulfonic acid relative to anhydroglucose units in the cellulose. The structure, composition, and functionality of the cellulose sulfate nanofibers were characterized by Fourier transform infrared spectroscopy, nuclear magnetic resonance spectroscopy, and elemental sulfur and thermogravimetric analysis techniques. The sulfonation reaction appears to occur first on lignin and hemicelluloses, but the pulping mechanism remains unclear. The best performing cellulose sulfate nanofiber sample exhibits the maximum ammonium adsorption capacity of 41.1 mg/g, which approaches the performance of the best synthetic adsorbents. Both the ammonium adsorption capacity and zeta potential values of the cellulose sulfate nanofibers remain about constant at low pH values (as low as 2.5). Although the cellulose sulfate nanofibers exhibit more robust ammonium adsorption performance than carboxylated nanofibers, a more facile and sustainable method to simultaneously pulp the lignocellulosic biomass and create sulfonic acid groups on the resulting nanocellulose surface needs to be further explored.

Author Contributions: Conceptualization, B.S.H.; Methodology, K.I.J., S.K.S. and B.S.H.; Validation, H.-Y.C.; Formal analysis, K.I.J.; Investigation, K.I.J., W.B., P.R.S., S.K.S. and B.S.H.; Data curation, K.I.J., W.B., P.R.S. and H.-Y.C.; Writing—original draft, K.I.J.; Writing—review & editing, M.M.A.-K., A.G.A. and B.S.H.; Supervision, B.S.H.; Project administration, B.S.H.; Funding acquisition, B.S.H. All authors have read and agreed to the published version of the manuscript.

Funding: This research received financial support of this work by two different National Science Foundation grants (NSF-PFI-TT-2140820 and NSF-DMR-2216585) and the Deanship of Scientific Research at Imam Mohammad Ibn Saud Islamic University (IMSIU) for funding through Research Partnership No. IMSIU-RP23052.

Data Availability Statement: Data are contained within the article.

Acknowledgments: The authors thank Hui Chen, Bernice Pham, and Min Liu for experimental assistance on various parts of the current study.

Conflicts of Interest: B.S.H. holds equity in and serves on the board of directors for SWFT Labs, Inc. Although a financial conflict of interest was identified for management based on the overall scope of the NSF-PFI-TT-2140820 project and its potential benefit to SWFT Labs, the research findings included in this work may not necessarily relate to the interests of SWFT Labs. The terms of this arrangement have been reviewed and approved by Stony Brook University in accordance with its policy on objectivity in research. The remaining authors declare that the research was conducted in the absence of any commercial or financial relationships that could be construed as a potential conflict of interest.

References

1. Isogai, A.; Saito, T.; Fukuzumi, H. TEMPO-oxidized cellulose nanofibers. *Nanoscale* **2011**, *3*, 71–85. [CrossRef] [PubMed]
2. Sharma, P.R.; Joshi, R.; Sharma, S.K.; Hsiao, B.S. A simple approach to prepare carboxycellulose nanofibers from untreated biomass. *Biomacromolecules* **2017**, *18*, 2333–2342. [CrossRef]
3. Jensen, D.; Weiss, J.; Rey, M.A.; Pohl, C.A. Novel weak acid cation-exchange column. *J. Chromatogr. A* **1993**, *640*, 65–71. [CrossRef]
4. Kunin, R.; Vassiliou, B. New deionization techniques based upon weak electrolyte ion exchange resins. *Ind. Eng. Chem. Process Des. Dev.* **1964**, *3*, 404–409. [CrossRef]
5. Chen, Y.G.; Sofińska-Chmiel, W.; Lv, G.Y.; Kołodyńska, D.; Chen, S.H. Application of modern research methods for the physicochemical characterization of ion exchangers. *Materials* **2021**, *14*, 7067. [CrossRef]
6. Basu, S.; Debnath, A.K. Chapter II—Main Equipment. In *Power Plant Instrumentation and Control Handbook*; Basu, S., Debnath, A.K., Eds.; Academic Press: Boston, MA, USA, 2015; pp. 39–146.
7. Davis, J.R.; Chen, Y.; Baygents, J.C.; Farrell, J. Production of acids and bases for ion exchange regeneration from dilute salt solutions using bipolar membrane electrodialysis. *ACSNF Sustain. Chem. Eng.* **2015**, *3*, 2337–2342. [CrossRef]
8. Meyers, P.; Gottlieb, L.; DeSilva, F. Lead removal by ion exchange. In *Recycling of Metals and Engineercd Materials*; John Wiley & Sons, Inc.: Hoboken, NJ, USA, 2013; pp. 675–685.
9. Luo, J.; Semenikhin, N.; Chang, H.; Moon, R.J.; Kumar, S. Post-sulfonation of cellulose nanofibrils with a one-step reaction to improve dispersibility. *Carbohydr. Polym.* **2018**, *181*, 247–255. [CrossRef]
10. Mestechkina, N.M.; Shcherbukhin, V.D. Sulfated polysaccharides and their anticoagulant activity: A review. *Appl. Biochem. Microbiol.* **2010**, *46*, 267–273. [CrossRef]
11. Zhang, K.; Brendler, E.; Geissler, A.; Fischer, S. Synthesis and spectroscopic analysis of cellulose sulfates with regulable total degrees of substitution and sulfation patterns via ^{13}C NMR and FT Raman spectroscopy. *Polymer* **2011**, *52*, 26–32. [CrossRef]

12. Strätz, J.; Liedmann, A.; Trutschel, M.-L.; Mäder, K.; Groth, T.; Fischer, S. Development of hydrogels based on oxidized cellulose sulfates and carboxymethyl chitosan. *Cellulose* **2019**, *26*, 7371–7382. [CrossRef]
13. Pingrey, B.; Hsieh, Y.-L. Sulfated cellulose nanofibrils from chlorosulfonic acid treatment and their wet spinning into high-strength fibers. *Biomacromolecules* **2022**, *23*, 1269–1277. [CrossRef] [PubMed]
14. Chen, X. Modeling of experimental adsorption isotherm data. *Information* **2015**, *6*, 14–22. [CrossRef]
15. Liu, X.; Chen, T.; Hu, Y.; Li, K.; Yan, L. Catalytic synthesis and antioxidant activity of sulfated polysaccharide from *Momordica charantia* L. *Biopolymers* **2014**, *101*, 210–215. [CrossRef] [PubMed]
16. Wohlert, M.; Benselfelt, T.; Wågberg, L.; Furó, I.; Berglund, L.A.; Wohlert, J. Cellulose and the role of hydrogen bonds: Not in charge of everything. *Cellulose* **2022**, *29*, 1–23. [CrossRef]
17. Malaspina, D.C.; Faraudo, J. Molecular insight into the wetting behavior and amphiphilic character of cellulose nanocrystals. *Adv. Colloid Interface Sci.* **2019**, *267*, 15–25. [CrossRef] [PubMed]
18. Kačuráková, M.; Capek, P.; Sasinková, V.; Wellner, N.; Ebringerová, A. FT-IR study of plant cell wall model compounds: Pectic polysaccharides and hemicelluloses. *Carbohydr. Polym.* **2000**, *43*, 195–203. [CrossRef]
19. Vane, C.H.; Drage, T.C.; Snape, C.E.; Stephenson, M.H.; Foster, C. Decay of cultivated apricot wood (*Prunus armeniaca*) by the ascomycete *Hypocrea sulphurea*, using solid state 13C NMR and off-line TMAH thermochemolysis with GC–MS. *Int. Biodeterior. Biodegrad.* **2005**, *55*, 175–185. [CrossRef]
20. Brebu, M.; Vasile, C. Thermal degradation of lignin—A review. *Cellul. Chem. Technol.* **2010**, *44*, 353.
21. Long, Y.; Yu, Y.; Chua, Y.W.; Wu, H. Acid-catalysed cellulose pyrolysis at low temperatures. *Fuel* **2017**, *193*, 460–466. [CrossRef]
22. Johnson, K.I.; Ilacas, G.; Das, R.; Chang, H.Y.; Sharma, P.R.; Dimkpa, C.O.; Hsiao, B.S. A circular solution to enhance the food-water nexus by nanocellulose technologies for ammonium recovery and reuse. *Sustain. Sci. Technol.* 2024; *under review*.
23. Widiastuti, N.; Wu, H.; Ang, H.M.; Zhang, D. Removal of ammonium from greywater using natural zeolite. *Desalination* **2011**, *277*, 15–23. [CrossRef]
24. Weatherley, L.R.; Miladinovic, N.D. Comparison of the ion exchange uptake of ammonium ion onto New Zealand clinoptilolite and mordenite. *Water Res.* **2004**, *38*, 4305–4312. [CrossRef] [PubMed]
25. Wang, Y.; Liu, S.; Xu, Z.; Han, T.; Chuan, S.; Zhu, T. Ammonia removal from leachate solution using natural Chinese clinoptilolite. *J. Hazard. Mater.* **2006**, *136*, 735–740. [CrossRef] [PubMed]
26. Vu, N.-T.; Do, K.-U. Insights into adsorption of ammonium by biochar derived from low temperature pyrolysis of coffee husk. *Biomass Convers. Biorefinery* **2021**, *13*, 2193–2205. [CrossRef]
27. Zheng, Y.; Liu, Y.; Wang, A. Fast removal of ammonium ion using a hydrogel optimized with response surface methodology. *Chem. Eng. J.* **2011**, *171*, 1201–1208. [CrossRef]

nanomaterials

MDPI

Article

The Effect of Copper Adsorption on Iron Oxide Magnetic Nanoparticles Embedded in a Sodium Alginate Bead

Michele Modestino [1], Armando Galluzzi [1], Marco Barozzi [2], Sabrina Copelli [2], Francesco Daniele [3], Eleonora Russo [4], Elisabetta Sieni [3,*], Paolo Sgarbossa [4], Patrizia Lamberti [5] and Massimiliano Polichetti [1,*]

1 Department of Physics, University of Salerno, Via Giovanni Paolo II 132, 84084 Fisciano, Italy; mmodestino@unisa.it (M.M.); agalluzzi@unisa.it (A.G.)

2 Science and High Technology Department, Insubria University, Via Valleggio 11, 22100 Como, Italy; marco.barozzi@uninsubria.it (M.B.); sabrina.copelli@uninsubria.it (S.C.)

3 Theoretical and Applied Science Department, Insubria University, Via O. Rossi 9, 21100 Varese, Italy; fdaniele@uninsubria.it

4 Industrial Engineering Department, Padova University, Via F. Marzolo 9, 35122 Padova, Italy; eleonora.russo.2@phd.unipd.it (E.R.); paolo.sgarbossa@unipd.it (P.S.)

5 Department of Information and Electrical Engineering and Applied Mathematics, University of Salerno, 84084 Fisciano, Italy; plamberti@unisa.it

* Correspondence: elisabetta.sieni@uninsubria.it (E.S.); mpolichetti@unisa.it (M.P.)

Abstract: The preparation and use of iron oxide magnetic nanoparticles for water remediation is a widely investigated research field. To improve the efficacy of such nanomaterials, different synthetic processes and functionalization methods have been developed in the framework of green chemistry to exploit their magnetic properties and adsorption capacity in a sustainable way. In this work, iron oxide magnetic nanoparticles embedded in cross-linked sodium alginate beads designed to clean water from metal ions were magnetically characterized. In particular, the effect of copper adsorption on their magnetic properties was investigated. The magnetic characterization in a DC field of the beads before adsorption showed the presence of a superparamagnetic state at 300 K—a state that was also preserved after copper adsorption. The main differences in terms of magnetic properties before and after Cu^{2+} adsorption were the reduction of the magnetic signal (observed by comparing the saturation magnetization) and a different shape of the blocking temperature distribution obtained by magnetization versus temperature measurements. The evaluation of the reduction in magnetization can be important from the application perspective since it can affect the efficiency of the beads' removal from the water medium after treatment.

Keywords: copper adsorption; sodium alginate beads; magnetic properties; magnetic nanomaterials

1. Introduction

The worldwide diffusion of industrialization has led to an increase in the presence of metals and metalloids in water. This is due to their use in processes like electrolysis, electroplating, and metal smelting, resulting in a contamination of the environment whenever the wastewater treatment methods are insufficient for their removal [1]. The effects of metals and metalloids, such as copper (Cu), iron (Fe), zinc (Zn), nickel (Ni), and arsenic (As), on human health are well known, and they should not be underestimated [2–4]. In the attempt to reduce as much as possible their concentration, if not to completely remove them from water bodies, many different technologies have been proposed and developed. Methods such as ion exchange, membrane filtration, and chemical precipitation are widely used, but their applications have disadvantages in terms of either economic and working costs or use of secondary chemicals [1,5]. Thus, new and sustainable methods based on non-environmentally impacting materials are necessary. In this framework, magnetic nanoparticles have been of great interest and a research focus due to a combination of unique magnetic properties and the possibility of functionalization [1,6–8]. Magnetic nanoparticles are ferromagnetic or antiferromagnetic materials that, as a result of their small dimension (with a diameter below c.a. 50 nm, depending on the material), can show superparamagnetic properties at room temperature. These properties are characterized by an absence of coercivity, as expected for a paramagnet, but with a magnetic signal with intensity comparable to ferromagnets [9–11]. Superparamagnetic properties are used in most applications concerning magnetic nanoparticles, from biomedical ones to wastewater remediation [12–16]. For these applications, the possibility of functionalizing the magnetic nanoparticles with an organic coating or embedding them in a polymer matrix is fundamental, giving rise to the possibility of exploiting the properties of the organic component, such as the biocompatibility of metal ions complexation. In fact, the use of magnetic nanoparticles for wastewater remediation has been extensively considered thanks to the possibility of choosing proper coatings for metal and metalloid adsorption [8]. The advantages of this application are the simplicity and high efficiency of the adsorption process, as well as the possibility of easily removing the nanoadsorbents after purification to regenerate and reutilize them [5,7,17]. With respect to the incorporation in polymers, this opens the possibility of preparing a wealth of hybrid materials, with the nanoparticles imparting useful magnetic properties. One example is the incorporation of magnetic nanoparticles in alginates [18,19]. The polysaccharide of natural origin and its modified or functionalized derivatives can easily form cross-linked hydrogels with good mechanical properties that, as small beads, can be used in wastewater remediation for their high adsorption capacity [20]. The advantages connected to the use of these materials are, again, low cost and the possibility of recovery and regeneration, particularly if they are integrated with magnetic nanoparticles, enabling the easy separation from water. In this work, sodium alginate beads with two types of incorporated iron oxide magnetic nanomaterials (namely, magnetite nanoparticles, MNPs, and graphene oxide decorated with MNPs) were prepared. Those with only MNPs were magnetically characterized and compared, before and after copper(II) ions adsorption, with the pristine magnetic nanoparticles. The intent is to evaluate how the adsorption of copper ions influences the magnetic properties of the embedded nanoparticles.

2. Materials and Methods

All reagents and solvents were purchased from Sigma-Aldrich, St. Louis, MO, USA, unless otherwise indicated, and used without further purification. Magnetite nanoparticles (MNPs) were prepared according to the protocols described in the previous publications of the same authors [21,22] and magnetically analyzed as a reference.

2.1. Alginate Beads Preparation

The preparation of nanoadsorbents is outlined in Figure 1: the iron oxide MNPs and the graphene oxide decorated with MNPs (GONP) were prepared by coprecipitation from an alkaline solution, as reported elsewhere [21,22], and included in alginate beads, as outlined below. The composite adsorbents were obtained in the form of spherical hydrogel beads with a diameter around 4 mm. Figure 1 shows the confocal microscope (Olympus IX51, manufactured by Olympus Scientific Solutions, part of Evident Corporation, Tokyo, Japan) images (Figure 1a) and TEM (Transmission Electron Microscope, model TECNAI

FEI G2, FEI company, Hillsboro, OR, USA) images (Figure 1b,c) of the nanoparticles in the beads with a detail of the graphene oxide sheet decorated with nanoparticles (Figure 1c).

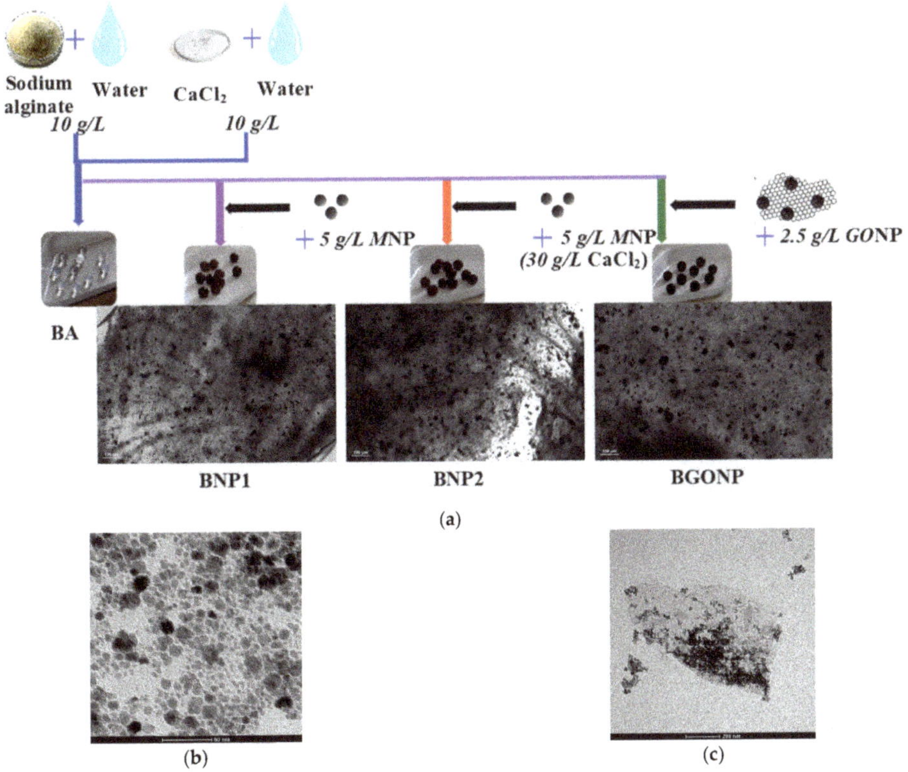

Figure 1. (**a**) The preparation scheme of the adsorbent beads. From left to right: BA (pure alginate beads), BNP1 (alginate beads with embedded MNPs), BNP2 (alginate beads with embedded MNPs and a higher concentration of hardener), BGONP (alginate beads with embedded GONPs), along with a 40× picture of their section at a confocal microscope (with white bar scale 100 μm long). TEM images of the magnetic nanomaterials used in the preparation of the beads (**b**) BNP1 (the NPs in BNP2 were similar and are not shown), and (**c**) BGONP with detail of graphene oxide sheet, respectively (scale impressed in the figure).

Depending on both the presence and type of magnetic nanomaterial filler, four types of beads were prepared and tested:

- Pure alginate beads (sample BA);
- Alginate beads with MNPs at different hardener (calcium chloride) solution concentrations (samples BNP1 and BNP2);
- Alginate beads with GONP (named BGONP).

Samples BNP1 and BNP2 were prepared following the method described in [21] and herein briefly resumed:

(1) Preparation of the sodium alginate solutions (10.0 g/L) with suspended MNPs (5.0 g/L for BNP1 and BNP2, 2.5 g/L for BGONP) by ultrasonication.

(2) Drop-wise addition of each suspension in a cross-linking solution, obtained by dissolving either 10.0 g/L (for BA, BNP1, and BGONP) or 30.0 g/L (for BNP2) of calcium

chloride ($CaCl_2$) in deionized water. During the addition, the Ca^{2+} solution was kept under constant agitation.

(3) Collection of the formed hydrated beads from the solution, followed by washing four times with deionized water, and storage in deionized water prior to use.

In particular, in BNP1 and BNP2, the same type of magnetic nanoparticles was embedded (the TEM in Figure 1b shows the NPs embedded in BNP1, and Figure 1c shows the GO sheet decorated with nanoparticle embedded in BGONP).

2.2. Experimental Procedure for Sample Preparation

According to the authors' previous work [21,22], different beads were tested on water samples containing Cu^{2+} ions obtained by diluting copper(II) sulphate pentahydrate ($CuSO_4 \cdot 5H_2O$, 99%) in milli-Q water. The beads loaded with copper have been prepared by adding 10 beads in 4 mL glass vials, where 2 mL of a solution at a defined copper concentration was added. The beads have been mixed for 24 h, and then, the solution was analyzed to establish the final Cu^{2+} concentration. The copper content on the beads was found by considering the difference between the initial and final copper found in the samples.

2.3. Spectrophotometric Determination of Metal Load

Metal ion loads were estimated by measuring the concentration of copper(II) in solution via absorbance by UV-VIS spectrophotometry (Jasco V-560 Spectrophotometer, Jaco Europe s.r.l., Cremella, Italy) with optical glass cuvettes. The use of such an analytical technique was previously validated by comparing the measured ion concentrations with other widely used techniques, such as ICP (Inductively Coupled Plasma). For Cu^{2+}, results showed a deviation below 4%, therefore justifying the use of such a technique for the aim of the present work.

The absorbance was measured at the maximum peak corresponding to 780 nm [23]. A calibration curve was created using a linear regression through the Beer–Lambert law by measuring a series of Cu^{2+} stock solutions at 3.00–2.00–1.00–0.25 g/L (each measurement was repeated 3 times, and the average value was taken). The stock solutions were prepared by dissolving copper(II) sulphate pentahydrate in milli-Q water. The pH of the solution was not corrected and was measured at values between 4.5 and 5 in all the solutions. Figure 2 shows the calibration curve with the determination coefficient (R^2).

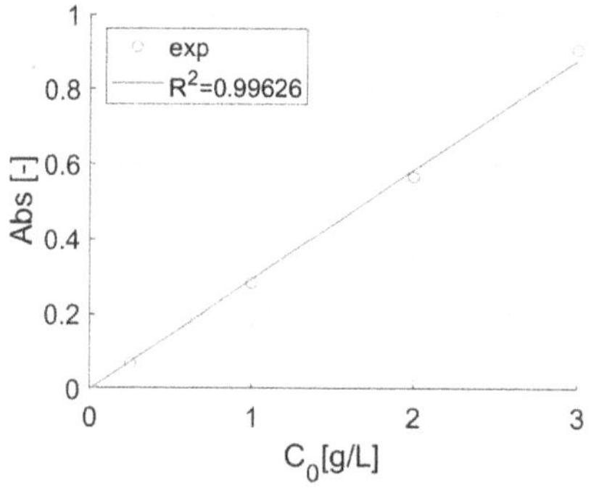

Figure 2. The calibration curve for copper (II) stock solutions.

The load tests were conducted according to the following protocol: a series of 4 mL vials were loaded with 2 mL of Cu^{2+} stock solution (0.25–1.00–2.00–3.00 g/L), 10 beads (of a single type per vial) and left for 24 h at 25 °C to ensure the equilibrium was reached.

After the test, the treated water was analyzed by spectrophotometry to determine the metal concentration, and the copper(II) load was calculated according to the following equation:

$$x_{load}\left[\frac{mg_{metal}}{g_{bead}}\right] = \frac{C_0 \cdot C_{eq}}{n \cdot m_{bead}} \cdot V \cdot 1000$$

where x_{load} [mg/g] is the amount of metal adsorbed per gram of bead. C_{eq} [g/L] is the equilibrium concentration of the metal in solution after the treatment, C_0 [g/L] is the metal starting concentration, n is the number of beads per vial, m_{bead} [g] is the average mass of each bead (determined from Table 1), V [L] is the volume of solution in the vial.

Table 1. Wet and dried mass of the different beads.

	BNP1		BNP2		BGONP		BA	
State	Wet	Dry	Wet	Dry	Wet	Dry	Wet	Dry
N. of Beads	27	27	29	29	34	34	20	20
Average Weight	23.4 mg	0.7 mg	23.1 mg	0.7 mg	22.6 mg	0.6 mg	15.9 mg	0.91 mg
Std. Dev.	2.5 mg	0.1 mg	1.6 mg	0.1 mg	3.6 mg	0.1 mg	0.91 mg	0.060 mg
Std. Dev. (%)	10.5%	12.4%	6.80%	13.1%	16.0%	13.4%	5.74%	12.0%

After treatment, the beads were isolated and dried prior to undergoing the magnetic measurements.

2.4. Magnetism Analysis

The magnetic characterization of the samples was performed by means of a Physical Properties Measurement System (PPMS) by Quantum Design equipped with a Vibrating Sample Magnetometer (VSM). The PPMS has a superconductor magnet that reaches a DC magnetic field of 90,000 Oe and can perform measurements in a temperature range from 1.9 K to 400 K. The samples were characterized by studying the magnetization as a function of both the applied DC field, M(H), and temperature, M(T). The M(H)s were measured at 300 K and 5 K by varying the applied field from 0 Oe to 90,000 Oe, then by looping it from 90,000 Oe to −90,000 Oe and back to 90,000 Oe. The M(T)s were measured with an applied DC field of 100 Oe and by varying temperature from 5 K to 300 K. Before both measurements, the trapped field in the superconductor magnet was reduced by following the procedure described in reference [24]. The M(T)s were measured with both Zero Field Cooling (ZFC) and Field Cooling (FC) protocols. For the ZFC protocol, the temperature was reduced after field zeroing, then a magnetic field of 100 Oe was applied, and the magnetization was measured by increasing the temperature to 300 K. Once 300 K was reached, to follow the FC protocol, the magnetization was acquired by reducing the temperature to 5 K without turning off the applied field.

3. Results

By applying an easy and reproducible preparation procedure, three types of nanocomposite adsorbent beads based on cross-linked alginate loaded with magnetite nanoparticles or MNP-decorated graphene oxide were prepared. They were compared with pure cross-linked alginate beads in the adsorption of copper(II) ions in aqueous solution to assess the effect of metal adsorption on the magnetic properties of the material.

In order to properly assess the equilibrium load of all the beads, the net mass of adsorbing material had to be determined. Since the produced alginate beads were in the form of a hydrogel, composed of up to 95% water, a sample of 20 to 40 beads per type was taken and lyophilized for 24 h. Table 1 reports the mass of wet vs. dry beads. The average dried mass was used to evaluate the load of copper(II) per gram of bead.

3.1. Isotherms Characterization

Batch isotherms were developed according to the following protocol: 5 beads were placed in a vial with 2 mL of copper(II) sulphate solution at 0.25, 1.00, 2.00, and 3.00 g/L (with respect to copper), and stirred for 1 h in a platform rocker at controlled temperature (25, 35, 45 °C). After adsorption was finished, the final solution was collected at once and sent to the spectrophotometer for analysis. From the data, isotherms according to the Langmuir model were developed. Both unbuffered solutions (pH between 4.5 and 5, due to copper(II) being a conjugated acid of a weak base, 3 repetitions for each concentration), and pH 3 (corrected with HCl, 2 repetitions for each concentration) were analyzed. Temperature dependence was studied at pH equal to 3. Table 2 reports the results of the maximum load Q_{max} (expressed as mg_{Cu}/g_{bead}) according to the Langmuir adsorption model.

Table 2. Maximum load determined according to the Langmuir model for each case studied (R^2 for each isotherm is reported in brackets).

	pH = 3			pH = Unbuffered
	25 °C	35 °C	45 °C	25 °C
Sample	Langmuir max load Q_{max} (mg/g_{bead})			
BNP1	264 (R^2 = 0.8949)	197 (R^2 = 0.9105)	164 (R^2 = 0.7589)	220 (R^2 = 0.9620)
BNP2	262 (R^2 = 0.7855)	156 (R^2 = 0.9062)	152 (R^2 = 0.8982)	276 (R^2 = 0.8549)
BGONP	218 (R^2 = 0.7667)	155 (R^2 = 0.8041)	141 (R^2 = 0.7589)	354 (R^2 = 0.9015)
BA	248 (R^2 = 0.8754)	159 (R^2 = 0.8553)	170 (R^2 = 0.8735)	338 (R^2 = 0.9206) 348 (R^2 = 0.9262)

Figure 3 reports both the experimental results and the fitting of the Langmuir isotherms. According to the achieved results, for each bead analyzed in buffered conditions at pH 3, higher temperatures show a general decrease in the maximum load, a behavior consistent with most chemical and physical adsorption isotherm models. This is also observed in other adsorbents bearing protonated carboxylic acid groups [25] in which the release of the exchangeable protons caused the desorption of Cu(II). The importance of the protonation–deprotonation equilibrium is also proved by observing the effect of pH, which generally shows an increase in the maximum load with unbuffered solutions. In fact, at acidic pH, the carboxylic groups are partially protonated, limiting their coordination ability/ionic interaction toward the copper cations. BGONP and BA showed the maximum increase in load, so we added adsorption tests at 5 g/L of copper(II) to ensure a proper estimation of the maximum load under equilibrium conditions. This is due to the fact that graphene oxide can contribute to the adsorption of copper ions thanks to the oxygenated functional groups in its structure.

3.2. Copper(II) Load on Beads

The equilibrium load of copper(II) on the four types of adsorbent beads was measured by spectrophotometric determination of the metal concentration in solution after 24 h of contact time with a selected number of beads. The results at different Cu^{2+} starting concentrations are reported in Table 3 in terms of the concentration of metal ions adsorbed per gram of beads. It can be deduced from Table 3 that the load determined on the beads is compatible with the maximum load found with the isotherms developed.

Copper(II) exhibited a comparable affinity with BNP1 and BNP2, as previously observed [21], resulting in similar loads, lower than pure alginate (BA), probably due to the reduced amount of alginate and, thus, of exchangeable divalent cations. However, BGONP displayed a similar load to BA thanks to the contribution of graphene oxide as an adsorbent co-filler. The loaded bead of type BNP1 was extracted from the solution with

3.00 g/L of Cu^{2+} concentration after treatment, washed with milli-q water, and dried for the magnetic characterization.

Table 3. Starting concentration and copper load on each sample of beads.

	Starting Cu^{2+} Concentration [g/L]			
	3.00	2.00	1.00	0.25
Sample	X_{load} of Cu^{2+} [mg/g_{beads}]			
BNP1	180	134	115	20.2
BNP2	197	155	105	17.3
BGONP	300	240	147	86.4
BA	295	227	170	90.1

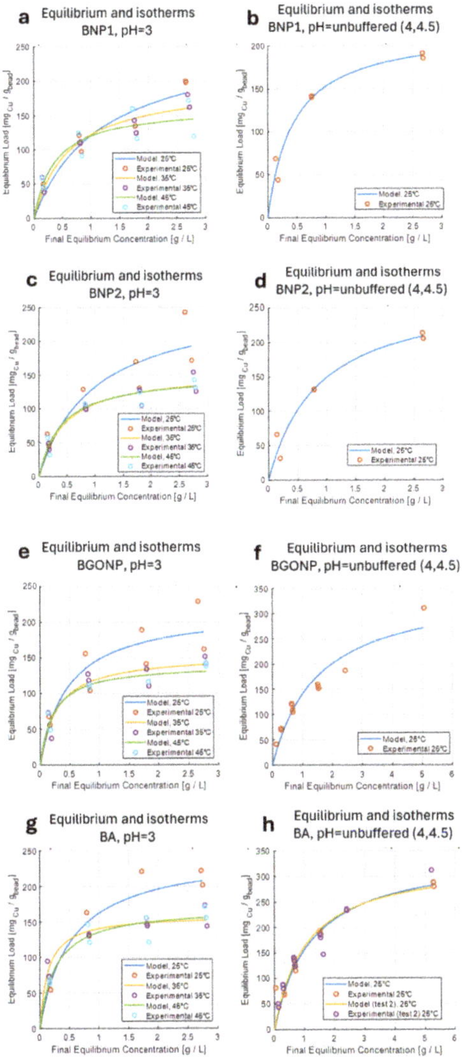

Figure 3. Adsorption equilibrium isotherms for copper under different conditions. (**a**) BNP1, pH 3; (**b**) BNP1, unbuffered pH; (**c**) BNP2, pH 3; (**d**) BNP2, unbuffered pH; (**e**) BGONP, pH 3; (**f**) BGONP, unbuffered pH; (**g**) BA, pH 3; (**h**) BA, unbuffered pH.

3.3. Magnetic Characterization

The curves obtained by performing the M(H) measurements on the powder of Fe_3O_4 magnetic nanoparticles at 5 K and 300 K are reported in Figure 4. The value of the mass magnetization at 90,000 Oe (M_s) was about 70 emu/g at 300 K and about 78 emu/g at 5 K. These values are comparable to the ones reported in the literature for samples of Fe_3O_4 nanoparticles [14,19,26]. The increase in mass magnetization at 5 K and the change in coercivity field from a value below 2 Oe at 300 K to a value about 250 Oe at 5 K suggested the presence of a superparamagnetic behavior at the highest temperatures [11]. In fact, despite the absence of a coercive field, which is expected for a paramagnet, the shape of the loop at 300 K showed a rapid increase in magnetization below a field intensity of 10,000 Oe, as expected for the signal of a ferromagnet.

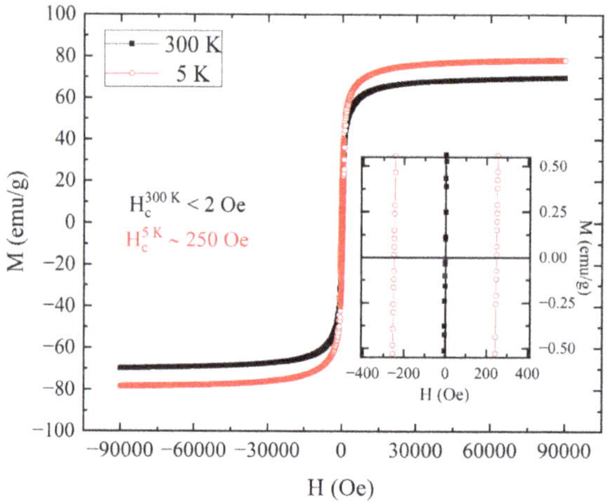

Figure 4. M(H) loops performed at 5 K (red open circles) and 300 K (black closed squares) for the powder of Fe_3O_4 magnetic nanoparticles. The inset shows the data near to zero field. From this, it is possible to evaluate the coercive field at 300 K as low as 2 Oe and the coercive field at 5 K of about 250 Oe.

To better investigate the transition from a blocked state to a superparamagnetic behavior for the powder of Fe_3O_4 magnetic nanoparticles, the study of magnetization versus temperature with 100 Oe of applied field was performed, and the curves obtained are reported in Figure 5. The inset reveals that the ZFC curve reaches its maximum (T_{Max}) at a temperature of 205 K. This T_{Max} represents the mean blocking temperature for the sample of magnetic nanoparticles, indicating a transition to a superparamagnetic behavior [27,28]. In fact, as observed in Figure 5, at temperatures below $T_{Max,}$ the ZFC and FC curves were completely separated, as expected for a sample of magnetic nanoparticles in a blocked state. For temperatures above T_{Max}, the separation between ZFC and FC curves suggested that not all the nanoparticles were in a superparamagnetic state [28]. The two curves had an overlap at near room temperature, indicating that the sample had a reversible behavior. This was an indication of the complete transition to the superparamagnetic state, and it is in accordance with the absence of hysteretic behavior at 300 K (see inset of Figure 4).

The same magnetic characterizations were performed on a bead BNP1 containing the same magnetic nanoparticles shown in Figures 4 and 5. To compare the magnetic properties of the pristine nanoparticles and their alginate composite beads, the sample BNP1 was chosen. This would allow for highlighting the effect of the embedment in the polysaccharide matrix. The M(H) loops obtained for the dried bead at 300 K and

5 K are reported in Figure 6. The mass magnetization was obtained by normalizing the magnetic moment with the mass of magnetic nanoparticles embedded in the sodium alginate. Considering the mass proportion between magnetic nanoparticles and sodium alginate determined for BNP1 in [21], the mass of the former was estimated to be one-third of the total dry mass of the bead. The first observation that can be made from Figure 6 is that the saturation magnetization at 300 K was lower than that at 5 K. Furthermore, from the inset, which showed the data around zero field, it was possible to evaluate the coercivity of the sample. In particular, the coercive field at 5 K was about 280 Oe, while the coercive field at 300 K was lower than 2 Oe. This reduction in coercivity suggests that magnetic nanoparticles retained their superparamagnetic behavior also when they were embedded in the bead. Despite this, there was a difference in the mass magnetization at 90,000 Oe, both at 5 K and 300 K for the two samples. From Figure 6, the mass magnetization reached a maximum of 54 emu/g at 300 K and of 62 emu/g at 5 K. Comparing these values with those measured in Figure 4 for the powder of magnetic nanoparticles (70 emu/g at 300 K and 78 emu/g at 5 K), it was possible to see a difference of around 16 emu/g at each temperature (reduction of 23% at 300 K and of 21% at 5 K). The mass ratio of 1/3 between magnetic nanoparticles and the total mass of the bead, used to evaluate the mass magnetization, was obtained considering the proportion between magnetic nanoparticles and sodium alginate in the synthesis process of the bead. For this reason, the one-third fraction is an average value but could slightly change when considering a single bead. This could explain the difference between the M_s of the pure powder of magnetic nanoparticles and the bead. Another possible explanation could be the presence of an alternating surface anisotropy due to the incorporation of nanoparticles in the sodium alginate matrix. Despite this, the value obtained for M_s was comparable to that reported in the literature for similar types of beads [19] produced for wastewater remediation.

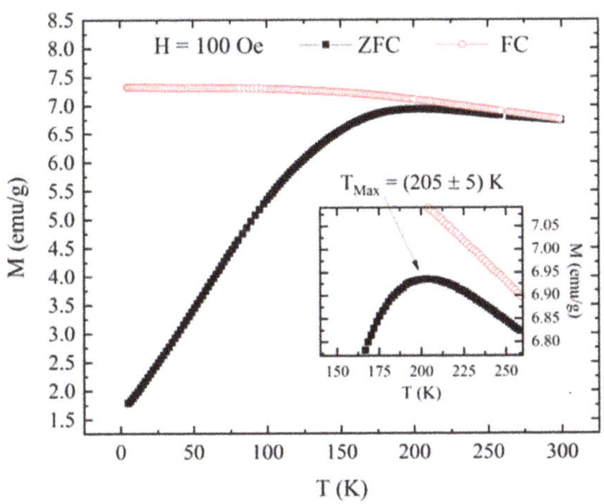

Figure 5. Magnetization versus temperature measured with an applied field of 100 Oe in the ZFC (black closed squares) and FC protocol (red open circles) for a powder of Fe_3O_4 magnetic nanoparticles. The inset shows the behavior around the peak of the ZFC curve, showing a maximum associated temperature of 205 K.

Focusing on this latter application, the magnetic properties of a bead of BNP1type dried after adsorbing copper ions (in a 3.00 g/L Cu^{2+} solution) were analyzed by performing the M(H) measurements (Figure 7). To maximize the metal effect on the adsorbent's magnetic signal, the

BNP1 sample with the highest copper load was selected for this characterization. From the M(H) shown in Figure 7, it was possible to observe that the superparamagnetic behavior at 300 K remained after the adsorption of copper ions. In fact, the coercive field changed from about 245 Oe at 5 K to a value below 2 Oe at 300 K. This difference was perfectly in accordance with the behavior observed in Figures 4 and 6. Despite this, the magnetic responses showed a difference in terms of signal intensity. In fact, as it can be seen in Figure 7, the maximum of mass magnetization for the bead after adsorption was 16 emu/g at 300 K and 21 emu/g at 5 K, with a difference of more than 40 emu/g with the data reported in Figure 6 (a reduction of 74% at 300 K and 65% at 5 K). One possibility to explain the change in the signal is to consider that the value of mass used to calculate mass magnetization in Figure 7 was higher than that used for Figure 6. In fact, the mass magnetization in Figure 7 was calculated by considering one-third of the mass of the dried bead after copper adsorption, exactly as in the case of a dried bead before adsorption; however, in this case (after the adsorption), the total mass of the bead considered the presence of not only sodium alginate and magnetic nanoparticles but also copper ions. So, there was an error in the estimation of the mass associated with magnetic nanoparticles due to the presence of copper. In fact, the average mass of a group of 20 beads before copper adsorption and of 20 beads after copper adsorption differed by around 0.5 mg. However, even rescaling the mass used for normalization in Figure 7 to the one used for the bead before adsorption (Figure 6), M_s would be 21 emu/g at 300 K and 27 emu/g at 300 K. Thus, the difference in the case of bead before adsorption would be more than 30 emu/g (a reduction of 55% at 300 K and 47% at 5 K). Another possibility to explain this change in the signal could be to consider the signal associated with the presence of copper inside the bead. Considering that the mass susceptibility of bulk copper at 300 K was -0.0830×10^{-6} emu/(Oe g) [29] and evaluating the signal associated with the estimated mass of 0.5 mg of copper at 300 K with an applied field of 90,000 Oe, the contribution of copper would be about -4×10^{-6} emu and about -0.017 emu/g after the normalization with mass used for data in Figure 7. From this estimation, a reduction of about 0.03% of M_s for the bead after copper adsorption, compared to that before copper adsorption, is expected, but it is too low to explain the difference between the data in Figures 6 and 7. After these evaluations, the difference in signal intensity obtained by comparing M(H) for the bead before and after copper adsorption cannot be explained by considering only the presence of copper ions or an underestimation of the mass of the magnetic component.

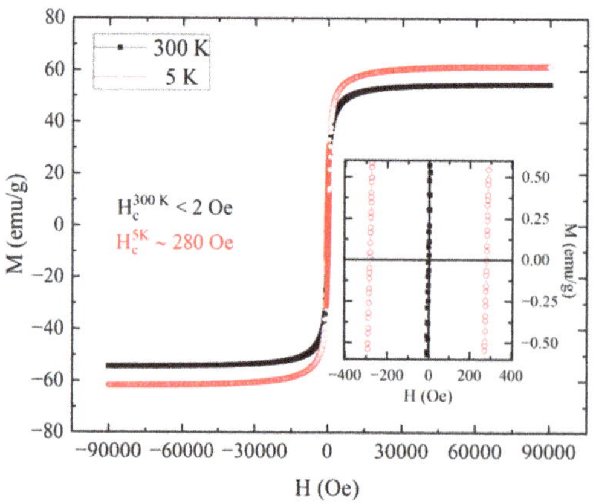

Figure 6. M(H) loops performed at 5 K (red open circles) and 300 K (black closed squares) for the BNP1 bead. The inset shows the data near to zero field. From this, it is possible to evaluate the coercive field at 300 K as 2 Oe and the coercive field at 5 K of about 280 Oe.

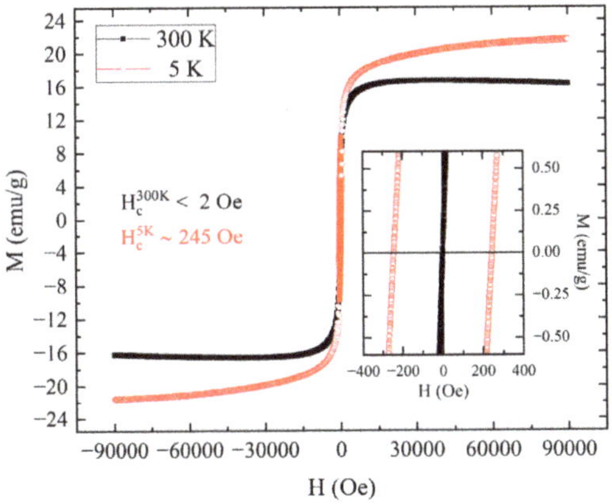

Figure 7. Magnetization versus field loops performed at 5 K (red open circles) and 300 K (black closed squares) for the BNP1 bead after copper ions adsorption. The inset shows the data near to zero field. From this, it is possible to evaluate the coercive field at 300 K as low as 2 Oe and the coercive field at 5 K of about 245 Oe.

Copper adsorption also has another effect, which can be observed in the behavior of magnetization at high fields at 300 K. In Figure 7, the magnetization associated with the BNP1 sample before adsorption shows a maximum around 58,000 Oe and then decreases with the increase in the field. This behavior is also observed for the sample of BNP1 after copper adsorption (see Figure 7). However, the maximum has reached around 37,000 Oe. After this field value, magnetization starts to decrease until the maximum field is reached. This can be explained by considering the diamagnetic behavior of copper, which is relevant when the signal of magnetic nanoparticles reduces its growth. In fact, this behavior is observed for the M(H) at 300 K, but not for the M(H) at 5 K, in which the magnetization does not reach a maximum in either case, before and after copper adsorption.

To investigate the signal reduction between Figures 6 and 7, the distribution of blocking temperature extracted from the M(T) curves, as the temperature derivative of the difference between ZFC and FC curves, was studied for all three samples [27,30]. In Figure 8, the comparison among the three blocking distributions is shown. As can be seen in Figure 8a, the presence of a peak at 16 K is evident for all three samples. This peak in the blocking temperature distribution has the physical meaning of the average blocking temperature for the samples [24,27,28,30,31]. The presence of this peak for all the samples underlines the fact that the microscopic properties of magnetic nanoparticles remained the same and they were not affected by either the incorporation in the bead or the presence of copper. This is more evident when considering Figure 8b, in which each distribution is normalized with the intensity of the peak at 16 K. As can be seen, the shapes of the three distributions were very similar, except for the intensities of the peaks. In fact, the distribution extracted for the powder of magnetic nanoparticles showed two peaks, the first one at 16 K, as previously mentioned, and the second one at the temperature of 65 K. The presence of double peaks in the distribution of blocking temperature has different possible explanations in the literature, connected to either different interactions between nanoparticles present in the sample [24,32] or a different dimension distribution for them [33]. After the incorporation of the beads, the intensity of the peak at 65 K was reduced, but the peak at 16 K remained, as stated before. This determines a relative increase in the peak amplitude at 16 K compared to that at 65 K (Figure 8a). The reduction in the intensity of the peak at 16 K is in line with the loss of magnetic signal observed in the M(H) measurements in Figure 6. In fact, considering the mass magnetization from Figures 3, 5 and 6 at H = 100 Oe, the field used to perform the M(T)s, it is possible to observe a reduction of 27% in the signal between the powder of MNPs and BNP1 and a reduction of 75% in the signal between the powder of MNPs and BNP1 after copper adsorption. Comparing the signals in Figure 8a, it is possible to observe a similar trend for the peak at 16 K with a reduction of 20% between the powder of MNPs (black curve) and BNP1 before copper adsorption (red curve) and a reduction of 73% between the same samples after copper adsorption (blue curve). This trend is not the same for the peak at 65 K, since the signal reduction is 42% before copper adsorption and 80% after copper adsorption. Thus, the intensity variation of the peak at 65 K is more pronounced than that at 16 K for BNP1 after copper adsorption. This result also gives the information that the change in blocking temperature distribution is due to a reduction of the peak at 65 K. The point now is to understand what can determine a change in the distribution of magnetic nanoparticles' properties in the bead. Following the results of Germanos et al. [26], copper adsorption in a bead of calcium alginate can change the spatial distribution of incorporated magnetic nanoparticles. Through a study of the beads via EDS, the authors observed that magnetic nanoparticles were perfectly distributed in the bead before copper adsorption, but after the adsorption, magnetic nanoparticles started to move from the center of the bead to its side regions. This not only changed the spatial concentration of magnetic nanoparticles, but it could also explain the change in the blocking distribution due to possible changes in MNPs interaction and aggregation. Considering these observations, the changes observed in Figure 8b after copper adsorption are in line with a variation of the MNPs configuration in the bead.

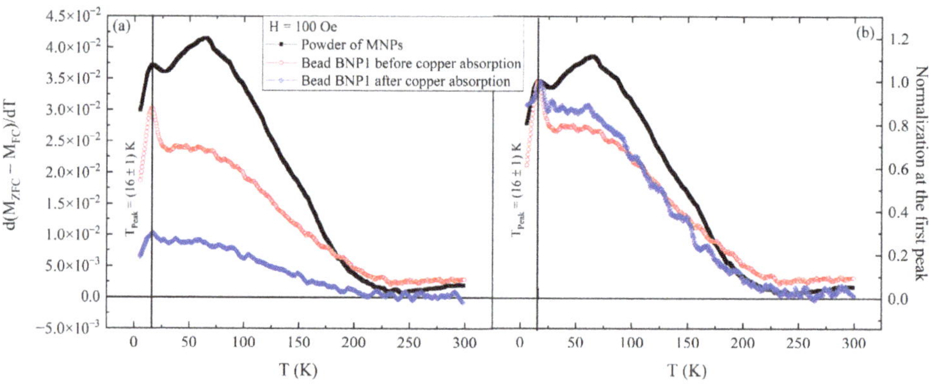

Figure 8. (**a**) Blocking temperature distribution extracted from the M(T) curves. The figure shows the comparison between the distributions obtained for the powder of Fe_3O_4 magnetic nanonoparticles (black closed squares), the dried BNP1 bead (red open circles), and the dried BNP1 bead after copper adsorption (blue crossed diamonds). Each distribution shows a peak at 16 K. On the right (**b**), the same curves are compared after normalization with the intensity of the peak at 16 K.

4. Conclusions

The magnetic measurements performed on a powder of Fe_3O_4 nanoparticles confirmed its superparamagnetic behavior at room temperature. This can be observed from either the change in coercivity measured at both 5 K and 300 K or the analysis of the M(T) measurement. The presence of a superparamagnetic state also remained when the nanoparticles were used in the synthesis process of a cross-linked sodium alginate bead. In particular, the saturation magnetization extracted from the M(H) measurement was lower than that measured for the pure MNPs powder, although of the same order of magnitude, confirming the possibility of using the bead in water remediation applications. The same characterization was performed on a bead dried after the adsorption of copper ions from a water solution. The variation in coercivity between the M(H) at 5 K and the M(H) at 300 K suggested that the superparamagnetic behavior still remained on the bead after copper adsorption. However, the saturation magnetization was somewhat reduced compared to that measured for the bead without copper adsorption and for the powder of magnetic nanoparticles. However, there is a significant reduction in the magnetic signal after the metal ion absorption. This reduction cannot be explained by considering the effect of the copper magnetic signal, and it can suggest a change in magnetic nanoparticle spatial configuration in the bead. This result was supported by the comparison of the blocking temperature distributions obtained from the M(T) measurements. The magnetization reduction is important from the perspective of applications because it can affect the efficiency of bead removal from the water medium. Starting with these results, the possibility of using magnetic signal measurements of the beads as a nondestructive determination of the adsorbent loading during use can be studied further.

Author Contributions: Conceptualization, A.G. and P.S.; methodology, P.S., M.B., E.S., P.L. and M.P.; software, M.B.; validation, F.D., M.M., and M.B.; formal analysis, S.C.; investigation, F.D., M.M., M.B., and E.R.; data curation, M.M. and M.B.; writing—original draft preparation, M.M. and E.S.; writing—review and editing, P.S. and S.C.; supervision, M.P. and P.L.; project administration, E.S. and P.L. All authors have read and agreed to the published version of the manuscript.

Funding: This research received no external funding.

Data Availability Statement: Data are contained within the article.

Acknowledgments: The authors thank Miriana Arletti and Francesca Malgrati for their contribution to the development of isotherms. P.S. thanks the UNIPD-DII BIRD2023 project "HYPURE-Hybrid 3D Printed Water Purification Systems for Emerging Pollutants" for financial support. This work was partially supported by the EU under the Italian National Recovery and Resilience Plan (NRRP) of NextGenerationEU, Project Title UPWARD CUP E53D23014490001—Grant Assignment Decree No. 1383 of 01/09/2023, by partnership on "Telecommunications of the Future" (PE00000001-program "RESTART") and from the Horizon Europe MSCA programme under FLORIN project (GA n° 101086142).

Conflicts of Interest: The authors declare no conflicts of interest.

References

1. Shrestha, R.; Ban, S.; Devkota, S.; Sharma, S.; Joshi, R.; Tiwari, A.P.; Kim, H.Y.; Joshi, M.K. Technological trends in heavy metals removal from industrial wastewater: A review. *J. Environ. Chem. Eng.* **2021**, *9*, 105688. [CrossRef]
2. Järup, L. Hazards of heavy metal contamination. *Br. Med. Bull.* **2003**, *68*, 167–182. [CrossRef] [PubMed]
3. Jaishankar, M.; Tseten, T.; Anbalagan, N.; Mathew, B.B.; Beeregowda, K.N. Toxicity, mechanism and health effects of some heavy metals. *Interdiscip. Toxicol.* **2014**, *7*, 60–72. [CrossRef] [PubMed]
4. Ali, H.; Khan, E.; Ilahi, I. Environmental Chemistry and Ecotoxicology of Hazardous Heavy Metals: Environmental Persistence, Toxicity, and Bioaccumulation. *J. Chem.* **2019**, *2019*, 6730305. [CrossRef]
5. Saleh, T.A.; Mustaqeem, M.; Khaled, M. Water treatment technologies in removing heavy metal ions from wastewater: A review. *Environ. Nanotechnol. Monit. Manag.* **2022**, *17*, 100617. [CrossRef]
6. Al-Jabri, M.T.K.; Devi, M.G.; Al Abri, M. Synthesis, characterization and application of magnetic nanoparticles in the removal of copper from aqueous solution. *Appl. Water Sci.* **2018**, *8*, 223. [CrossRef]
7. Liosis, C.; Papadopoulou, A.; Karvelas, E.; Karakasidis, T.E.; Sarris, I.E. Heavy Metal Adsorption Using Magnetic Nanoparticles for Water Purification: A Critical Review. *Materials* **2021**, *14*, 7500. [CrossRef]
8. Xue, S.; Xiao, Y.; Wang, G.; Fan, J.; Wan, K.; He, Q.; Gao, M.; Miao, Z. Adsorption of heavy metals in water by modifying Fe_3O_4 nanoparticles with oxidized humic acid. *Colloids Surf. A Physicochem. Eng. Asp.* **2021**, *616*, 126333. [CrossRef]
9. Bean, C.P.; Livingston, J.D. Superparamagnetism. *J. Appl. Phys.* **1959**, *30*, S120–S129. [CrossRef]

10. Blundell, S. *Magnetism in Condensed Matter*; Oxford Master Series in Condensed Matter Physics, No. 4; Reprint; Oxford University Press: Oxford, UK, 2014; ISBN 978-0-19-850591-4.
11. Narlikar, A.V.; Fu, Y.Y. (Eds.) *The Oxford Handbook of Nanoscience and Technology*; Oxford University Press: New York, NY, USA, 2010; ISBN 978-0-19-953304-6.
12. Pankhurst, Q.A.; Connolly, J.; Jones, S.K.; Dobson, J. Applications of magnetic nanoparticles in biomedicine. *J. Phys. Appl. Phys.* **2003**, *36*, R167–R181. [CrossRef]
13. Kalia, S.; Kango, S.; Kumar, A.; Haldorai, Y.; Kumari, B.; Kumar, R. Magnetic polymer nanocomposites for environmental and biomedical applications. *Colloid Polym. Sci.* **2014**, *292*, 2025–2052. [CrossRef]
14. Maldonado-Camargo, L.; Unni, M.; Rinaldi, C. Magnetic Characterization of Iron Oxide Nanoparticles for Biomedical Applications. In *Biomedical Nanotechnology*; Petrosko, S.H., Day, E.S., Eds.; Methods in Molecular Biology; Springer: New York, NY, USA, 2017; Volume 1570, pp. 47–71. ISBN 978-1-4939-6838-1. [CrossRef]
15. Ali, A.; Shah, T.; Ullah, R.; Zhou, P.; Guo, M.; Ovais, M.; Tan, Z.; Rui, Y. Review on Recent Progress in Magnetic Nanoparticles: Synthesis, Characterization, and Diverse Applications. *Front. Chem.* **2021**, *9*, 629054. [CrossRef] [PubMed]
16. Frey, N.A.; Peng, S.; Cheng, K.; Sun, S. Magnetic nanoparticles: Synthesis, functionalization, and applications in bioimaging and magnetic energy storage. *Chem. Soc. Rev.* **2009**, *38*, 2532. [CrossRef] [PubMed]
17. Bhateria, R.; Singh, R. A review on nanotechnological application of magnetic iron oxides for heavy metal removal. *J. Water Process Eng.* **2019**, *31*, 100845. [CrossRef]
18. Lin, Z.; Zhang, Y.; Chen, Y.; Qian, H. Extraction and recycling utilization of metal ions (Cu^{2+}, Co^{2+} and Ni^{2+}) with magnetic polymer beads. *Chem. Eng. J.* **2012**, *200–202*, 104–112. [CrossRef]
19. De Castro Alves, L.; Yáñez-Vilar, S.; Piñeiro-Redondo, Y.; Rivas, J. Efficient Separation of Heavy Metals by Magnetic Nanostructured Beads. *Inorganics* **2020**, *8*, 40. [CrossRef]
20. Gao, X.; Guo, C.; Hao, J.; Zhao, Z.; Long, H.; Li, M. Adsorption of heavy metal ions by sodium alginate based adsorbent-a review and new perspectives. *Int. J. Biol. Macromol.* **2020**, *164*, 4423–4434. [CrossRef]
21. Russo, E.; Sgarbossa, P.; Gelosa, S.; Copelli, S.; Sieni, E.; Barozzi, M. Adsorption of Heavy Metal Ions on Alginate-Based Magnetic Nanocomposite Adsorbent Beads. *Materials* **2024**, *17*, 1942. [CrossRef]
22. Barozzi, M.; Copelli, S.; Russo, E.; Sgarbossa, P.; Lavagnolo, M.C.; Sandon, A.; Morosini, C.; Sieni, E. Implementation of Magnetic Nanostructured Adsorbents for Heavy Metals Separation from Textile Wastewater. *Sustainability* **2022**, *14*, 11785. [CrossRef]
23. Jancsó, G. Effect of D and 18O isotope substitution on the absorption spectra of aqueous copper sulfate solutions. *Radiat. Phys. Chem.* **2005**, *74*, 168–171. [CrossRef]
24. Modestino, M.; Galluzzi, A.; Sarno, M.; Polichetti, M. The Effect of a DC Magnetic Field on the AC Magnetic Properties of Oleic Acid-Coated Fe_3O_4 Nanoparticles. *Materials* **2023**, *16*, 4246. [CrossRef]
25. Ji, C.; Wu, D.; Lu, J.; Shan, C.; Ren, Y.; Li, T.; Lv, L.; Pan, B.; Zhang, W. Temperature regulated adsorption and desorption of heavy metals to A-MIL-121: Mechanisms and the role of exchangeable protons. *Water Res.* **2021**, *189*, 116599. [CrossRef] [PubMed]
26. Germanos, G.; Youssef, S.; Abboud, M.; Farah, W.; Lescop, B.; Rioual, S. Diffusion and agglomeration of iron oxide nanoparticles in magnetic calcium alginate beads initiated by copper sorption. *J. Environ. Chem. Eng.* **2017**, *5*, 3727–3733. [CrossRef]
27. Bruvera, I.J.; Mendoza Zélis, P.; Pilar Calatayud, M.; Goya, G.F.; Sánchez, F.H. Determination of the blocking temperature of magnetic nanoparticles: The good, the bad, and the ugly. *J. Appl. Phys.* **2015**, *118*, 184304. [CrossRef]
28. Modestino, M.; Galluzzi, A.; Sarno, M.; Polichetti, M. Study of the Coercivity Field in a Sample of $NiFe_2O_4$-OA Nanoparticles at Different Temperatures: Comparing the Different Processes to Evaluate TB by DC Measurements. In Proceedings of the 2023 IEEE Nanotechnology Materials and Devices Conference (NMDC), Paestum, Italy, 22–25 October 2023; pp. 259–263. [CrossRef]
29. Bowers, R. Magnetic Susceptibility of Copper Metal at Low Temperatures. *Phys. Rev.* **1956**, *102*, 1486–1488. [CrossRef]
30. Micha, J.S.; Dieny, B.; Régnard, J.R.; Jacquot, J.F.; Sort, J. Estimation of the Co nanoparticles size by magnetic measurements in Co/SiO_2 discontinuous multilayers. *J. Magn. Magn. Mater.* **2004**, *272–276*, E967–E968. [CrossRef]
31. Polichetti, M.; Modestino, M.; Galluzzi, A.; Pace, S.; Iuliano, M.; Ciambelli, P.; Sarno, M. Influence of citric acid and oleic acid coating on the dc magnetic properties of Fe_3O_4 magnetic nanoparticles. *Mater. Today Proc.* **2020**, *20*, 21–24. [CrossRef]
32. Urian, Y.A.; Atoche-Medrano, J.J.; Quispe, L.T.; León Félix, L.; Coaquira, J.A.H. Study of the surface properties and particle-particle interactions in oleic acid-coated Fe_3O_4 nanoparticles. *J. Magn. Magn. Mater.* **2021**, *525*, 167686. [CrossRef]
33. Goya, G.F.; Lima, E.; Arelaro, A.D.; Torres, T.; Rechenberg, H.R.; Rossi, L.; Marquina, C.; Ibarra, M.R. Magnetic Hyperthermia with Fe_3O_4 Nanoparticles: The Influence of Particle Size on Energy Absorption. *IEEE Trans. Magn.* **2008**, *44*, 4444–4447. [CrossRef]

nanomaterials

MDPI

Article

A 3D-Printed Portable UV and Visible Photoreactor for Water Purification and Disinfection Experiments

Nelson Castro [1,2], Joana M. Queirós [1,2,3], Dinis C. Alves [1,3], Margarida M. Macedo Fernandes [1,4,5], Senetxu Lanceros-Méndez [1,6,7,*] and Pedro M. Martins [2,3,*]

1 Physics Centre of Minho and Porto Universities (CF-UM-UP) and LaPMET—Laboratory of Physics for Materials and Emergent Technologies, University of Minho, 4710-057 Braga, Portugal; nelson.castro@inl.int (N.C.); id10031@alunos.uminho.pt (J.M.Q.); dinisalves@fisica.uminho.pt (D.C.A.); margaridafernandes@cmems.uminho.pt (M.M.M.F.)

2 International Iberian Nanotechnology Laboratory (INL), 4715-330 Braga, Portugal

3 Centre of Molecular and Environmental Biology, University of Minho, 4710-057 Braga, Portugal

4 IB-S—Institute for Research and Innovation on Bio-Sustainability, University of Minho, 4710-057 Braga, Portugal

5 Center for Microelectromechanical Systems (CMEMS) and LABBELS—Associate Laboratory, University of Minho, 4800-058 Guimarães, Portugal

6 BCMaterials, Basque Center for Materials, Applications and Nanostructures, UPV/EHU Science Park, 48940 Leioa, Spain

7 IKERBASQUE, Basque Foundation for Science, 48009 Bilbao, Spain

* Correspondence: senentxu.lanceros@bcmaterials.net (S.L.-M.); pamartins@bio.uminho.pt (P.M.M.)

Abstract: Water scarcity and contamination are urgent issues to be addressed. In this context, different materials, techniques, and devices are being developed to mitigate contemporary and forthcoming water constraints. Photocatalysis-based approaches are suitable strategies to address water contamination by degrading contaminants and eliminating microbes. Photoreactors are usually designed to perform photocatalysis in a scalable and standardised way. Few or none have been developed to combine these characteristics with portability, flexibility, and cost effectiveness. This study reports on designing and producing a portable (490 g), low-cost, and multifunctional photoreactor that includes adjustable radiation intensity and two types of wavelengths (UV-A and visible), including combined agitation in a compact mechanism produced through 3D printing technology. The mechanical, electrical, and optical subsystems were designed and assembled into a robust device. It is shown that it is possible to apply radiations that can reach 65 mW/cm^2 and 110 mW/cm^2 using the installed visible and UV LEDs and apply mechanical agitation up to 200 rpm, all under a ventilated system. Regarding functionality, the photoreactor proof of concept indicated the ability to degrade ~80% and 30% ciprofloxacin under UV and visible irradiation of TiO_2 and Ag/TiO_2 nanoparticles. The device also showed the ability to eliminate *E. coli* bacteria, recurring to radiation set-ups and nanoparticles. Therefore, the originally designed and constructed photoreactor concept was characterised and functionally validated as an exciting and flexible device for lab-scaled or outdoor experiments, assuring standardised and comparable results.

Keywords: portable photoreactor; photocatalysis; ciprofloxacin; water purification; 3D printing

1. Introduction

Water contamination is a planetary-scale problem driven by several factors, such as industrial activities, agricultural practices, population density, and societal habits, that demands urgent action to reduce its harmful effects on aquatic organisms and human health. The United Nations members have identified clean water as the sixth goal of the Sustainable Development Goals. The World Health Organisation estimates that approximately 800 thousand people die yearly from illnesses related to contaminated water consumption [1,2].

In this critical context, there are increasing efforts to develop new methods, materials, and devices that can efficiently decontaminate water [3].

As a result, research on advanced oxidation processes (AOPs) for water purification has gained significant attention as an alternative and complement to conventional water treatment methods. These innovative techniques facilitate the breakdown of various organic contaminants by generating highly reactive species [4–7].

AOPs have paid much attention to the photocatalytic processes to degrade organic contaminants in water matrixes [8–10]. The photocatalyst more commonly used is titanium dioxide (TiO_2), owing to its advantageous properties, such as low toxicity, abundance, cost effectiveness, and chemical stability [11]. To efficiently perform photocatalytic degradation processes in a well-controlled, scalable, and reproducible way, it is necessary to design and develop photoreactors. These devices should allow the photons from a radiation source to be in contact with the photocatalysts (e.g., titanium dioxide (TiO_2)) and the contaminant in the water matrix under continuous stirring for homogenisation, allowing for initiation and maintaining the photocatalytic degradation process [12].

Typically, the photocatalytic reactors presented in the literature can be divided into two main types: a reactor with the photocatalyst (e.g., TiO_2 nanoparticles) in suspension and a configuration with immobilised photocatalysts. The main difference between those is the catalyst placement: in suspension reactors, the catalyst is suspended in a reactant solution; in the other, a catalyst is immobilised in the reactor walls or on a substrate (e.g., zeolites, polymer), where the reactant solution flows through [13]. The former presents higher efficiencies, as the nanoparticle's entire surface area is available for radiation harvesting and interaction with the organic contaminants in suspension, and it is usually the first stage that allows the evaluation of newly produced photocatalysts. On the other hand, it demands post-processing for nanoparticle removal to prevent secondary pollution, which makes decontamination more expensive [14]. The latter type of reactor immobilises nanoparticles that show significant photocatalytic degradation efficiencies into a substrate. This enables their reuse and the reduction in post-processing steps and costs. It also prevents secondary pollution with the trade-off of lower photocatalytic efficiency due to decreasing the nanoparticle's active contact area [14].

Photoreactors with very different configurations have been produced; for instance, an immobilised solution involving a tubular photoreactor based on cellulose acetate monolithic/TiO_2 thin films has been irradiated with simulated or natural solar light, removing approximately 100% of Cr in the solution [15]. Another immobilised solution developed a fixed-bed reactor utilising 12% wt. TiO_2-coated alumina monoliths incorporated in the space between a centrally deployed UV lamp and the reactor wall to degrade 1,8-Diazabicyclo[5,4,0]undec-7-ene, obtaining 100% conversion [16].

Regarding suspension reactors, Khawla Azalok et al. photodegraded 98% of metronidazole in 60 min, using TiO_2 as a catalyst under UV irradiation [17]. Further, García-Muñoz et al. used a batch suspension reactor with TiO_2 as a catalyst to degrade polystyrene nanobeads (PS); this work showed the removal of approximately 100% of 20 ppm PS 6 h after the beginning of the experiment [18].

Additionally, water disinfection applications have also been reported, with the inactivation of 92.6% of Escherichia coli [19] after 120 min of visible irradiation over polystyrene spheres with nitrogen-doped TiO_2 particles and 100% using 0.6% wt. Ag/TiO_2 nanoparticles [20].

To evaluate photoreactors and among the many possible contaminants, one category of pollutants deserves particular attention: the contaminants of emergent concern (CECs). These substances can be pharmaceuticals, personal care products, or pesticides that are not regularly monitored in the environment. However, they can potentially pose negative ecological and health consequences for humans [21,22], and conventional wastewater treatment techniques cannot effectively remove these CECs [23].

Further, the aquatic environment provides a thriving environment for microorganisms such as enteric bacteria, viruses, protozoa, and parasitic worms [24,25]. Public health authorities often prioritise pollutants that cause immediate illness and the spread of diseases, particularly regarding non-potable wastewater recycling [26]. Nevertheless, it has been demonstrated that traditional disinfection methods, such as chlorine disinfection, are inadequate and have led to resistant bacterial strains, posing a significant health risk in water-related applications [27].

Based on the mentioned state of the art, regarding both photoreactors for contaminant degradation and disinfection, there is a vast diversity of devices whose control of parameters such as radiation intensity, type of radiation, container (volume), and agitation are difficult to vary, control and, therefore, make them reproducible, aiming to create comparable scientific data. In this context of photocatalytic assays, there is a demand for new and standardised approaches that allow for straightforward comparisons between results.

In this work, the focus was on the development of a new concept of a portable photoreactor for lab-scale photocatalytic assays under UV or visible radiation with an articulated stirrer connected to a motor. The prototype parts were produced by FDM (Fused Deposition Modelling) 3D printing, and the circuits were designed for PCB (Printed Circuit Board) construction. The constructed photoreactor functionalities were validated regarding its photocatalytic and disinfection properties using well-known photocatalysts (TiO_2 and Ag/TiO_2) to evaluate a model contaminant's (ciprofloxacin—CIP) photocatalytic degradation and efficiency against *E. coli*, both under UV and visible LED irradiation installed in the photoreactor. The suitability of this reactor for the assessed functionalities was proved, and the ability to standardise and tune experiments was equally reassured.

2. Materials and Methods

2.1. Photoreactor Design and Fabrication

- Mechanical subsystem

The system was designed as a small portable actuation device of 144 × 83 × 149.5 mm (L × W × H), which enables LED irradiation to 50 mm diameter beakers used with the contaminated solutions to be tested with UVA or visible radiation.

For this purpose, to reduce power consumption and improve light transmission efficiency, light actuation was applied from the bottom. However, this architecture hinders a magnetic stirrer at the bottom of the beaker, requiring a mechanical stirrer placed at the top. In Figure 1, the schematic representation of the device is shown where the stirrer attached to the motor was designed to be articulated, allowing for its insertion or removal from the reaction beaker (100 mL) that contains the contaminated solution and the photocatalysis (e.g., titanium dioxide nanoparticles). Furthermore, the stirrer is easily withdrawn and washed at the end of each use.

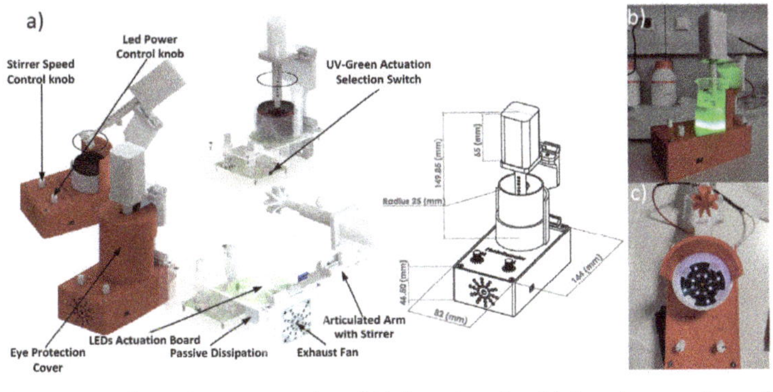

Figure 1. A schematic representation of (**a**) the external and internal perspectives of the mechanical appearance of the 3D-printed photoreactor fully assembled, and all device measurements are in mm. The real pictures represent the (**b**) lateral and (**c**) superior perspectives of the photoreactor displaying VIS (green) and UV (purple) radiation turned on.

A removable printed cover protects the user's eye from the UVA and visible irradiation. A side switch provides the control with three positions: no light actuation, visible light, or UVA light. The user interface is a simple control of the stirrer rpm (rotations per minute) and light intensity, which relies on the PWM (Pulse Width Modulation) signal manipulation, which actuates the inner electronics. The mechanical parts were designed using SolidWorks 2017® as CAD software, which enabled the possibility of 3D printing them using a Prusa® iMK3 printer with PLA (polylactic acid filament—a Prusament reference from Prusa, Prague, Czech Republic). Further, a conventional 200 rpm, small DC motor (FIT0492 from BDFROBOT) was installed and screwed to the printed structure to apply mechanical stirring to the features of the device. The final assembled device weighs 490 g, which perfectly matched the intentions of developing a lightweight device that allows for easy transportation (e.g., field experiments) and high practicability and flexibility (e.g., lab scale assays) for the photocatalytic and disinfection assessment set-ups.

- Electrical subsystem

The electrical design of the photoreactor is summarised in Figure 2 with the simplified circuit diagrams. Two potentiometers (RK1191124001 from Alps Alpine, Tokyo, Japan) are instrumented with a reference resistor (standard surface mount 0805 sizes) and the respective DC voltage variation, received by an LTC6992 from Analog Devices (Wilmington, MA, USA) and transduced into a PWM signal accordingly. The transduction happens the same way for both system components; however, the following circuits are different: in Figure 2a, a MOSFET receives the signal in the gate pin, which enables the power drawing from the source by the motor (maximum 200 rpm) and thus actuating the stirrer mechanically coupled; in Figure 2b, a dedicated LED driver integrated circuit with reference 172946001 from Wurth Electronics (Yokohama, Japan), receives the PWM signal and accordingly applies a controlled current throughout the LED array PCB.

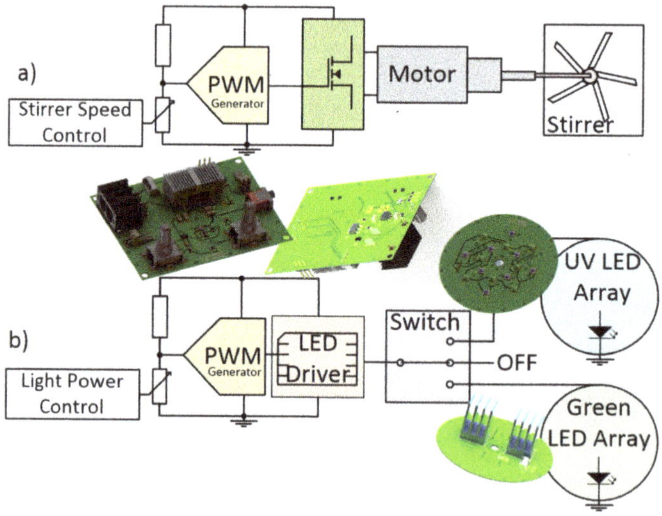

Figure 2. Simplified circuit diagrams for the control of (**a**) the motor stirrer by a PWM signal being applied to a field-effect transistor followed by the motor, and (**b**) an LED driver receives the PWM signal and applies a current stream to the LED array.

However, transforming electrical power into light generates heat in the system, requiring both passive and active heatsink components. The heat dissipation is critical to maintaining the stability of the electrical part, but also because the system's main structure is 3D printed in PLA, which could melt in the vicinities of the heatsinks and LEDs.

Thus, as shown in Figure 3, the heatsink for the LED driver was based on encapsulation DA-T263-401E-TR from Ohmite (Warrenville, IL, USA), and for the LED array, on a PCB holding ATS-54150W-C1-R0 from Advanced Thermal Solutions (Norwood, MA, USA).

Figure 3. Schematic representation of the photoreactor PCB's disposition according to passive and active heatsink systems.

Regarding active heat dissipation, a 12 V 40 mm conventional fan (CFM-4010B-180-296-22 from CUI Devices, Lake Oswego, OR, USA) and frontal and back air entrances were installed for air circulation around the passive heatsinks. Furthermore, the system uses only one of the two arrays of LEDs to keep power dissipation at lower levels.

- Irradiation subsystems

The eight UV and eight visible LEDs were installed, separated by approximately 15 mm between them. The LEDs' voltage was tuned by dividing 30 V by eight LEDs, yielding 3.75 V for each. The wavelength of the LEDs matches the typical semiconductor catalyst's absorbance peaks in the UVA region, namely 365 nm (SST-10-UV from Luminus, Sunnyvale, CA, USA). The effective radiation source range depends on the photocatalyst; for TiO_2, the light absorption is in the ultraviolet range, up to 400 nm [28]. The ultraviolet radiation spectrum is divided into three zones depending on the wavelength: long wave (UV-A) from 315 to 400 nm, middle wave (UV-B) from 280 to 315 nm, and short wave (UV-C) from 200 to 280 nm [29]. Traditional tube technologies for UV-B and UV-C radiation have drawbacks, such as the need for expensive high voltages, fragility, and the use of mercury. This radiation is also associated with skin and eye problems [30–32]. In this context, to make the process cheap, practical, and safe in this work, LED-based UV-A technologies emitting a radiation of 365 nm are adequate and readily commercially available [33]. Concerning visible range radiation, the LED peak of 535 nm (XPEBGR-L1-0000-00G03 from Cree XLamp XP-E2) was installed to allow for the assessment of the ability of nanocatalysts (e.g., Ag/TiO_2) to absorb the visible range radiation to perform photocatalysis. The device's structural design provides up to 450 mA flow in the LED arrays, with a controlled PWM of 800 kHz up to 80% (hardware limited).

The photocatalytic degradation of ciprofloxacin and the disinfection processes are driven by irradiation. Therefore, the LED disposition was carefully studied to allow for UV and visible irradiation through the device and simultaneously allow for uniform and efficient photocatalytic and disinfection processes. The four-pointed star arrangement for UV and visible LEDs (Figure 4a) allows for a uniform irradiation of the bottom of the beaker, assuring that solutions containing the nanocatalysts and the contaminant are entirely and equally irradiated. It is equally important to quantify photo actuation from the device. The light actuation of both LED arrays was measured at a constant 5 mm distance from the bottom to the beaker using an S425C surface absorber from Thorlabs, Figure 4b.

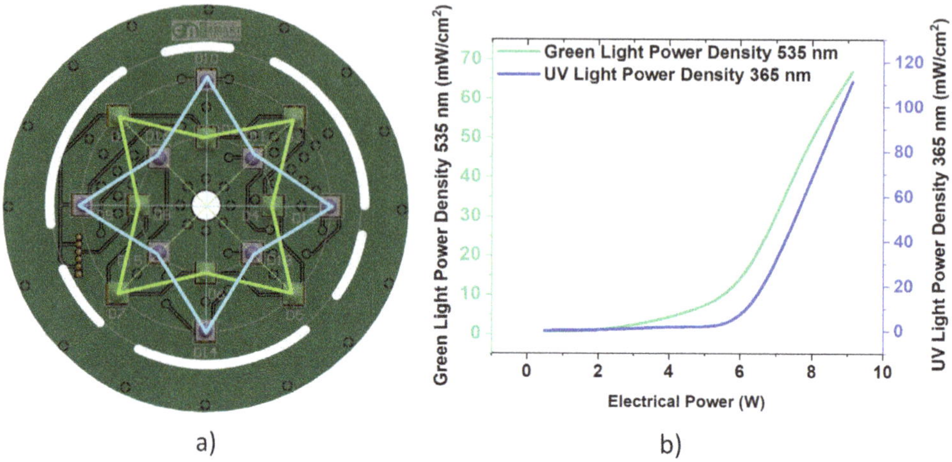

Figure 4. Light power is produced from both LED arrays according to the electrical power applied. (**a**) Mechanical disposition of both types of LED with UV and Visible light intertwined for a similar light actuation disposition and (**b**) Ratio of light intensity for each Wavelength type according to applied power.

The electrical power drawn by the LEDs is controlled by the PWM signal in the LED driver set by the potentiometer position, and a similar approach controls the motor rpm. Regarding the light actuation results, it is realised that below 5 W of electrical power applied to a single array of LEDs yields no significant light irradiation (less than 10 mW/cm^2) for both collections. Concerning UV radiation (365 nm), 6 W produces ≈7.5 mW/cm^2 power density, increasing to ≈110 mW/cm^2 with 9 W of electrical power. Additionally, the visible light (535 nm) produced by applying 6 W of electrical power is ≈14 mW/cm^2 and reaches 66 mW/cm^2 for 9 W. It is relevant to stress that the electrical power of the LEDs was in part limited by the electric-to-light transformation, presenting plenty of heat loss. In a miniaturised and portable device, electrical power/radiation intensity was sacrificed to maintain heat dissipation at secure levels.

2.2. Functional Performance Evaluation of the Photoreactor

After the complete assemblage of all components, functional tests to evaluate the photocatalytic and disinfection features of the device were performed. Figure 5 emphasises the versatility of the developed photoreactor, which allows for tuning the type of radiation (UV-A or visible range), its intensity, and agitation rate in a standardised, reproducible, comparable manner usually lacking in photocatalytic assessments.

In this way, many applications, such as the photocatalytic degradation of contaminants (e.g., ciprofloxacin) or microbial disinfection (e.g., *E. coli*), can be investigated with the proposed multifunctional system. Photocatalytic nanoparticles (TiO_2 and Ag/TiO_2) were employed during both functional tests. It is essential to highlight that the nanocatalysts used in these photocatalytic tests were already synthesised, characterised, and applied to remove contaminants and bacteria [9,34,35]. The role of these materials in this work is to validate the developed photoreactor concept. The portability allows for experiments in the field or laboratory, and the flexibility enables tuning parameters such as radiation type, intensity, agitation, and exposure time, as demonstrated in Figure 5.

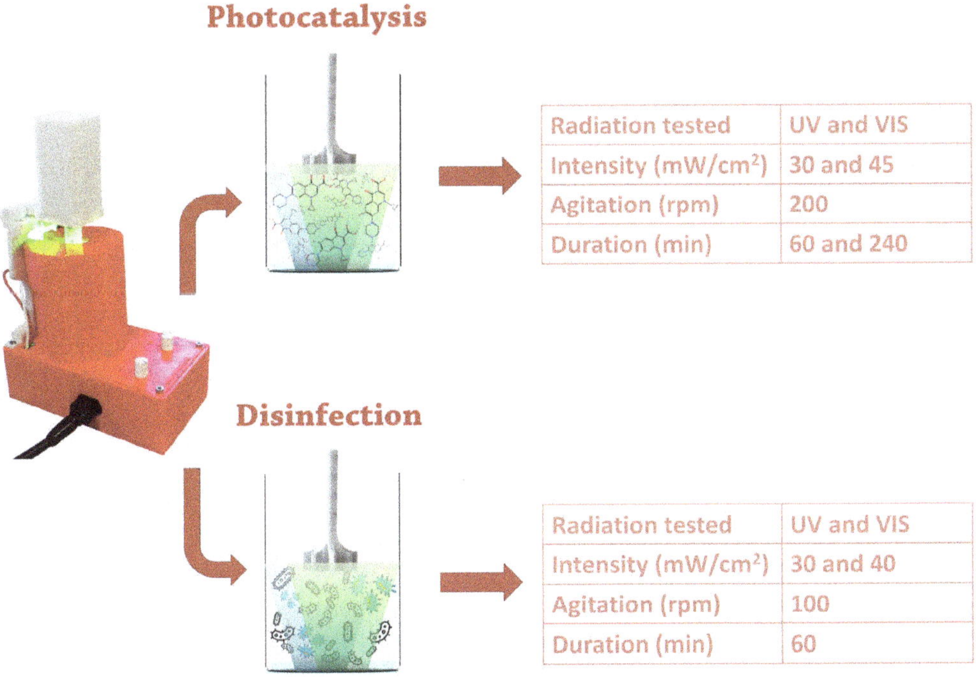

Figure 5. A schematic representation of the functional tests to evaluate the photocatalytic degradation of ciprofloxacin and the disinfection of *E. coli* with the designed and assembled photoreactor. Tables list the experimental conditions employed in each assessment.

2.2.1. Photocatalytic Assays

Commercial titanium dioxide nanoparticles (TiO_2), with a surface area ranging from 35 to 65 m^2/g, were acquired from Evonik Industries AG. The Ag/TiO_2 nanocomposites were produced by the deposition–precipitation method, as reported in [35]. Ciprofloxacin (Mw = 331.34 g/mol) with a maximum absorption of 277 nm was purchased from Sigma-Aldrich (St. Louis, MO, USA).

The photocatalytic activity of TiO_2 and Ag/TiO_2 was evaluated by monitoring the degradation of ciprofloxacin under ultraviolet (UV) and visible (VIS) LED radiation. A standard solution of CIP (5 mg L^{-1}, pH = 2.9) was prepared. Then, to 50 mL of this solution, 50 mg of the catalyst was added and kept under agitation in the dark to reach adsorption/desorption equilibrium. Subsequently, the solution was exposed to UV radiation to test the photocatalytic activity of TiO_2 nanoparticles. Similarly, the solution was exposed to VIS radiation to test the photocatalytic activity of Ag/TiO_2 nanoparticles. The photocatalytic degradation efficiencies of CIP were analysed by measuring the maximum absorption peak (277 nm) over time using an Infinite M Plex spectrophotometer (TECAN). The degradation fits the Langmuir–Hinshelwood model, expressed by Equation (1):

$$\frac{C}{C_0} = e^{-kt} \tag{1}$$

where C_0 and C are the initial concentration of the pollutant and the concentration at time (t), respectively, and k is the first-order rate constant of the reaction (min^{-1}) [36].

2.2.2. Antimicrobial Assays

Gram-negative Escherichia coli ATCC® 8739™ were purchased from the American Type Culture Collection (LGC Standards S.L.U, Barcelona, Spain). Bacterial pre-inoculum was prepared using a single colony from the corresponding stock bacterial culture, resuspending it in nutrient broth (NB), and then incubating overnight at 37 °C and 110 rpm. After 20 h, the bacteria were harvested by centrifugation at 4500 rpm for 5 min and resuspended in NaCl 0.9% (*w*/*v*) twice. The *E. coli* culture optical density (OD) was adjusted to OD = 0.26, measured at 600 nm, giving rise to a working inoculum of approximately 1×10^7 colony-forming units (CFU) per mL.

The bactericidal activity was assessed according to the standard shake flask method (ASTM-E2149-01) with some modifications as an adaptation to the photoreactor set-up. This method provides quantitative data for measuring the reduction rate in the number of *E. coli* colonies formed, converted to the average colony-forming units per millilitre of buffer solution in the flask (CFU/mL). E. coli eradication was assessed using the developed photoreactor with two different radiation sources.

The UV radiation was tested in the presence of the bacterial inoculum at 37 °C for one hour (the control was the same bacterial inoculum without applying UVs) without any particles to evaluate the effect of UV radiation on the bacteria viability. The VIS (green) radiation was also tested by mixing the solution with 0.5 mg/mL of Ag/TiO_2 nanoparticles, previously synthesised and tested in other studies [34,35], to assess the synergy between the VIS radiation and the particles for efficient bacterial killing. The final volume of the solutions was 40 mL to fit the photoreactor. After one hour of incubation, the solutions subjected to the UV and green irradiation and respective controls were serially diluted (1:10) in sterile buffer solution, plated on a plate count NB agar, and further incubated at 37 °C for 24 h to determine the number of surviving bacteria. Antimicrobial activity is the percentage of bacteria log reduction calculated as the ratio between the number of surviving bacteria after and before the treatment (Equation (2)):

$$Bacteria\ reduction\ (\%) = \frac{(A - B)}{A} \times 100 \tag{2}$$

where *A* and *B* are the average number of bacteria before and after contact with the samples. The results were further expressed as a $\log_{10}$ reduction by calculating the $\log_{10}$ of bacteria reduction. All antibacterial data represent mean values ± SD (n = 3).

3. Results and Discussion

3.1. Photocatalytic Assays

CIP degradation tests were performed under UV and VIS LED irradiation for 60 and 240 min to validate the developed photoreactor and its practicability in photocatalytic assays (Figure 6).

The analysis of the photocatalytic experiments indicates negligible CIP adsorption by nanoparticles (~0%), suggesting a reliance on LED irradiation for pollutant removal. The contaminated solution was irradiated with UV and VIS LEDs in the photoreactor to estimate CIP photolysis. The results, shown in Figure 6a,b, show minimal degradation rates of approximately 3% and 0% of CIP under UV and VIS light, respectively.

Afterwards, using TiO_2 nanoparticles in contact with CIP solution under UV irradiation yielded a degradation efficiency of 78% and an apparent degradation rate constant of 0.024 min^{-1}. Regarding the photocatalytic results obtained with Ag/TiO_2 nanoparticles under VIS visible radiation, 31% of CIP in the solution was degraded with an apparent reaction rate of 0.002 min^{-1} obtained by applying Equation (1). These results indicate that irradiance energy plays an essential role in pollutant degradation. Therefore, as UV radiation is more energetic than visible light, it makes the valence electrons more easily excited so they to interact with the molecular species in the water, generating higher concentrations of reactive oxidative species (ROS). Therefore, this difference in degradation efficiency between UV and visible light tests can be attributed to the higher concentration of ROS generated by the former [37]. CIP was used as a model contaminant since its photocatalytic degradation process is already well known, with several works focused on the photocatalytic condition's optimisation, photocatalytic degradation mechanism, and the toxicity of the generated by-products, among many others [9,38,39].

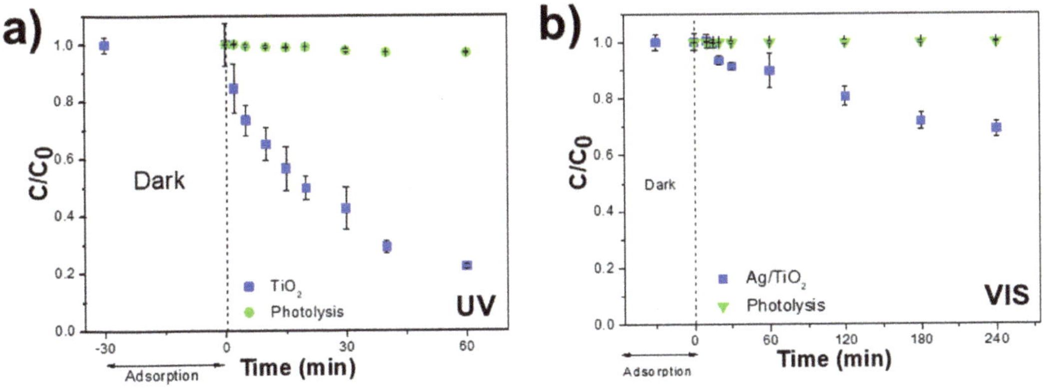

Figure 6. Photocatalytic degradation of CIP (5 mg L^{-1}; pH ≈ 3; V_{CIP} = 50 mL) in aqueous solution under (**a**) UV radiation with TiO_2-P25 (m_{TiO2} = 50 mg) particles for 60 min and (**b**) visible radiation with Ag/TiO_2-P25 ($m_{Ag/TiO2\text{-}P25}$ = 50 mg) nanoparticles for 240 min.

Despite the interesting photocatalytic results, the focus of this work was the validation of the design and the functional concept of the photoreactor, indicating that the bottom irradiation is efficient (intensity and spatial distribution), both from the standpoint of the photocatalytic degradation process, which was efficient, and in terms of the photoreactor stability, by avoiding excessive heating despite the prolonged irradiation (i.e., 1 and 4 h), and finally validating the agitation using an articulated mechanical stirrer. Additionally, it was possible to certify that the mechanical agitation provided is perfectly adjustable and in a range that allows for maintaining adequate nanoparticle dispersion in the CIP solution.

Table 1 lists different studies focused on applying photocatalytic materials based on TiO_2 with varying radiation sources for the degradation of antibiotics in an aqueous solution. Due to the remarkably different experimental conditions, such as pH, contaminant concentration, and radiation intensity/wavelength, direct comparisons cannot be made. This is one of the significant drawbacks of the photocatalytic experiments performed in the scientific community; a vast and infinite combination of experimental conditions hinders any direct comparison of the obtained photocatalytic degradation results. Nonetheless, the information is helpful for contextualisation purposes.

The antibiotic degradation rates (k) related to the different materials and degradation conditions in Table 1 range between 0.02 and 0.001 min^{-1}. Through an analysis of Table 1, this work presents a superior performance with similar experimental conditions. It is crucial to underline that most of the present works involve extensive, mono radiation, only UV or visible, non-portable photoreactors with higher irradiation intensities, yielding higher photocatalytic activities. The present photoreactor shows efficiencies interestingly similar to other systems in this context. But this device offers many features that others lack, such as UV and VIS radiation, adjustable agitation, portability, and standardised efficiencies that allow for direct comparisons if assessments are performed using the same conditions. Also, the idea of this photoreactor is not to compete with commercial devices in the market but to be a cost-effective, 3D-printable, and portable device that allows lab-scale (in lab or field) assessments to be performed before upscaled experiments.

Table 1. Works applying TiO_2-based materials to degrade different contaminants, the degradation reaction rate (k), and the type of radiation used.

Reactor Type	Photocatalytic Material	Contaminant	Degradation Efficiency (%)	Degradation Rate k (min^{-1})	Radiation Source	Reference
Slurry	TiO_2	Ciprofloxacin	78	0.024	8 UV LEDs (365 nm)	Present work
Immobilised	TiO_2	Ciprofloxacin	75	0.009	Irradiance 100 Wm^{-2} 365 nm	[40]
Slurry	TiO_2	Metronidazole	98	0.022	UV lamp (9 W)	[17]
Slurry	TiO_2	Roxithromycin	99	0.008	6 lamps (8 W each) at 350 nm	[41]
Slurry	Ag/TiO_2	Ciprofloxacin	31	0.002	8 VIS LEDs (535 nm)	Present work
Immobilised	TiO_2/AuNRs 650°	Nalidixic acid	65	0.001	QTH lamp (0.22 W/cm^2~2 SUN	[42]
Immobilised	TiO_2/GO	Ciprofloxacin	91	0.002	300 W xenon lamp with a 420 nm	[5]

3.2. Antimicrobial Assays

The photoreactor was also tested on an *E. coli* bacterium, a microorganism commonly found in wastewater treatment facilities and usually a concern due to the risk of discharge into drinking water [43]. Thus, an alternative method to eliminate or inactivate these faecal bacteria through photocatalytic processes was tested in the developed reactor. Two different radiation sources were used: UV radiation directly on a bacterial inoculum to test the ability of UV irradiation to eliminate bacteria without the presence of nanoparticles, and VIS radiation on bacteria inoculated with 0.5 mg/mL of Ag/TiO_2 to induce bactericidal effects through the activation of the ROS in these NPs due to the presence of Ag. The UV radiation has a negligible impact on reducing *E. coli* colonies (0.5 log_{10} reduction). At the same time, the NPs alone, without light, induced a 1.5 log_{10} reduction (Figure 7a) and represented colonies forming units (CFU) on Petri dishes in sequential dilutions (Figure 7b). The VIS green light from the photoreactor enhanced the bactericidal effect, yielding an impressive 8 log_{10} reduction, thus eliminating this bacterium. The VIS radiation applies wavelengths near the plasmon resonance of silver, which promotes the formation of ROS and changes the overall plasmon resonance of the Ag/TiO_2 NPs, thus making it highly effective in killing bacteria [35]. During disinfection assays, the photoreactor showed irradiation stability and a suitable refrigeration system that prevented the heating of the contaminated water in the beaker. If the refrigeration was inefficient, temperature variations could have occurred and interfered with bacteria viability—cell death could be a consequence of the temperature and not the nanomaterial or irradiation.

This approach may represent an essential breakthrough in the treatment of contaminated water. Usually, the effluents are disinfected using oxidative processes, where chlorine is the most commonly used disinfectant [44]. Ozonation and UV irradiation approaches are gaining significant attention and have been increasingly applied [45,46]. UV LEDs have been directly used in wastewater treatment plants to inactivate bacteria. It has been found that a 280 nm wavelength LED was effective in eliminating *E. coli*, inducing a 4 log reduction [47]. In another study at 190 nm wavelength, pulsed UV irradiation was applied, generating an 11 log_{10} reduction rate [48].

Nevertheless, these wavelengths are also considered harmful to humans; thus, alternatives employing UV photoreactors have been explored. A 3D-printed UV reactor was developed, in which, similarly to our photoreactor, the lamps do not come in contact with the water, thus avoiding fouling processes and inducing a 6.9 log_{10} reduction in *E. coli* [49]. Despite the exciting results, this work aimed not to develop profound studies on optimising the disinfection process or its mechanism; several other studies focused on this topic [35,50,51]. Instead, the primary goal of the antimicrobial assays was to demonstrate the practicability and versatility of the developed photoreactor in disinfection studies. The benefit of the developed systems is very similar to the advantage enunciated in the CIP photocatalytic degradation tests, the possibility of adjusting radiation type and intensity, the agitation rate, reproducibility, and standardisation of the obtained results. This is particularly relevant for lab-scaled assays where the first assessments require standardisation to allow for conclusive results and straightforward comparisons between different antimicrobial efficiencies.

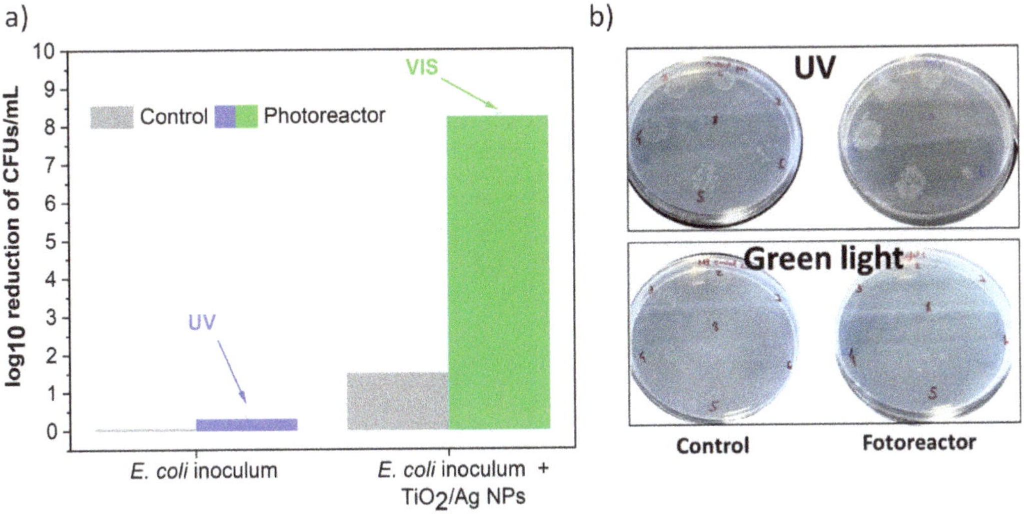

Figure 7. (**a**) The antimicrobial effect of the photocatalytic nanoparticles evaluated in the photoreactor applied to an *E. coli* inoculum using UV and green light, measured in a $\log_{10}$ reduction in CFUs; (**b**) Petri dishes displaying colonies that grew in nutrient broth agar.

4. Conclusions

Water remediation is a relevant area in the framework of the sustainable goals identified by the United Nations. Humankind's future relies on our ability to develop new materials and devices to reestablish the water and natural environments to their original state. The possibility of having a multifunctional, portable, and efficient photoreactor will certainly leverage all the studies and outcomes in this area as it allows for the creation of experimental conditions that can be accurately controlled and reproduced in any laboratory or field using this device.

In this work, we report on a new photoreactor based on concepts of portability, the possibility of using UV or visible radiation, incorporated agitation, and stability. The device was assembled by creating all the electronics, irradiation set-up, mechanical agitation arm, and an external FDM 3D-printed case.

The portable device permits radiation intensities in an adjustable range between 7.5 and 110 mW/cm^2 for UV radiation and 14 to 66 mW/cm^2 for visible radiation, permitting studies regarding the radiation impact in photocatalytic and disinfection processes. Also, the agitation and refrigeration proved suitable and stable during lengthy irradiation assessments, which are paramount in the context of persistent contaminants.

Concerning the application assessments and the device proof of concept, the photocatalytic photoreactor proved to degrade 30% and 80% of ciprofloxacin under UV and visible radiation and eliminate *E. coli* efficiently.

Compared to other photoreactors, this aesthetic device offers a simple, lightweight, and multifunctional approach with open-loop control, portability, and miniaturisation. It is convenient for swift laboratory photocatalytic assays and cost-effective actuation as a starting point for more standardised photocatalytic and disinfection experiments. The application of this photoreactor in a new lab or field combined with new photocatalysts and different water matrixes will be explored in the future based on the versatility of this device.

Author Contributions: Conceptualization, N.C. and P.M.M.; Methodology, N.C., J.M.Q., M.M.M.F. and P.M.M.; Software, N.C., J.M.Q. and D.C.A.; Validation, N.C., J.M.Q., M.M.M.F., S.L.-M. and P.M.M.; Formal analysis, M.M.M.F., S.L.-M. and P.M.M.; Investigation, N.C., J.M.Q., D.C.A. and M.M.M.F.; Resources, S.L.-M.; Data curation, N.C., J.M.Q., D.C.A. and M.M.M.F.; Writing—original draft, N.C., J.M.Q., D.C.A., M.M.M.F. and P.M.M.; Writing—review & editing, N.C., M.M.M.F., S.L.-M. and P.M.M.; Visualization, N.C., D.C.A. and P.M.M.; Supervision, S.L.-M. and P.M.M.; Project administration, S.L.-M.; Funding acquisition, S.L.-M. All authors have read and agreed to the published version of the manuscript.

Funding: The authors thank the Fundação para a Ciência e Tecnologia (FCT) for financial support under the framework of Strategic Funding UIDB/04650/2020, UIDB/04050/2020, and UID/FIS/04650/2020. Joana M. Queirós and P. M. Martins thank the FCT for the fellowship 2021.08822.BD and contract DOI 10.54499/2020.02802.CEECIND/CP1600/CT0017. Dinis C. Alves and M. M. Fernandes thank the fellowship 009/TT/2023 and the contract CEECINST/00018/2021. This study formed part of the Advanced Materials programme and was supported by MCIN with funding from European Union NextGenerationEU (PRTR-C17.I1) as well as by IKUR Strategy under the collaboration agreement between Ikerbasque Foundation and Fundación BCMaterials on behalf of the Department of Education of the Basque Government.

Data Availability Statement: Data are contained within the article.

Conflicts of Interest: The authors declare no conflict of interest.

References

1. Prüss-Ustün, A.; Bartram, J.; Clasen, T.; Colford, J.M., Jr.; Cumming, O.; Curtis, V.; Bonjour, S.; Dangour, A.D.; De France, J.; Fewtrell, L.; et al. Burden of disease from inadequate water, sanitation and hygiene in low- and middle-income settings: A retrospective analysis of data from 145 countries. *Trop. Med. Int. Health* **2014**, *19*, 894–905. [CrossRef]
2. Prüss-Ustün, A.; Wolf, J.; Bartram, J.; Clasen, T.; Cumming, O.; Freeman, M.C.; Gordon, B.; Hunter, P.R.; Medlicott, K.; Johnston, R. Burden of disease from inadequate water, sanitation and hygiene for selected adverse health outcomes: An updated analysis with a focus on low- and middle-income countries. *Int. J. Hyg. Environ. Health* **2019**, *222*, 765–777. [CrossRef]
3. Sathish, T.; Ahalya, N.; Thirunavukkarasu, M.; Senthil, T.S.; Hussain, Z.; Haque Siddiqui, M.I.; Panchal, H.; Kumar Sadasivuni, K. A comprehensive review on the novel approaches using nanomaterials for the remediation of soil and water pollution. *Alex. Eng. J.* **2024**, *86*, 373–385. [CrossRef]
4. Comninellis, C.; Kapalka, A.; Malato, S.; Parsons, S.A.; Poulios, I.; Mantzavinos, D. Advanced oxidation processes for water treatment: Advances and trends for R&D. *J. Chem. Technol. Biotechnol.* **2008**, *83*, 769–776. [CrossRef]
5. Wang, J.; Zhuan, R. Degradation of antibiotics by advanced oxidation processes: An overview. *Sci. Total Environ.* **2020**, *701*, 135023. [CrossRef] [PubMed]
6. Xia, L.; Zhang, K.; Wang, X.; Guo, Q.; Wu, Y.; Du, Y.; Zhang, L.; Xia, J.; Tang, H.; Zhang, X.; et al. 0D/2D Schottky junction synergies with 2D/2D S-scheme heterojunction strategy to achieve uniform separation of carriers in 0D/2D/2D quasi CN-QDs/TCN/ZnIn2S4 towards photocatalytic remediating petroleum hydrocarbons polluted marine. *Appl. Catal. B Environ.* **2023**, *325*, 122387. [CrossRef]
7. Feng, H.; Yu, J.; Tang, J.; Tang, L.; Liu, Y.; Lu, Y.; Wang, J.; Ni, T.; Yang, Y.; Yi, Y. Enhanced electro-oxidation performance of FeCoLDH to organic pollutants using hydrophilic structure. *J. Hazard. Mater.* **2022**, *430*, 128464. [CrossRef] [PubMed]
8. Mozia, S. Photocatalytic membrane reactors (PMRs) in water and wastewater treatment. A review. *Sep. Purif. Technol.* **2010**, *73*, 71–91. [CrossRef]
9. Zheng, F.; Queirós, J.M.; Martins, P.M.; de Luis, R.F.; Fidalgo-Marijuan, A.; Vilas-Vilela, J.L.; Lanceros-Méndez, S.; Reguera, J. Au-sensitised TiO_2 and ZnO nanoparticles for broadband pharmaceuticals photocatalytic degradation in water remediation. *Colloids Surf. A Physicochem. Eng. Asp.* **2023**, *671*, 131594. [CrossRef]
10. Xia, L.; Sun, Z.; Wu, Y.; Yu, X.-F.; Cheng, J.; Zhang, K.; Sarina, S.; Zhu, H.-Y.; Weerathunga, H.; Zhang, L.; et al. Leveraging doping and defect engineering to modulate exciton dissociation in graphitic carbon nitride for photocatalytic elimination of marine oil spill. *Chem. Eng. J.* **2022**, *439*, 135668. [CrossRef]
11. Guo, Q.; Zhou, C.; Ma, Z.; Yang, X. Fundamentals of TiO_2 Photocatalysis: Concepts, Mechanisms, and Challenges. *Adv. Mater.* **2019**, *31*, 1901997. [CrossRef]
12. Braham, R.J.; Harris, A.T. Review of Major Design and Scale-up Considerations for Solar Photocatalytic Reactors. *Ind. Eng. Chem. Res.* **2009**, *48*, 8890–8905. [CrossRef]
13. Wetchakun, K.; Wetchakun, N.; Sakulsermsuk, S. An overview of solar/visible light-driven heterogeneous photocatalysis for water purification: TiO_2- and ZnO-based photocatalysts used in suspension photoreactors. *J. Ind. Eng. Chem.* **2019**, *71*, 19–49. [CrossRef]
14. Martins, P.M.; Miranda, R.; Marques, J.; Tavares, C.J.; Botelho, G.; Lanceros-Mendez, S. Comparative efficiency of TiO_2 nanoparticles in suspension vs. immobilization into P(VDF–TrFE) porous membranes. *RSC Adv.* **2016**, *6*, 12708–12716. [CrossRef]

15. Marinho, B.A.; Cristóvão, R.O.; Djellabi, R.; Loureiro, J.M.; Boaventura, R.A.R.; Vilar, V.J.P. Photocatalytic reduction of Cr(VI) over TiO_2-coated cellulose acetate monolithic structures using solar light. *Appl. Catal. B Environ.* **2017**, *203*, 18–30. [CrossRef]
16. Ochuma, I.J.; Osibo, O.O.; Fishwick, R.P.; Pollington, S.; Wagland, A.; Wood, J.; Winterbottom, J.M. Three-phase photocatalysis using suspended titania and titania supported on a reticulated foam monolith for water purification. *Catal. Today* **2007**, *128*, 100–107. [CrossRef]
17. Azalok, K.A.; Oladipo, A.A.; Gazi, M. Hybrid MnFe-LDO–biochar nanopowders for degradation of metronidazole via UV-light-driven photocatalysis: Characterization and mechanism studies. *Chemosphere* **2021**, *268*, 128844. [CrossRef]
18. García-Muñoz, P.; Allé, P.H.; Bertoloni, C.; Torres, A.; de la Orden, M.U.; Urreaga, J.M.; Dziurla, M.-A.; Fresno, F.; Robert, D.; Keller, N. Photocatalytic degradation of polystyrene nanoplastics in water. A methodological study. *J. Environ. Chem. Eng.* **2022**, *10*, 108195. [CrossRef]
19. Sacco, O.; Vaiano, V.; Rizzo, L.; Sannino, D. Photocatalytic activity of a visible light active structured photocatalyst developed for municipal wastewater treatment. *J. Clean. Prod.* **2018**, *175*, 38–49. [CrossRef]
20. van Grieken, R.; Marugán, J.; Sordo, C.; Martínez, P.; Pablos, C. Photocatalytic inactivation of bacteria in water using suspended and immobilized silver-TiO_2. *Appl. Catal. B Environ.* **2009**, *93*, 112–118. [CrossRef]
21. Rosenfeld, P.E.; Feng, L.G.H. 16—Emerging Contaminants. In *Risks of Hazardous Wastes*; Rosenfeld, P.E., Feng, L.G.H., Eds.; William Andrew Publishing: Boston, MA, USA, 2011; pp. 215–222. [CrossRef]
22. Salimi, M.; Esrafili, A.; Gholami, M.; Jonidi Jafari, A.; Rezaei Kalantary, R.; Farzadkia, M.; Kermani, M.; Sobhi, H.R. Contaminants of emerging concern: A review of new approach in AOP technologies. *Environ. Monit. Assess.* **2017**, *189*, 414. [CrossRef] [PubMed]
23. Deblonde, T.; Cossu-Leguille, C.; Hartemann, P. Emerging pollutants in wastewater: A review of the literature. *Int. J. Hyg. Environ. Health* **2011**, *214*, 442–448. [CrossRef] [PubMed]
24. Sanganyado, E.; Gwenzi, W. Antibiotic resistance in drinking water systems: Occurrence, removal, and human health risks. *Sci. Total Environ.* **2019**, *669*, 785–797. [CrossRef]
25. Cabral, J.P.S. Water Microbiology. Bacterial Pathogens and Water. *Int. J. Environ. Res. Public Health* **2010**, *7*, 3657–3703. [CrossRef] [PubMed]
26. Chahal, C.; van den Akker, B.; Young, F.; Franco, C.; Blackbeard, J.; Monis, P. Chapter Two—Pathogen and Particle Associations in Wastewater: Significance and Implications for Treatment and Disinfection Processes. In *Advances in Applied Microbiology*; Sariaslani, S., Michael Gadd, G., Eds.; Academic Press: Cambridge, MA, USA, 2016; pp. 63–119.
27. Luo, L.-W.; Wu, Y.-H.; Yu, T.; Wang, Y.-H.; Chen, G.-Q.; Tong, X.; Bai, Y.; Xu, C.; Wang, H.-B.; Ikuno, N.; et al. Evaluating method and potential risks of chlorine-resistant bacteria (CRB): A review. *Water Res.* **2021**, *188*, 116474. [CrossRef]
28. Peiris, S.; de Silva, H.B.; Ranasinghe, K.N.; Bandara, S.V.; Perera, I.R. Recent development and future prospects of TiO_2 photocatalysis. *J. Chin. Chem. Soc.* **2021**, *68*, 738–769. [CrossRef]
29. Eskandarian, M.R.; Choi, H.; Fazli, M.; Rasoulifard, M.H. Effect of UV-LED wavelengths on direct photolytic and TiO_2 photocatalytic degradation of emerging contaminants in water. *Chem. Eng. J.* **2016**, *300*, 414–422. [CrossRef]
30. Lui, G.Y.; Roser, D.; Corkish, R.; Ashbolt, N.J.; Stuetz, R. Point-of-use water disinfection using ultraviolet and visible light-emitting diodes. *Sci. Total Environ.* **2016**, *553*, 626–635. [CrossRef]
31. Sliney, D.H.; Stuck, B.E. A Need to Revise Human Exposure Limits for Ultraviolet UV-C Radiation†. *Photochem. Photobiol.* **2021**, *97*, 485–492. [CrossRef]
32. Pasupuleti, K.S.; Chougule, S.S.; Jung, N.; Yu, Y.-J.; Oh, J.-E.; Kim, M.-D. Plasmonic Pt nanoparticles triggered efficient charge separation in TiO_2/GaN NRs hybrid heterojunction for the high performance self-powered UV photodetectors. *Appl. Surf. Sci.* **2022**, *594*, 153474. [CrossRef]
33. Ferreira, L.C.; Fernandes, J.R.; Rodríguez-Chueca, J.; Peres, J.A.; Lucas, M.S.; Tavares, P.B. Photocatalytic degradation of an agro-industrial wastewater model compound using a UV LEDs system: Kinetic study. *J. Environ. Manag.* **2020**, *269*, 110740. [CrossRef] [PubMed]
34. Martins, P.; Kappert, S.; Nga Le, H.; Sebastian, V.; Kühn, K.; Alves, M.; Pereira, L.; Cuniberti, G.; Melle-Franco, M.; Lanceros-Méndez, S. Enhanced Photocatalytic Activity of Au/TiO_2 Nanoparticles against Ciprofloxacin. *Catalysts* **2020**, *10*, 234. [CrossRef]
35. Salazar, H.; Martins, P.M.; Santos, B.; Fernandes, M.M.; Reizabal, A.; Sebastián, V.; Botelho, G.; Tavares, C.J.; Vilas-Vilela, J.L.; Lanceros-Mendez, S. Photocatalytic and antimicrobial multifunctional nanocomposite membranes for emerging pollutants water treatment applications. *Chemosphere* **2020**, *250*, 126299. [CrossRef] [PubMed]
36. Ayati, A.; Ahmadpour, A.; Bamoharram, F.F.; Tanhaei, B.; Mänttäri, M.; Sillanpää, M. A review on catalytic applications of Au/TiO_2 nanoparticles in the removal of water pollutant. *Chemosphere* **2014**, *107*, 163–174. [CrossRef]
37. Vaiano, V.; Sacco, O.; Sannino, D.; Ciambelli, P. Photocatalytic removal of spiramycin from wastewater under visible light with N-doped TiO_2 photocatalysts. *Chem. Eng. J.* **2015**, *261*, 3–8. [CrossRef]
38. Silva, A.R.; Martins, P.M.; Teixeira, S.; Carabineiro, S.A.C.; Kuehn, K.; Cuniberti, G.; Alves, M.M.; Lanceros-Mendez, S.; Pereira, L. Ciprofloxacin wastewater treated by UVA photocatalysis: Contribution of irradiated TiO_2 and ZnO nanoparticles on the final toxicity as assessed by Vibrio fischeri. *RSC Adv.* **2016**, *6*, 95494–95503. [CrossRef]
39. Zheng, F.; Martins, P.M.; Queirós, J.M.; Tavares, C.J.; Vilas-Vilela, J.L.; Lanceros-Méndez, S.; Reguera, J. Size Effect in Hybrid TiO_2:Au Nanostars for Photocatalytic Water Remediation Applications. *Int. J. Mol. Sci.* **2022**, *23*, 13741. [CrossRef] [PubMed]
40. Triquet, T.; Tendero, C.; Latapie, L.; Manero, M.-H.; Richard, R.; Andriantsiferana, C. TiO_2 MOCVD coating for photocatalytic degradation of ciprofloxacin using 365 nm UV LEDs—kinetics and mechanisms. *J. Environ. Chem. Eng.* **2020**, *8*, 104544. [CrossRef]

41. Kwiecień, A.; Krzek, J.; Żmudzki, P.; Matoga, U.; Długosz, M.; Szczubiałka, K.; Nowakowska, M. Roxithromycin degradation by acidic hydrolysis and photocatalysis. *Anal. Methods* **2014**, *6*, 6414–6423. [CrossRef]
42. Truppi, A.; Petronella, F.; Placido, T.; Margiotta, V.; Lasorella, G.; Giotta, L.; Giannini, C.; Sibillano, T.; Murgolo, S.; Mascolo, G.; et al. Gram-scale synthesis of UV–vis light active plasmonic photocatalytic nanocomposite based on TiO_2/Au nanorods for degradation of pollutants in water. *Appl. Catal. B Environ.* **2019**, *243*, 604–613. [CrossRef]
43. Anastasi, E.M.; Matthews, B.; Stratton, H.M.; Katouli, M. Pathogenic Escherichia coli found in sewage treatment plants and environmental waters. *Appl. Environ. Microbiol.* **2012**, *78*, 5536–5541. [CrossRef]
44. Poyatos, J.M.; Muñio, M.M.; Almecija, M.C.; Torres, J.C.; Hontoria, E.; Osorio, F. Advanced Oxidation Processes for Wastewater Treatment: State of the Art. *Water Air Soil Pollut.* **2009**, *205*, 187. [CrossRef]
45. Hallmich, C.; Gehr, R. Effect of pre- and post-UV disinfection conditions on photoreactivation of fecal coliforms in wastewater effluents. *Water Res.* **2010**, *44*, 2885–2893. [CrossRef]
46. Yu, G.; Wang, Y.; Cao, H.; Zhao, H.; Xie, Y. Reactive Oxygen Species and Catalytic Active Sites in Heterogeneous Catalytic Ozonation for Water Purification. *Environ. Sci. Technol.* **2020**, *54*, 5931–5946. [CrossRef] [PubMed]
47. Silva, N.B.; Leonel, L.P.; Tonetti, A.L. UV-LED for Safe Effluent Reuse in Agriculture. *Water Air Soil Pollut.* **2020**, *231*, 343. [CrossRef]
48. Faghihzadeh, F.; Anaya, N.M.; Hadjeres, H.; Boving, T.B.; Oyanedel-Craver, V. Pulse UV light effect on microbial biomolecules and organic pollutants degradation in aqueous solutions. *Chemosphere* **2019**, *216*, 677–683. [CrossRef]
49. Younis, B.A.; Mahoney, L.; Palomo, N. A Novel System for Water Disinfection with UV Radiation. *Water* **2018**, *10*, 1275. [CrossRef]
50. Yang, Y.-Y.; Niu, C.-G.; Huang, D.-W.; Guo, H.; Feng, H.-P.; Li, L.; Liu, H.-Y.; Fan, Q.-Q.; Qin, M.-Z. Appropriate oxygen vacancies and Mo-N bond synergistically modulate charge transfer dynamics of $MoO3-x$/S-CN for superior photocatalytic disinfection: Unveiling synergistic effects and disinfection mechanism. *J. Hazard. Mater.* **2023**, *445*, 130481. [CrossRef] [PubMed]
51. Yang, Y.-Y.; Guo, H.; Huang, D.-W.; Li, L.; Liu, H.-Y.; Sui, L.; Wu, Q.; Zhu, J.-J.; Zhang, L.; Niu, C.-G. Simultaneously tuning oxygen reduction pathway and charge transfer dynamics toward sacrificial agent-free photocatalytic H_2O_2 production for in-situ water disinfection. *Chem. Eng. J.* **2024**, *479*, 147863. [CrossRef]

MDPI AG
Grosspeteranlage 5
4052 Basel
Switzerland
Tel.: +41 61 683 77 34

Nanomaterials Editorial Office
E-mail: nanomaterials@mdpi.com
www.mdpi.com/journal/nanomaterials

www.ingramcontent.com/pod-product-compliance
Lightning Source LLC
Chambersburg PA
CBHW040812250926
51850CB00007BA/1149